Lecture Notes in Computer Science 8716

Commenced Publication in 1973
Founding and Former Series Editors:
Gerhard Goos, Juris Hartmanis, and Jan van Leeuwen

Yannis Manolopoulos   Goce Trajcevski
Margita Kon-Popovska (Eds.)

Advances in Databases and Information Systems

18th East European Conference, ADBIS 2014
Ohrid, Macedonia, September 7-10, 2014
Proceedings

 Springer

Volume Editors

Yannis Manolopoulos
Aristotle University of Thessaloniki, Department of Informatics
Thessaloniki, Greece
E-mail: manolopo@csd.auth.gr

Goce Trajcevski
Northwestern University, EECS Department
Evanston, IL, USA
E-mail: goce@eecs.northwestern.edu

Margita Kon-Popovska
University Ss. Cyril and Methodius Skopje
Faculty of Computer Sciences and Engineering
Skopje, Macedonia
E-mail: margita.kon-popovska@finki.ukim.mk

ISSN 0302-9743   e-ISSN 1611-3349
ISBN 978-3-319-10932-9   e-ISBN 978-3-319-10933-6
DOI 10.1007/978-3-319-10933-6
Springer Cham Heidelberg New York Dordrecht London

Library of Congress Control Number: 2014946898

LNCS Sublibrary: SL 3 – Information Systems and Application,
incl. Internet/Web and HCI

Typesetting: Camera-ready by author, data conversion by Scientific Publishing Services, Chennai, India

Printed on acid-free paper

Springer is part of Springer Science+Business Media (www.springer.com)

Preface

This volume contains a selection of the papers presented at the 18th East European Conference on Advances in Databases and Information Systems (ADBIS 2014), held during September 7–10, 2014, in Ohrid, Republic of Macedonia.

The ADBIS series of conferences aims to provide a forum for the presentation and dissemination of research on database theory, development of advanced DBMS technologies, and their advanced applications. ADBIS 2014 continued the ADBIS series held every year in different countries of Europe, beginning in St. Petersburg (1997), Poznan (1998), Maribor (1999), Prague (2000), Vilnius (2001), Bratislava (2002), Dresden (2003), Budapest (2004), Tallinn (2005), Thessaloniki (2006), Varna (2007), Pori (2008), Riga (2009), Novi Sad (2010), Vienna (2011), Poznan (2012), and Genoa (2013). The conferences are initiated and supervised by an international Steering Committee consisting of representatives from Armenia, Austria, Bulgaria, Czech Republic, Estonia, Finland, Germany, Greece, Hungary, Israel, Italy, Latvia, Lithuania, Poland, Russia, Serbia, Slovakia, Slovenia, and Ukraine.

The program of ADBIS 2014 included keynotes, research papers, a tutorial session entitled "Online Social Networks Analytics - Communities and Sentiment Detection" by Athena Vakali, a doctoral consortium, and thematic workshops. The conference attracted 82 paper submissions from 34 different countries representing all the continents (Algeria, Australia, Austria, Bangladesh, Bosnia and Herzegovina, Brazil, China, Croatia, Czech Republic, Estonia, France, Germany, Greece, Hungary, Italy, Japan, Lebanon, Lithuania, Macedonia, Netherlands, Poland, Romania, Russian Federation, Singapore, Slovakia, Slovenia, Spain, Switzerland, Tunisia, Turkey, UK, USA, Vietnam) with 210 authors. After a rigorous reviewing process by the members of the international Program Committee consisting of 115 reviewers from 34 countries, the 26 papers included in this LNCS proceedings volume were accepted as full contributions.

Moreover, the Program Committee selected 15 more papers to be accepted as short contributions which, in addition to the three selected papers from the doctoral consortium, and eight papers from three workshops, are published in a companion volume entitled *New Trends in Databases and Information Systems 2* in the Springer series *Advances in Intelligent Systems* and Computing. All papers were evaluated by at least three reviewers and most of them by four to five reviewers. The selected papers span a wide spectrum of topics in the database field and related technologies, tackling challenging problems and presenting inventive and efficient solutions. In this volume, they are organized in eight sections: (1) Data Models and Query Languages; (2) Data Warehousing; (3) Query and Data-Flow Optimization; (4) Information Extraction and Integration; (5) Spatial, Temporal and Streaming Data; (6) Data Mining and

Knowledge Discovery; (7) Data Organization and Physical Issues; (8) Data and Business Processes.

Three keynote lecturers were invited and they gave talks on timely aspects pertaining to the theme of the conference, namely, Maarten de Rijke (University of Amsterdam, Netherlands), Minos Garofalakis (Technical University of Crete in Chania, Greece), and João Gama (University of Porto, Portugal). The volume also includes an invited paper for the conference keynote talk from Minos Garofalakis.

ADBIS 2014 strived to create conditions for more experienced researchers to share their knowledge and expertise with the young researchers participating in the doctoral consortium. In addition, the following three workshops associated with the ADBIS conference were co-located with the main conference:

- Third Workshop on GPUs in Databases (GID), organized by Witold Andrzejewski (Poznan University of Technology), Krzysztof Kaczmarski (Warsaw University of Technology), and Tobias Lauer (Jedox).
- Third Workshop on Ontologies Meet Advanced Information Systems (OAIS) organized by Ladjel Bellatreche (LIAS/ENSMA, Poitiers) and Yamine Aït Ameur (IRIT/ENSEIHT, Toulouse).
- First Workshop on Technologies for Quality Management in Challenging Applications (TQMCA) organized by Isabelle Comyn-Wattiau (CNAM, Paris), Ajantha Dahanayake (Prince Sultan University, Saudi Arabia), and Bernhard Thalheim (Christian Albrechts University).

Each workshop had its own international Program Committee. The accepted papers were published by Springer in the *Advances in Intelligent Systems and Computing series*.

The conference is supported by the President of the Republic of Macedonia, H.E. Dr. Gjorge Ivanov. We would like to express our gratitude to every individual who contributed to the success of ADBIS 2014. Firstly, we thank all the authors who submitted papers to the conference. However, we are also indebted to the members of the community who offered their time and expertise in performing various roles ranging from organizational to reviewing ones – their efforts, energy, and degree of professionalism deserve the highest commendations. Special thanks go to the Program Committee members, as well as to the external reviewers, for their support in evaluating the papers submitted to ADBIS 2014, ensuring the quality of the scientific program. Thanks also to all the colleagues involved in the conference organization, as well as the workshop organizers. A special thank you is due to the members of the Steering Committee and, in particular, its chair, Leonid Kalinichenko, for all their help and guidance. Finally, we thank Springer for publishing the proceedings containing invited and research papers in the LNCS series. The Program Committee work relied on EasyChair, and we thank its development team for creating and maintaining it; it offered great support throughout the different phases of the reviewing process. The conference would not have been possible without our supporters and sponsors: the Ministry of Information Society and Administration, Ss. Cyril and

Methodius University, Faculty of Computer Sciences and Engineering, ICT-ACT Association, and Municipality of Ohrid.

Last, but not least, we thank the participants of ADBIS 2014 for sharing their works and presenting their achievement, thus providing a lively, fruitful, and constructive forum, and giving us the pleasure of knowing that our work was purposeful.

September 2014   Margita Kon-Popovska
                                                         Yannis Manolopulos
                                                           Goce Trajcevski

Organization

General Chair

Margita Kon-Popovska   Ss. Cyril and Methodius University in Skopje,
Republic of Macedonia

Program Committee Co-chairs

Yannis Manolopoulos   Aristotle University of Thessaloniki, Greece
Goce Trajcevski   Northwestern University, USA

Workshop Co-chairs

Themis Palpanas   Paris Descartes University, France
Athena Vakali   Aristotle University of Thessaloniki, Greece

Doctoral Consortium Co-chairs

Nick Bassiliades   Aristotle University of Thessaloniki, Greece
Mirjana Ivanovic   University of Novi Sad, Serbia

Publicity Chair

Goran Velinov   Ss. Cyril and Methodius University in Skopje,
Republic of Macedonia

Website Chair

Vangel Ajanovski   Ss. Cyril and Methodius University in Skopje,
Republic of Macedonia

Proceedings Technical Editor

Ioannis Karydis   Department of Informatics, Ionian University
Corfu, Greece

Local Organizing Committee Chair

Goran Velinov   Ss. Cyril and Methodius University in Skopje,
Republic of Macedonia

Local Organizing Committee

Anastas Mishev   Ss. Cyril and Methodius University in Skopje,
                              Republic of Macedonia
Boro Jakimovski   Ss. Cyril and Methodius University in Skopje,
                              Republic of Macedonia
Ivan Chorbev   Ss. Cyril and Methodius University in Skopje,
                              Republic of Macedonia

Supporters

Ministry of Information Society and Administration
Ss. Cyril and Methodius University in Skopje
Faculty of Computer Sciences and Engineering
ICT-ACT Association
Municipality of Ohrid

Steering Committee

Leonid Kalinichenko, Russian Academy of Science, Russia (Chair)

Paolo Atzeni, Italy   Joris Mihaeli, Israel
Andras Benczur, Hungary   Tadeusz Morzy, Poland
Albertas Caplinskas, Lithuania   Pavol Navrat, Slovakia
Barbara Catania, Italy   Boris Novikov, Russia
Johann Eder, Austria   Mykola Nikitchenko, Ukraine
Theo Haerder, Germany   Jaroslav Pokornyv, Czech Republic
Marite Kirikova, Latvia   Boris Rachev, Bulgaria
Hele-Mai Haav, Estonia   Bernhard Thalheim, Germany
Mirjana Ivanovic, Serbia   Gottfried Vossen, Germany
Hannu Jaakkola, Finland   Tatjana Welzer, Slovenia
Mikhail Kogalovsky, Russia   Viacheslav Wolfengagen, Russia
Yannis Manolopoulos, Greece   Robert Wrembel, Poland
Rainer Manthey, Germany   Ester Zumpano, Italy
Manuk Manukyan, Armenia

Program Committee

Marko Bajec   University of Ljubljana, Slovenia
Mirta Baranovic   University of Zagreb, Croatia
Guntis Barzdins   University of Latvia, Latvia
Andreas Behrend   University of Bonn, Germany

Krzysztof Stencel	University of Warsaw, Poland
Leonid Stoimenov	University of Nis, Serbia
Panagiotis Symeonidis	Aristotle University of Thessaloniki, Greece
Amirreza Tahamtan	Vienna University of Technology, Austria
Ernest Teniente	Universitat Politècnica de Catalunya, Spain
Manolis Terrovitis	Institute for the Management of Information Systems, Greece
Bernhard Thalheim	Christian Albrechts University of Kiel, Germany
A. Min Tjoa	Vienna University of Technology, Austria
Ismail Toroslu	Middle East Technical University, Turkey
Juan Trujillo	University of Alicante, Spain
Traian Marius Truta	Northern Kentucky University, USA
Ozgur Ulusoy	Bilkent University, Turkey
Maurice Van Keulen	University of Twente, The Netherlands
Olegas Vasilecas	Vilnius Gediminas Technical University, Lithuania
Panos Vassiliadis	University of Ioannina, Greece
Jari Veijalainen	University of Jyvaskyla, Finland
Goran Velinov	Ss. Cyril and Methodius University in Skopje, Republic of Macedonia
Gottfried Vossen	Universität Münster, Germany
Boris Vrdoljak	University of Zagreb, Croatia
Fan Wang	Microsoft, USA
Gerhard Weikum	Max Planck Institute for Informatics, Germany
Tatjana Welzer	University of Maribor, Slovenia
Marek Wojciechowski	Poznan University of Technology, Poland
Robert Wrembel	Poznan University of Technology, Poland
Vladimir Zadorozhny	University of Pittsburgh, USA
Jaroslav Zendulka	Brno University of Technology, Czech Republic
Andreas Zuefle	Ludwig-Maximilians-Universität München, Germany

Additional Reviewers

Selma Bouarar	LIAS/ISAE-ENSMA, France
Kamel Boukhalfa	LSI/USTHB, Algiers
Ljiljana Brkić	University of Zagreb, Croatia
Jacek Chmielewski	Poznań University of Economics, Poland
Armin Felbermayr	Catholic University of Eichstätt-Ingolstadt, Germany
Flavio Ferrarotti	Software Competence Center Hagenberg (SCCH), Austria
Olga Gkountouna	National Technical University of Athens, Greece

Keynote Presentations

Querying Distributed Data Streams

Prof. Minos Garofalakis

Computer Science at the School of ECE
Technical University of Crete in Chania, Greec
Director of the Software Technology and Network Applications Laboratory
(SoftNet)
minos@softnet.tuc.gr

Effective big data analytics pose several difficult challenges for modern data management architectures. One key such challenge arises from the naturally streaming nature of big data, which mandates efficient algorithms for querying and analyzing massive, continuous data streams (i.e., data that are seen only once and in a fixed order) with limited memory and CPU-time resources. Such streams arise naturally in emerging large-scale event-monitoring applications; for instance, network-operations monitoring in large ISPs, where usage information from numerous sites needs to be continuously collected and analyzed for interesting trends. In addition to memory- and time-efficiency concerns, the inherently distributed nature of such applications also raises important communication-efficiency issues, making it critical to carefully optimize the use of the underlying network infrastructure. In this talk, we introduce the distributed data streaming model, and discuss recent work on tracking complex queries over massive distributed streams as well as new research directions in this space.

Challenges in Learning from Streaming Data

Prof. João Gama

LIAAD-INESC TEC, University of Porto,
Faculty of Economics, University Porto,
jgama@fep.up.pt

Nowadays, there are applications in which the data are modeled best not as persistent tables, but rather as transient data streams. In this article, we discuss the limitations of current machine learning and data mining algorithms. We discuss the fundamental issues in learning in dynamic environments such as continuously maintain learning models that evolve over time, learning and forgetting, concept drift, and change detection. Data streams produce a huge amount of data that introduce new constraints in the design of learning algorithms: limited computational resources in terms of memory, CPU power, and communication bandwidth. We present some illustrative algorithms, designed to take these constrains into account, for decision-tree learning, hierarchical clustering, and frequent pattern mining. We identify the main issues and current challenges that emerge in learning from data streams that open research lines for further developments.

Table of Contents

Information Extraction and Integration

Spatial, Temporal and Streaming Data

Data Mining and Knowledge Discovery

Data Organization and Physical Issues

Data and Business Processes

Querying Distributed Data Streams
(Invited Keynote Talk)

Minos Garofalakis

School of Electronic and Computer Engineering
Technical University of Crete
minos@softnet.tuc.gr

Abstract. Effective Big Data analytics pose several difficult challenges for modern data management architectures. One key such challenge arises from the naturally streaming nature of big data, which mandates efficient algorithms for querying and analyzing massive, continuous data streams (that is, data that is seen only once and in a fixed order) with limited memory and CPU-time resources. Such streams arise naturally in emerging large-scale event monitoring applications; for instance, network-operations monitoring in large ISPs, where usage information from numerous sites needs to be continuously collected and analyzed for interesting trends. In addition to memory- and time-efficiency concerns, the inherently distributed nature of such applications also raises important communication-efficiency issues, making it critical to carefully optimize the use of the underlying network infrastructure. In this talk, we introduce the distributed data streaming model, and discuss recent work on tracking complex queries over massive distributed streams, as well as new research directions in this space.

1 Introduction

Traditional data-management systems are typically built on a *pull-based paradigm*, where users issue one-shot queries to static data sets residing on disk, and the system processes these queries and returns their results. Recent years, however, have witnessed the emergence of a new class of large-scale event monitoring applications, that require the ability to efficiently process continuous, high-volume *streams* of data in real time. Examples include monitoring systems for IP and sensor networks, real-time analysis tools for financial data streams, and event and operations monitoring applications for enterprise clouds and data centers. As both the scale of today's networked systems, and the volumes and rates of the associated data streams continue to increase with no bound in sight, algorithms and tools for effectively analyzing them are becoming an important research mandate.

Large-scale stream processing applications rely on *continuous*, event-driven monitoring, that is, real-time tracking of measurements and events, rather than one-shot answers to sporadic queries. Furthermore, the vast majority of these applications are inherently *distributed*, with several remote monitor sites observing their local, high-speed data streams and exchanging information through a communication network. This distribution of the data naturally implies critical communication constraints that typically prohibit centralizing all the streaming data, due to either the huge volume of the data

Y. Manolopoulos et al. (Eds.): ADBIS 2014, LNCS 8716, pp. 1–10, 2014.

(e.g., in IP-network monitoring, where the massive amounts of collected utilization and traffic information can overwhelm the production IP network [11]), or power and bandwidth restrictions (e.g., in wireless sensornets, where communication is the key determinant of sensor battery life [25]). Finally, an important requirement of large-scale event monitoring is the effective support for tracking complex, *holistic queries* that provide a global view of the data by combining and correlating information across the collection of remote monitor sites. For instance, tracking aggregates over the result of a distributed join (the "workhorse" operator for combining tables in relational databases) can provide unique, real-time insights into the workings of a large-scale distributed system, including system-wide correlations and potential anomalies [6]. Monitoring the precise value of such holistic queries without continuously centralizing all the data seems hopeless; luckily, when tracking statistical behavior and patters in large scale systems, *approximate answers* (with reasonable approximation error guarantees) are often sufficient. This often allows algorithms to effectively tradeoff efficiency with approximation quality (e.g., using sketch-based stream approximations [6]).

Given the prohibitive cost of data centralization, it is clear that realizing sophisticated, large-scale distributed data-stream analysis tools must rely on novel algorithmic paradigms for processing local streams of data *in situ* (i.e., locally at the sites where the data is observed). This, of course, implies the need for intelligently decomposing a (possibly complex) global data-analysis and monitoring query into a collection of "safe" local queries that can be tracked independently at each site (without communication), while guaranteeing correctness for the global monitoring operation. This decomposition process can enable truly distributed, event-driven processing of real-time streaming data, using a *push-based paradigm*, where sites monitor their local queries and communicate only when some local query constraints are violated [6,29]. Nevertheless, effectively decomposing a complex, holistic query over the global collections of streams into such local constraints is far from straightforward, especially in the case of *non-linear* queries (e.g., joins) [29].

The bulk of early work on data-stream processing has focused on developing space-efficient, one-pass algorithms for performing a wide range of *centralized, one-shot computations* on massive data streams; examples include computing quantiles [21], estimating distinct values [18], and set-expression cardinalities [14], counting frequent elements (i.e., "heavy hitters") [4,9,26], approximating large Haar-wavelet coefficients [20], and estimating join sizes and stream norms [1,2,13]. Monitoring *distributed* data streams has attracted substantial research interest in recent years [5,27], with early work focusing on the monitoring of *single values*, and building appropriate models and filters to avoid propagating updates if these are insignificant compared to the value of simple *linear* aggregates (e.g., to the SUM of the distributed values). For instance, [28] proposes a scheme based on "adaptive filters" — that is, bounds around the value of distributed variables, which shrink or grow in response to relative stability or variability, while ensuring that the total uncertainty in the bounds is at most a user-specified bound. Still, in the case of linear aggregate functions, deriving local filter bounds based on a global monitoring condition is rather straightforward, with the key issue being how to intelligently distribute the available aggregate "slack" across all sites [3,7,22].

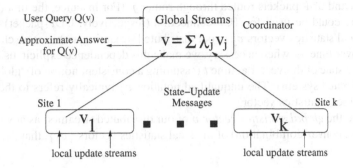

Fig. 1. Distributed stream processing architecture

In this talk, we focus on recently-developed algorithmic tools for effectively tracking a broad class of complex queries over massive, distributed data streams. We start by describing the key elements of a generic distributed stream-processing architecture and define the class of distributed query-tracking problems addressed in this talk, along with some necessary background material on *randomized sketching techniques* for data streams. We then give an overview of the *geometric method* for distributed threshold monitoring that lies at the core of our distributed query-tracking methodology, and discuss recent extensions to the basic geometric framework that incorporate sketches and local prediction models. Finally, we conclude with a brief discussion of new research directions in this space.

2 System Architecture

We consider a distributed-computing environment, comprising a collection of k *remote sites* and a designated *coordinator site*. Streams of data updates arrive continuously at remote sites, while the coordinator site is responsible for generating approximate answers to (possibly, continuous) user queries posed over the *unions* of remotely-observed streams (across all sites). Following earlier work in the area [3,6,7,12,28], our distributed stream-processing model does not allow direct communication between remote sites; instead, as illustrated in Figure 1, a remote site exchanges messages only with the coordinator, providing it with state information on its (locally-observed) streams. Note that such a hierarchical processing model is, in fact, representative of a large class of applications, including network monitoring where a central Network Operations Center (NOC) is responsible for processing network traffic statistics (e.g., link bandwidth utilization, IP source-destination byte counts) collected at switches, routers, and/or Element Management Systems (EMSs) distributed across the network.

Each remote site $j \in \{1, \ldots, k\}$ observes (possibly, several) local update streams that incrementally render a *local statistics vector* v_j capturing the current local state of the observed stream(s) at site j. As an example, in the case of IP routers monitoring the number of TCP connections and UDP packets exchanged between source and destination IP addresses, the local statistics vector v_j has 2×2^{64} entries capturing the up-to-date frequencies for specific (source, destination) pairs observed in TCP connections and UDP packets routed through router j. (For instance, the first (last) 2^{64} entries of v_j could be used for TCP-connection (respectively, UDP-packet) frequencies.) All local statistics vectors v_j in our distributed streaming architecture change dynamically over time — when necessary, we make this dependence explicit, using $v_j(t)$ to denote the state of the vector at time t (assuming a consistent notion of "global time" in our distributed system). The unqualified notation v_j typically refers to the *current* state of the local statistics vector.

We define the *global statistics vector* v of our distributed stream(s) as any *weighted average* (i.e., convex combination) of the local statistics vectors $\{v_j\}$; that is,

$$v = \sum_{j=1}^{k} \lambda_j v_j, \quad \text{where } \sum_j \lambda_j = 1 \text{ and } \lambda_j \geq 0 \text{ for all } j.$$

(Again, to simplify notation, we typically omit the explicit dependence on time when referring to the current global vector.) Our focus is on the problem of effectively answering user queries (or, functions) over the global statistics vector at the coordinator site. Rather than one-time query/function evaluation, we assume a continuous-querying environment which implies that the coordinator needs to *continuously maintain* (or, *track*) the answers to queries as the local update streams v_j evolve at individual remote sites. There are two defining characteristics of our problem setup that raise difficult algorithmic challenges for our query tracking problems:

• *The distributed nature and large volumes of local streaming data* raise important communication and space/time efficiency concerns. Naïve schemes that accurately track query answers by forcing remote sites to ship every remote stream update to the coordinator are clearly impractical, since they can impose an inordinate burden on the underlying communication infrastructure (especially, for high-rate data streams and large numbers of remote sites). Furthermore, the voluminous nature of the local data streams implies that effective streaming tools are needed at the remote sites in order to manage the streaming local statistics vectors in sublinear space/time. Thus, a practical approach is to adopt the paradigm of continuous tracking of *approximate* query answers at the coordinator site with strong guarantees on the quality of the approximation. This allows schemes that can effectively trade-off space/time/communication efficiency and query-approximation accuracy in a precise, quantitative manner.

• *General, non-linear queries/functions* imply fundamental and difficult challenges for distributed monitoring. For the case of linear functions, a number of approaches have been proposed that rely on the key idea of allocating appropriate *"slacks"* to the remote sites based on their locally-observed function values (e.g., [3,28,22]). Unfortunately, it is not difficult to find examples of simple *non-linear* functions on one-dimensional data, where it is basically impossible to make any assumptions about the value of the global function based on the values observed locally at the sites [29]. This renders conventional slack-allocation schemes inapplicable in this more general setting.

3 Sketching Continuous Data Streams

Techniques based on small-space pseudo-random *sketch* summaries of the data have proved to be very effective tools for dealing with massive, rapid-rate data streams in centralized settings [1,2,10,13,20]. The key idea in such sketching techniques is to represent a streaming frequency vector v using a much smaller (typically, randomized) *sketch* vector (denoted by $\mathtt{sk}(v)$) that (1) can be easily maintained as the updates incrementally rendering v are streaming by, and (2) provide probabilistic guarantees for the quality of the data approximation. The widely used AMS sketch (proposed by Alon, Matias, and Szegedy in their seminal paper [2]) defines i^{th} sketch entry $\mathtt{sk}(v)[i]$ as the random variable $\sum_k v[k] \cdot \xi_i[k]$, where $\{\xi_i\}$ is a family of four-wise independent binary random variables uniformly distributed in $\{-1, +1\}$ (with mutually-independent families used across different entries of the sketch). The key here is that, using appropriate pseudo-random hash functions, each such family can be efficiently constructed on-line in small (logarithmic) space [2]. Note that, by construction, each entry of $\mathtt{sk}(v)$ is essentially a *randomized linear projection* (i.e., an inner product) of the v vector (using the corresponding ξ family), that can be easily maintained (using a simple counter) over the input update stream. Another important property is the *linearity* of AMS sketches: Given two "parallel" sketches (built using the same ξ families) $\mathtt{sk}(v_1)$ and $\mathtt{sk}(v_2)$, the sketch of the union of the two underlying streams (i.e., the streaming vector $v_1 + v_2$) is simply the component-wise sum of their sketches; that is, $\mathtt{sk}(v_1 + v_2) = \mathtt{sk}(v_1) + \mathtt{sk}(v_2)$. This linearity makes such sketches particularly useful in *distributed* streaming settings [6].

The following theorem summarizes some of the basic estimation properties of AMS sketches for (centralized) stream query processing. (Throughout, the notation $x \in (y \pm z)$ is equivalent to $|x - y| \leq |z|$.) We use $f_{\mathrm{AMS}}()$ to denote the standard *AMS estimator function*, involving both averaging and median-selection operations over the components of the sketch-vector inner product [1,2]. Formally, each sketch vector can be conceptually viewed as a two-dimensional $n \times m$ array, where $n = O(\frac{1}{\epsilon^2})$, $m = O(\log(1/\delta))$ and ϵ, $1 - \delta$ denote the desired bounds on error and probabilistic confidence (respectively), and the AMS estimator function is defined as:

$$f_{\mathrm{AMS}}(\mathtt{sk}(v), \mathtt{sk}(u)) = \underset{i=1,\ldots,m}{\mathrm{median}}\{\frac{1}{n}\sum_{l=1}^{n}\mathtt{sk}(v)[l, i] \cdot \mathtt{sk}(u)[l, i]\}. \tag{1}$$

Theorem 1 ([1,2]). *Let $\mathtt{sk}(v)$ and $\mathtt{sk}(u)$ denote two parallel sketches comprising $O(\frac{1}{\epsilon^2} \log(1/\delta))$ counters, built over the streams v and u. Then, with probability at least $1 - \delta$, $f_{\mathrm{AMS}}(\mathtt{sk}(v), \mathtt{sk}(u)) \in (v \cdot u \pm \epsilon\|v\|\|u\|)$. The processing time required to maintain each sketch is $O(\frac{1}{\epsilon^2} \log(1/\delta))$ per update.*

Thus, AMS sketch estimators can effectively approximate *inner-product queries* $v \cdot u = \sum_i v[i] \cdot u[i]$ over streaming data vectors and tensors. Such inner products naturally map to *join and multi-join aggregates* when the the vectors/tensors capture the frequency distribution of the underlying join attribute(s) [13]. Furthermore, they can capture several other interesting query classes, including range and quantile queries [19], heavy hitters and top-k queries [4], and approximate histogram and wavelet representations [8,20]. An interesting special case is that of the (squared) L_2 *norm* (or, *self-join*) query (i.e., $u = v$): Theorem 1 implies that the AMS estimator $f_{\mathrm{AMS}}(\mathrm{sk}(v), \mathrm{sk}(v))$ (or, simply $f_{\mathrm{AMS}}(\mathrm{sk}(v))$) is within ϵ relative error of the true squared L_2 norm $\|v\|^2 = \sum_k (v[k])^2$; that is, $f_{\mathrm{AMS}}(\mathrm{sk}(v)) \in (1 \pm \epsilon)\|v\|^2$. To provide ϵ relative-error guarantees for the general inner-product query $v \cdot u$, Theorem 1 can be applied with error bound $\epsilon' = \epsilon(v \cdot u)/(\|v\|\|u\|)$, giving a total sketching space requirement of $O(\frac{\|v\|^2\|u\|^2}{\epsilon^2(v \cdot u)^2} \log(1/\delta))$ counters [1].

4   The Geometric Method

Sharfman et al. [29] consider the fundamental problem of *distributed threshold monitoring*; that is, determine whether $f(v) < \tau$ or $f(v) > \tau$, for a given (general) function $f()$ over the global statistics vector and a fixed threshold τ. Their key idea is that, since it is generally impossible to connect the locally-observed values of $f()$ to the global value $f(v)$, one can employ geometric arguments to monitor the *domain* (rather than the range) of the monitored function $f()$. More specifically, assume that at any point in time, each site j has informed the coordinator of some prior state of its local vector v_j^p; thus, the coordinator has an estimated global vector $e = \sum_{j=1}^k \lambda_j v_j^p$. Clearly, the updates arriving at sites can cause the local vectors v_j to drift too far from their previously reported values v_j^p, possibly leading to a violation of the τ threshold. Let $\Delta v_j = v_j - v_j^p$ denote the local *delta vector* (due to updates) at site j, and let $u_j = e + \Delta v_j$ be the *drift vector* from the previously reported estimate at site j. We can then express the current global statistics vector v in terms of the drift vectors:

$$v = \sum_{j=1}^k \lambda_j (v_j^p + \Delta v_j) = e + \sum_{j=1}^k \lambda_j \Delta v_j = \sum_{j=1}^k \lambda_j (e + \Delta v_j).$$

That is, the current global vector is a convex combination of drift vectors and, thus, guaranteed to lie somewhere within the convex hull of the delta vectors around e. Figure 2 depicts an example in $d = 2$ dimensions. The current value of the global statistics vector lies somewhere within the shaded convex-hull region; thus, as long as the convex hull does not overlap the inadmissible region (i.e., the region $\{v \in \mathbb{R}^2 : f(v) > \tau\}$ in Figure 2), we can guarantee that the threshold has not been violated (i.e., $f(v) \leq \tau$).

The problem, of course, is that the Δv_j's are spread across the sites and, thus, the above condition cannot be checked locally. To transform the global condition into a local constraint, we place a d-dimensional *bounding ball* $B(c, r)$ around each local delta vector, of radius $r = \frac{1}{2}\|\Delta v_j\|$ and centered at $c = e + \frac{1}{2}\Delta v_j$ (see Figure 2). It can be shown that the union of all these balls completely covers the convex hull of the drift vectors [29]. This observation effectively reduces the problem of monitoring the global statistics vector to the local problem of each remote site monitoring the ball around its local delta vector.

More specifically, given the monitored function $f()$ and threshold τ, we can partition the d-dimensional space into two sets $V = \{v : f(v) > \tau\}$ and $\overline{V} = \{v : f(v) \leq \tau\}$. (Note that these sets can be arbitrarily complex, e.g., they may comprise multiple disjoint regions of $\mathbb{R}^d$.) The basic protocol is now quite simple: Each site monitors its

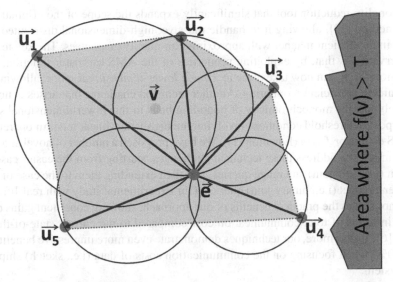

Fig. 2. Estimate vector e, delta vectors $\Delta v(p_i)$ (arrows out of e), convex hull enclosing the current global vector v (dotted outline), and bounding balls $B(e + \frac{1}{2}\Delta v_j, \frac{1}{2}\|\Delta v_j\|)$

delta vector Δv_j and, with each update, checks whether its bounding ball $B(e + \frac{1}{2}\Delta v_j, \frac{1}{2}\|\Delta v_j\|)$ is *monochromatic*, i.e., all points in the ball lie within the same region (V or $\overline{V}$). If this is not the case, we have a *local threshold violation*, and the site communicates its local Δv_j to the coordinator. The coordinator then initiates a *synchronization process* that typically tries to resolve the local violation by communicating with only a subset of the sites in order to "balance out" the violating Δv_j and ensure the monochromicity of all local bounding balls [29]. In the worst case, the delta vectors from all k sites are collected, leading to an accurate estimate of the current global statistics vector, which is by definition monochromatic (since all bounding balls have 0 radius).

In more recent work, Sharfman et al. [23] demonstrate that their geometric monitoring method can employ properties of the function and the data to guide the choice of a global *reference point* and local *bounding ellipsoids* for defining the local constraints. Furthermore, they show that the local bounding balls/ellipsoids defined by the geometric method are actually special cases of a more general theory of *Safe Zones (SZs)*, which can be broadly defined as *convex subsets of the admissible region* of a threshold query. It is not difficult to see that, as long as the local drift vectors stay within such a SZ, the global vector is guaranteed (by convexity) to be within the admissible region of the query [23].

5 Extensions: Sketches and Prediction Models

In more recent work [15], we have proposed novel query-tracking protocols that exploit the combination of the geometric method of Sharfman et al. [23,29] for monitoring general threshold conditions over distributed streams and AMS sketch estimators for querying massive streaming data [1,2,13]. The sketching idea offers an effective streaming dimensionality-reduction tool that significantly expands the scope of the original geometric method [29], allowing it to handle massive, high-dimensional distributed data streams in an efficient manner with approximation-quality guarantees. The key technical observation is that, by exploiting properties of the AMS estimator function, geometric monitoring can now take place in a *much lower-dimensional space*, allowing for communication-efficient monitoring. Another technical challenge that arises is how to effectively test the monochromicity of bounding balls in this lower-dimensional space with respect to threshold conditions involving the highly non-linear median operator in the AMS estimator $f_{AMS}()$ (Equation (1)). We have proposed a number of novel algorithmic techniques to address these technical challenges, starting from the easier cases of L_2-norm (i.e., self-join) and range queries, and then extending them to the case of general inner-product (i.e., binary-join) queries. Our experimental study with real-life data sets demonstrates the practical benefits of our approach, showing consistent gains of up to 35% in terms of total communication cost compared to the current state-of-the-art method [6]; furthermore, our techniques demonstrate even more impressive benefits (of over 100%) when focusing on the communication costs of data (i.e., sketch) shipping in the system.

In other recent work [16,17], we have proposed a novel combination of the geometric method with *local prediction models* for describing the temporal evolution of local data streams. (The adoption of prediction models has already been proven beneficial in terms of bandwidth preservation in distributed settings [6].) We demonstrate that prediction models can be incorporated in a very natural way in the geometric method for tracking general, non-linear functions; furthermore, we show that the initial geometric monitoring method of Sharfman et al. [23,29] is only a special case of our, more general, prediction-based geometric monitoring framework. Interestingly, the mere utilization of local predictions is not enough to guarantee lower communication overheads even when predictors are quite capable of describing local stream distributions. We establish a theoretically solid monitoring framework that incorporates conditions that can lead to fewer contacts with the coordinator. We also develop a number of mechanisms, along with extensive probabilistic models and analysis, that relax the previously introduced framework, base their function on simpler criteria, and yield significant communication benefits in practical scenarios.

6   Future Directions

We have discussed some basic, recently-proposed algorithmic tools for the difficult problem of tracking complex queries over distributed data streams. Continuous distributed streaming is a vibrant, rapidly evolving field of research, and a community of researchers has started forming around theoretical, algorithmic, and systems issues in the area [27] Naturally, there are several promising directions for future research. First, the single-level hierarchy model (depicted in Figure 1) is simplistic and also introduces a single point of failure (i.e., the coordinator). Extending the model to general hierarchies is probably not that difficult (even though effectively distributing the error bounds across the internal hierarchy nodes can be challenging [6]); however, extending the ideas to general, scalable distributed architectures (e.g., P2P networks) raises several theoretical and practical challenges. Second, while most of the proposed algorithmic tools have been prototyped and tested with real-life data streams, there is still a need for real system implementations that also address some of the key systems questions that arise (e.g., what functions and query language to support, how to interface to real users and applications, and so on). We have already started implementing some of the geometric monitoring ideas using Twitter's Storm/λ-architecture, and exploiting these ideas for large-scale, distributed Complex Event Processing (CEP) in the context of the FERARI project (www.ferari-project.eu). Finally, from a more foundational perspective, there is a need for developing new models and theories for studying the complexity of such *continuous distributed computations*. These could build on the models of *communication complexity* [24] that study the complexity of distributed *one-shot* computations, perhaps combined with very relevant ideas from information theory (such as distributed source coding).

Acknowledgements. This work was partially supported by the European Commission under ICT-FP7-FERARI (Flexible Event Processing for Big Data Architectures), www.ferari-project.eu.

References

1. Alon, N., Gibbons, P.B., Matias, Y., Szegedy, M.: Tracking Join and Self-Join Sizes in Limited Storage. In: Proc. of the 18th ACM Symposium on Principles of Database Systems, Philadeplphia, Pennsylvania (May 1999)
2. Alon, N., Matias, Y., Szegedy, M.: The Space Complexity of Approximating the Frequency Moments. In: Proc. of the 28th Annual ACM Symposium on the Theory of Computing, Philadelphia, Pennsylvania, pp. 20–29 (May 1996)
3. Babcock, B., Olston, C.: Distributed Top-K Monitoring. In: Proc. of the 2003 ACM SIGMOD Intl. Conference on Management of Data, San Diego, California (June 2003)
4. Charikar, M., Chen, K., Farach-Colton, M.: Finding Frequent Items in Data Streams. In: Widmayer, P., Triguero, F., Morales, R., Hennessy, M., Eidenbenz, S., Conejo, R. (eds.) ICALP 2002. LNCS, vol. 2380, pp. 693–703. Springer, Heidelberg (2002)
5. Cormode, G., Garofalakis, M.: Streaming in a connected world: querying and tracking distributed data streams. In: SIGMOD (2007)
6. Cormode, G., Garofalakis, M.: Approximate Continuous Querying of Distributed Streams. ACM Transactions on Database Systems 33(2) (June 2008)
7. Cormode, G., Garofalakis, M., Muthukrishnan, S., Rastogi, R.: Holistic Aggregates in a Networked World: Distributed Tracking of Approximate Quantiles. In: Proc. of the 2005 ACM SIGMOD Intl. Conference on Management of Data, Baltimore, Maryland (June 2005)
8. Cormode, G., Garofalakis, M., Sacharidis, D.: Fast Approximate Wavelet Tracking on Streams. In: Ioannidis, Y., et al. (eds.) EDBT 2006. LNCS, vol. 3896, pp. 4–22. Springer, Heidelberg (2006)
9. Cormode, G., Muthukrishnan, S.: What's Hot and What's Not: Tracking Most Frequent Items Dynamically. In: Proc. of the 22nd ACM Symposium on Principles of Database Systems, San Diego, California, pp. 296–306 (June 2003)
10. Cormode, G., Muthukrishnan, S.: An improved data stream summary: The count-min sketch and its applications. In: Latin American Informatics, pp. 29–38 (2004)
11. Cranor, C., Johnson, T., Spatscheck, O., Shkapenyuk, V.: Gigascope: A Stream Database for Network Applications. In: Proc. of the 2003 ACM SIGMOD Intl. Conference on Management of Data, San Diego, California (June 2003)

12. Das, A., Ganguly, S., Garofalakis, M., Rastogi, R.: Distributed Set-Expression Cardinality Estimation. In: Proc. of the 30th Intl. Conference on Very Large Data Bases, Toronto, Canada (September 2004)
13. Dobra, A., Garofalakis, M., Gehrke, J., Rastogi, R.: Processing Complex Aggregate Queries over Data Streams. In: Proc. of the 2002 ACM SIGMOD Intl. Conference on Management of Data, Madison, Wisconsin, pp. 61–72 (June 2002)
14. Ganguly, S., Garofalakis, M., Rastogi, R.: Processing Set Expressions over Continuous Update Streams. In: Proc. of the 2003 ACM SIGMOD Intl. Conference on Management of Data, San Diego, California (June 2003)
15. Garofalakis, M., Keren, D., Samoladas, V.: Sketch-based Geometric Monitoring of Distributed Stream Queries. In: Proc. of the 39th Intl. Conference on Very Large Data Bases, Trento, Italy (August 2013)
16. Giatrakos, N., Deligiannakis, A., Garofalakis, M., Sharfman, I., Schuster, A.: Prediction-based Geometric Monitoring over Distributed Data Streams. In: Proc. of the 2012 ACM SIGMOD Intl. Conference on Management of Data (June 2012)
17. Giatrakos, N., Deligiannakis, A., Garofalakis, M., Sharfman, I., Schuster, A.: Distributed Geometric Query Monitoring using Prediction Models. ACM Transactions on Database Systems 39(2) (2014)
18. Gibbons, P.B.: Distinct Sampling for Highly-Accurate Answers to Distinct Values Queries and Event Reports. In: Proc. of the 27th Intl. Conference on Very Large Data Bases, Roma, Italy (September 2001)
19. Gilbert, A.C., Kotidis, Y., Muthukrishnan, S., Strauss, M.J.: How to Summarize the Universe: Dynamic Maintenance of Quantiles. In: Proc. of the 28th Intl. Conference on Very Large Data Bases, Hong Kong, China, pp. 454–465 (August 2002)
20. Gilbert, A.C., Kotidis, Y., Muthukrishnan, S., Strauss, M.J.: One-pass wavelet decomposition of data streams. IEEE Transactions on Knowledge and Data Engineering 15(3), 541–554 (2003)
21. Greenwald, M.B., Khanna, S.: Space-Efficient Online Computation of Quantile Summaries. In: Proc. of the 2001 ACM SIGMOD Intl. Conference on Management of Data, Santa Barbara, California (May 2001)
22. Keralapura, R., Cormode, G., Ramamirtham, J.: Communication-efficient distributed monitoring of thresholded counts. In: Proc. of the 2006 ACM SIGMOD Intl. Conference on Management of Data, Chicago, Illinois, pp. 289–300 (June 2006)
23. Keren, D., Sharfman, I., Schuster, A., Livne, A.: Shape-Sensitive Geometric Monitoring. IEEE Transactions on Knowledge and Data Engineering 24(8) (August 2012)
24. Kushilevitz, E., Nisan, N.: Communication Complexity. Cambridge University Press, Cambridge (1997)
25. Madden, S.R., Franklin, M.J., Hellerstein, J.M., Hong, W.: The Design of an Acquisitional Query Processor for Sensor Networks. In: Proc. of the 2003 ACM SIGMOD Intl. Conference on Management of Data, San Diego, California (June 2003)
26. Manku, G.S., Motwani, R.: Approximate Frequency Counts over Data Streams. In: Proc. of the 28th Intl. Conference on Very Large Data Bases, Hong Kong, China, pp. 346–357 (August 2002)
27. NII Shonan Workshop on Large-Scale Distributed Computation, Shonan Village, Japan (January 2012), http://www.nii.ac.jp/shonan/seminar011/.
28. Olston, C., Jiang, J., Widom, J.: Adaptive Filters for Continuous Queries over Distributed Data Streams. In: Proc. of the 2003 ACM SIGMOD Intl. Conference on Management of Data, San Diego, California (June 2003)
29. Sharfman, I., Schuster, A., Keren, D.: A geometric approach to monitoring threshold functions over distributed data streams. In: Proc. of the 2006 ACM SIGMOD Intl. Conference on Management of Data, Chicago, Illinois, pp. 301–312 (June 2006)

Towards a Normal Form for Extended Relations Defined by Regular Expressions

András Benczúr and Gyula I. Szabó

Eötvös Loránd University, Faculty of Informatics,
Pázmány Péter sétány, 1/C, Budapest, 1118 Hungary
abenczur@inf.elte.hu, gyula@szaboo.de

Abstract. XML elements are described by XML schema languages such as a DTD or an XML Schema definition. The instances of these elements are semi-structured tuples. We may think of a semi-structure tuple as a sentence of a formal language, where the values are the terminal symbols and the attribute names are the nonterminal symbols. In our former work [13] we introduced the notion of the extended tuple as a sentence from a regular language generated by a grammar where the nonterminal symbols of the grammar are the attribute names of the tuple. Sets of extended tuples are the extended relations. We then introduced the dual language, which generates the tuple types allowed to occur in extended relations. We defined functional dependencies (regular FD - RFD) over extended relations. In this paper we rephrase the RFD concept by directly using regular expressions over attribute names to define extended tuples. By the help of a special vertex labeled graph associated to regular expressions the specification of substring selection for the projection operation can be defined. The normalization for regular schemas is more complex than it is in the relational model, because the schema of an extended relation can contain an infinite number of tuple types. However, we can define selection, projection and join operations on extended relations too, so a lossless-join decomposition can be performed.

1 Introduction

XML has evolved to become the de-facto standard format for data exchange over the World Wide Web. XML was originally developed to describe and present individual documents, it has also been used to build databases. Our original motivation for the introduction of the regular relational data model [13] was to find a good representation of the XML ELEMENT type declaration. The instances of a given element type in an XML document can be considered as a collection of data of complex row types. The set of attribute names in the row types are the element names occurring in the DTD declaration of the element. In the case of recursive regular expression in the element declaration, there are possibly infinite number of different row types for the element instances. The same attribute name may occur several times in a type instance. This leads to the problem of finding a formal way to define the projection operator, similar to the relational

Y. Manolopoulos et al. (Eds.): ADBIS 2014, LNCS 8716, pp. 11–24, 2014.

algebra, on the syntactical structure of the data type. That is necessary to define the left and right side of a functional dependency. We defined the attribute sequence by a traversal on the vertex labeled graph associated to the regular expression of the DTD. This form is also good to define attribute subsequences for the projection operator, for the selection operator and for equijoin operator. Set operations can be extended in a straightforward way, so this leads to the full extension of relational algebra operators. Using the extension of projection and equijoin (or natural join) the join dependency can be defined in the same way as in the relational model.

Motivation. Our previous model [13] could be effectively used for handling functional dependencies (FD). In the relational model FDs offer the basis for normalization (e.g. BCNF), to build non-redundant, well-defined database schema. But our model cannot handle the join operation among instances (that is used to secure lossless join decomposition) because the projection of a schema according to a set of nodes or two joined schemas would not necessarily leads to a new, valid schema. We need an improved model for regular data bases. To denote a regular language we can use regular expressions, our actual model bases upon a graph representation for regular expressions. This model is more redundant than our last one, but it is capable for handling database schema normalization.

Contributions. The main contribution of this paper is the concept of extended relations over the graph representation for regular expressions. We rephrase regular functional dependencies and also define regular join dependencies that constrain extended relations. We determine the schema of an extended relation as $(IN, \ldots, OUT)$ traversals on the graph representation for a given regular expression. We apply the classical Chase algorithm to a counterexample built on this graph. In this way, we show that the logical implication is decidable for this class of functional dependencies.

2   Related Work

As far as we know, each XML functional dependency (XFD) concept involves regular expressions or regular languages. Arenas and Libkin [2] prove different complexities for logical implication concerning their tree tuples XFD model according to the involved regular expressions. They prove quadratic time complexity in case of simple regular expressions. Our new model represents all possible instances of the regular expression at the same time and so it differs from theirs.

The notion data words has been introduced by Bouyer et al. in [4], based upon finite automata of Kaminski et al. [8]. Data words are pairs of a letter from a finite alphabet and a data from an infinite domain. Our concept differs substantially from data words: we assign data values (selected from infinite domains) to letters (from a finite alphabet) after generating a sentence by a regular expression. For data words letters and data values are processed together. Libkin and Vrgo[9] define regular expression for data words. They analyze the complexity of the main decision problems (nonemptiness, membership) for these regular expressions. Their model is similar to ours but our point of view is differs from theirs: we view finite subsets of the set of data words and specify dependencies over them.

3   Extended Relations

Let us start with the definition of extended relation given by a regular language.

Definition 1 (Extended Relation for Regular Types). *Let L be a regular language over the set of attribute names U. Let $w = w_1 \ldots w_n \in L$ a sentence, then we say that w is a regular tuple type over U. Let $dom_u; u \in U$ be sets of data values, then $\{(w_1 : a_1, \ldots, w_n : a_n) \,|\, a_i \in dom_{w_i}\}$ is the set of possible tuples of type w. A finite subset of these tuples is an instance of the regular relation. We say that the set of these tuple types for all $w \in L$ compose the schema of a regular relation based on L.*

We have introduced the notion of the extended relation [13] for a regular language associated with its dual language. The sentences of the dual language are either the concatenated nonterminals used by generating a regular sentence or the states of the accepting automaton, visited during the acception process. Equivalently, the dual language can be given by a vertex labeled graph with a unique IN and OUT node as start and end nodes. The vertex labels along each traversal on this graph (from IN to OUT) represent a schema for the extended relation (we get the sentences of the regular language by valuation). As said in Sect. 1 the dual language model cannot handle the join operation among instances because two joined schemas would not necessarily realize a new, valid schema.

We need an improved model, based upon a suitable graph representation for regular expressions. In our new model we use regular expressions over attribute names to directly define regular relational schemes (e.g. DTD element descriptions), and create the corresponding tuples by valuation (picking data values from suitable domains) similarly to relational databases. In the next Section we present a finite graph representation for the sentences denoted by a regular expression. This graph representation should support node-selection for the projection operation.

4   Graph Representation for Regular Expressions

Definition 2 (Regular Expression Syntax). *Let Σ be a finite set of symbols (alphabet), then a regular expression RE over Σ (denoted by RE_Σ, or simply RE, if Σ is understood from the context) is recursively defined as follows:*
$RE ::= 0\,|\,1\,|\,\alpha\,|\,RE + RE\,|\,RE \circ RE\,|\,RE^*\,|\,RE^?$
where $\alpha \in \Sigma$

For a given regular expression RE we denote the set of alphabet symbols appearing in RE by $[RE]$.

There are efficient constructions of finite state automaton from a regular expression [16,7,5]. The classical algorithm of Berry and Sethi [3] constructs efficiently a DFA from a regular expression if all symbols are distinct.

Berry-Sethi's algorithm constructs a deterministic automaton with at most a quadratic number of transitions [11] and in quadratic computing time (inclusive of marking and unmarking symbols) [6] with respect to the size of the input regular expression (the number of its symbols).

Example 1. Let $G\left(\{S, A, B\}, \{a, b\}, S, P\right)$ be a regular grammar, where $P = \{S \Rightarrow a\,S, S \Rightarrow b\,S, S \Rightarrow a\,A, A \Rightarrow b B, B \Rightarrow a\}$.

The regular expression $RE = (a + b)^* a\, b\, a$ generates the regular language L(G) too. Fig. 1 shows the graph of the non-deterministic FSA constructed by the Berry-Sethi algorithm (BSA). The nodes represent the states of the automaton: they are distinct. Each node complies with a symbol in the regular expression (small letters), they are not distinct after unmarking. We assign the ingoing edge symbol to each node (capital letter) as vertex label. The language, generated by the vertex labels of the visited nodes, is equivalent with the dual language iff the symbols in the regular expression are distinct.

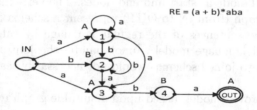

Fig. 1. Graph of the automaton for Example 1 constructed by BSA

As shown in Example 1, for a given regular expression RE we can construct a vertex labeled connected digraph G(RE), with a unique source (IN) and a unique sink (OUT) which represents RE so that that regular language denoted by RE consists of the $(IN, \ldots, OUT)$ traversals on G(RE). This graph is not too large (the number of its vertices equals to the number of symbols in RE), but it is not optimal for our aims because the different $(IN, \ldots, OUT)$ traversals have mostly common subpaths. We need another construction for the graph representation of regular expressions with disjoint $(IN, \ldots, OUT)$ traversals (regardless of IN and OUT). We will construct a graph from vertices picked from a suitably large symbol set Γ. We assume that $\{IN, OUT\} \subseteq \Gamma$ and by picking a node $v \in \Gamma$ we remove it from Γ. The vertices IN and OUT get the labels IN and OUT, respectively. We denote the empty traversal (IN, OUT) by $\bowtie$.

Algorithm 1. *Construction of the Graph-Representation for a regular expression.*

Input: regular expression RE (built from the alphabet Σ),
Output: vertex labeled digraph $G(RE)=(V,E)$ representing RE.

1. *if $RE = 0$ or $RE = 1$, then $V = \{IN, OUT\}$ and $E = \{(IN, OUT)\}$.*

2. *if $RE = A, A \in \Sigma$, then we pick a node $v \in \Gamma$, set $V = \{IN, OUT, v\}$, and $E = \{(IN, v), (v, OUT)\}$. We label the node v with A.*

3. *if RE_1 and RE_2 are regular expressions, then $G(RE_1 + RE_2)$ will be formed by uniting the IN and OUT nodes of $G(RE_1)$ and $G(RE_2)$, respectively.*

4. *if RE_1 and RE_2 are regular expressions, then in order to build the graph $G(RE_1 \circ RE_2)$ we first rename the OUT node of $G(RE_1)$ and the IN node of $G(RE_2)$ to $JOIN$ (Fig. 2), then unite them and connect all "left" paths with all "right" paths while eliminating the $JOIN$ node (Fig. 3).*

5. *if RE is a regular expression, then $G(RE^?) = G(RE) \cup (IN, OUT)$.*

6. *if RE is a regular expression, then in order to build the graph $G(RE^*)$ we first pick a node $v \in \Gamma$, then we create the graph $G^*(RE) = G(RE) \cup \{v\}$ (It means that $V^* = V \cup \{v\}$, the node v gets the special label $STAR$). Let us denote $\{a_1, \ldots, a_n\}$ the nodes with ingoing edge from IN and $\{z_1, \ldots, z_n\}$ the nodes with outgoing edge to OUT, respectively. Let us create the graph $G_{IN}(RE, STAR) = \cup_{i=1}^n (v, a_i)$ and the graph $G_{OUT}(RE, STAR) = \cup_{i=1}^n (z_i, v)$, respectively. Then $G(RE^*) = G^*(RE) \cup G_{IN}(RE, STAR) \cup G_{OUT}(RE, STAR) \cup (IN, STAR) \cup (STAR, OUT)$.*

Theorem 1. *The $(IN, \ldots, OUT)$ traversals on the graph representation $G(RE)$ for the regular expression RE constructed by Alg. 1 generate exactly the regular language $L(RE)$.*

Proof. Algorithm 1 constructs the representation graph so that each elementary step of building a regular expression (Def. 2) will be covered. With induction by the length of the expression and using the regular expression semantics we yield the result.

Example 2. The graph representation for the regular expression $RE = (a + b)^*$ $a\,b\,a$ constructed by the Alg. 1 may be seen on Fig. 4. This graph represents the same regular language as its counterpart on Fig. 1 but it consists of disjoint traversals.

RE = (A + B + C) · (D + E)

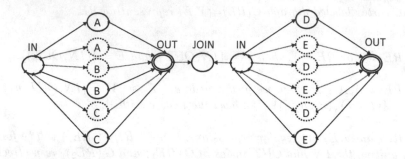

Fig. 2. First joining step of concatenation for two RE graphs

RE = (A + B + C) · (D + E)

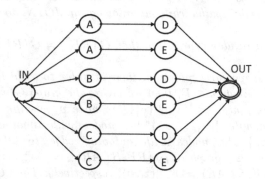

Fig. 3. Eliminating the JOIN node from the concatenation of two RE graphs

Regular expressions present a compact form for specifying regular languages. We look at the sentences of this regular language as types (schemas) of complex value tuples. We can represent these types as IN-OUT traversals on a graph constructed from the symbols in the regular expression. We say that this graph is the schemagraph for the regular expression.

Definition 3 (Schemagraph for Regular Expression). *Let RE be a regular expression built from the alphabet Σ. We say that the graph G is the schemagraph for the regular expression RE (denoted by $G(RE)$) iff*

1. *is a directed, (not necessarily strongly) connected graph,*
2. *has a unique source (IN) and a unique sink (OUT),*
3. *fulfills $OutDegree(IN) = InDegree(OUT)$,*
4. *for any two $P_A = (IN, A_1, \ldots, A_n, OUT)$, $P_B = (IN, B_1, \ldots, B_m, OUT)$ is true that $\{IN, OUT\} \subseteq P_A \cap P_B$, and if $v \in P_A \cap P_B$, then $label(v) \in \{IN, OUT, STAR\}$,*

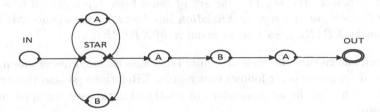

Fig. 4. Vertex labeled RE representation graph for Example 2 constructed by Alg. 1

5. *is vertex-labeled with a single symbol for each node,*
6. *each cycle of the graph involves a vertex with label STAR, this is the start and end node of the cycle,*
7. *each vertex v with label STAR fulfills OutDegree $(v) = $ InDegree (v),*
8. *the set of vertex-labels is the set $[RE] \cup \{IN, OUT, STAR\}$,*
9. *the labels of vertices visited by an $(IN, \ldots, OUT)$ walk on $G(RE)$ set up a string generated by RE (the labels IN, OUT, STAR will be ignored). Each symbol string denoted by RE can be obtained in this way.*

We say that an $(IN, \ldots, OUT)$ walk on G is a traversal on G. We denote the set of traversals on G by $T(G)$. The item (9) of Def. 3 states that for a regular expression RE $L(RE) = T(G(RE))$.

Lemma 1. *If RE is a regular expression, then the graph $G(RE)$ generated by Alg. 1 is a schemagraph for RE.*

Proof. Starting with an empty regular expression a structural recursion by Alg. 1 gives the result. For instance, an empty RE fulfills (3) of Def. 3 and each step of Alg. 1 preserves this attribute of the graph.

Definition 4 (Schema Foundation Graph). *We say that a graph G complying with features 1-7 from Def. 3 is a schema foundation graph. We denote the set of vertex-labels for G by $Lab(G)$.*

Lemma 2. *If G is a schema foundation graph, then there exists a regular expression RE so that $G(RE) = G$ and $L(RE) = T(G)$ and $[RE] = Lab(G)$.*

5   Relational Algebra for Regular XRelation

Definition 5 (Regular XRelation for Regular Expressions). *Let RE be a regular expression and let G be a schemagraph for RE, moreover, let $w = (IN, v_1, \ldots, v_n, OUT) \in T(G)$ be a traversal on G. Let $dom_U; U \in [RE]$ be sets of data values, then $\{(v_1 : a_1, \ldots, v_n : a_n) | a_i \in dom_{v_i}\}$ is a tuple of type w. We say that a finite set of these tuples is a table instance of type w, and w is the type (schema) of the table instance. A regular relational instance, e.g. I, is a finite* *set of table instances. The schema of a relational instance is the set of types of its table instances. We say that the set of these tuple types for all $w \in T(G)$ compose the schema of a regular XRelation based on RE. We denote this regular XRelation by $XR(RE)$, so I is an instance of $XR(RE)$.*

It is well known that the class of regular languages is closed under union, intersection and complement. It follows that regular XRelations possess these closure properties. That is, the set operations of relational databases are applicable for XRelations.

Let RE_1 and RE_2 be regular expressions and let I_1 and I_2 be regular relational instances for the regular XRelations $XR(RE_1)$ and $XR(RE_2)$, respectively.

Union. The union of the schemagraphs $G(RE_1) \cup G(RE_2)$ is a schemagraph too ($= G(RE_1 + RE_2)$). That is, the union $XR(RE_1) \cup XR(RE_2)$ of regular XRelations is again an XRelation and its regular instances have the form $I_1 \cup I_2$.

Intersection. The intersection $XR(RE_1) \cap XR(RE_2)$ of regular XRelations is again an XRelation and its regular instances have the form $I_1 \cap I_2$.

Difference. The set difference of two regular instances I_1 and I_2 for the regular XRelation $XR(RE_1)$ is also a regular instance for it.

5.1 Projection

Definition 6 (Node-selection). *Let RE be a regular expression and let $G(V,E)$ be a schemagraph for RE. We say that a subset $X \subset V$ is a node-selection over G iff $IN \in X$ and $OUT \in X$. If X is a node-selection, then we denote by $\overline{X}$ the complementer node-selection for X, defined by $\overline{X} = V \setminus X \cup \{IN, OUT\}$.*

Remark 1. Definition 6 presents a rigid method for fixing the scope of the projecting window. If the selected nodes belong to a cycle, then the selection chooses all occurrences from a given transversal. A more flexible selection method can be realized on an extended graph. We may add a a given number of walks (as new nodes and edges) for any (or all) cycles and select nodes on the new graph. E.g., if the RE involves the (sub)expression $(ABC)^*$, the original graph contains the nodes a, b, c (labeled with A, B, C, respectively), and the edges $(a, b), (b, c), (c, a)$. The node-selection of $\{a, b\}$ selects the labels $ABAB$ from the traversal which repeats twice the cycle. The extended graph (with two cycles) would give the new nodes and edges

$(a_1, b_1), (b_1, c_1), (c_1, a_2)$
$(a_2, b_2), (b_2, c_2)$

The labels on vertices are $ABCABC$. We can select, for instance, the nodes a_1, b_1, a_2 which brings ABA. No selection on the original graph can produce this result.

Definition 7 (Projection). *Let $G(V, E)$ be a schemagraph and let X be a node-selection over G. Let $E[X] = \{(a_1, a_n) \,|\, a_1, a_n \in X; a_2, \ldots, a_{n-1} \notin X\}$, $(a_1, a_2, \ldots, a_{n-1}, a_n) \in P(G)$, where $P(G)$ is the set of paths for G.*

We say that $G[X] = (V \setminus X \cup \{IN, OUT\}, E[X])$ is the projection graph of G onto X.

Lemma 3. *If G is a schemagraph for the regular expression RE and X a node-selection over G, then $G[X]$ is a schema foundation graph.*

Proof. $G[X]$ is the result of deleting the complement of the subgraph X from the schemagraph G and re-connecting during the deletion disconnected vertices of X. Clearly, the features 1-7 from Def. 3 of the schemagraph will be preserved. For instance, a traversal (an $(IN, \ldots, OUT)$ walk) on G will be either deleted (it contains no vertex from X) or preserved (perhaps reconnected), so the attribute (3) of Def. 3 will be preserved.

Let $RE[X]$ be a regular expression complying with the schema foundation graph $G[X]$, then we say that $RE[X]$ is the projection of RE onto X. ($Lab(X) \subseteq [RE]$, but different vertices in X can have the same label).

Definition 8 (Projection of Schema). *Let G be a schemagraph of the regular expression RE and let X be a node-selection over G.*

Let $w = (v_0, v_1, \ldots, v_n, v_{n+1}); v_0 = IN, v_{n+1} = OUT$ be a traversal on G (w is a type for RE). We denote by $w[X]$ the projection of w to X, defined as follows: $w[X] = (v_0, v_{i_1}, \ldots, v_{i_k}, v_{n+1}); v_r \in X$ for $r \in \{i_1, \ldots, i_k\}$ and $v_r \notin X$ otherwise.

$w[X]$ is either a traversal on G or its re-connected edges belong to $G[X]$, so:

Lemma 4. *If G is a schemagraph for the regular expression RE and X a node-selection over G and w is schema for RE, then $w[X]$ is a traversal on $G[X]$.*

Definition 9 (Projection of Instance). *Let RE be a regular expression and let XR be a regular XRelation based on RE and let I be a table instance for XR with $type(I) = w$. The projection of I to X, denoted by $\pi_X(I)$, is the set of tuples $\{t[X] \,|\, t \in I; type(t[X]) = w[X]\}$ (that is, $t[X]$ is the subsequence of constants from t according to the subsequence $w[X]$ in w).*

Definition 10 (Functional Dependency). *Let G be a schemagraph of the regular expression RE and let X, Y be node-selections over G. The regular relational instance I satisfies the functional dependency (XRFD) $X \to Y$ if for any to tuples $t_1, t_2 \in I$ with $type(t_1) = w_1$ and $type(t_2) = w_2$, whenever $w_1[X] = w_2[X]$ and $t_1[X] = t_2[X]$, then $w_1[Y] = w_2[Y]$ and $t_1[Y] = t_2[Y]$.*

Example 3. Let $\mathbb{R} = (R_1, \ldots, R_n)$ be a relational database schema. The regular expression $RE = (R_1|R_2|\ldots|R_n)$ (if $R_i = (a, b, c, d, e)$ then for the regular expression we use the concatenation $abcde$ of the attributes), then the schemagraph for RE consists of parallel, linear $(IN, \ldots, OUT)$ traversals. Each relational functional dependency over $\mathbb{R}$ can be defined on the schemagraph using Def. 10, with the restriction that both participant node-selections will be located on the same $(IN, \ldots, OUT)$ path.

5.2 Natural Join

Definition 11 (Disjunctive Natural Join). *Let G_1, G_2 be schemagraphs for the regular expressions RE_1, RE_2 and let X_1, X_2 be node-selections over G_1, G_2, respectively, so that $G_1[X_1] = G_2[X_2] = G$. Let $w_1 \in T(G_1)$ and $w_2 \in T(G_2)$ so that $w_1[X_1] = w_2[X_2] = w$. Let $(A, B) \in G$, then $(A, x, B) \in G_1$ and $(A, y, B) \in G_2$ for some paths x and y, respectively, so that AxB and AyB are subsequences of w_1 and w_2, respectively. Let I_1 and I_2 be table instances for the regular XRelations $XR(RE_1)$ and $XR(RE_2)$, respectively, so that $type(I_1) = w_1$ and $type(I_2) = w_2$, then we say that w_1 and w_2 (and also I_1 and I_2) can be joined. We define $I = I_1 \bowtie I_2$ as a (disjunctive joined) regular relational instance, for which if $t \in I$, then there exist $t_1 \in I_1, t_2 \in I_2$ so that $t_1[X_1] = t_2[X_2]$, then $t[u] = t_1[u] \,|\, u \in (IN, \dots, A)$ and $t[u] = t_2[u] \,|\, u \in (B, \dots, OUT)$. Moreover, let $t[A \bowtie B] = \{t[ApB] \,|\, ApB \in P(G_1) \cup P(G_2)\}$, then $t[A \bowtie B] = \{t_1[AxB] \cup t_2[AyB] \,|\, t_1[A] = t_2[A], t_1[B] = t_2[B]\}$.*

Remark 2. If $w_1[X_1] = w_2[X_2] = \bowtie$, then the disjunctive join of the two table instances I_1 and I_2 will be $I = I_1 \bowtie I_2 = I_1 \cup I_2$, moreover, $schema(I) = \{w_1, w_2\}$. The same is true for the special case $[X_1] = [X_2] = \emptyset$ as well.

Remark 3. We have defined the join operator for two table instances joined on two single attributes. We can extend this definition to joining two table instances on any number (or a single one) of attributes. We can also extend this definition to joining any (finite) number of table instances in a natural way.

Example 4. The disjunctive natural join of table instances means in fact union for the background regular expressions. Let $RE_1 = A \circ X \circ Y \circ B$ and $RE_2 = A \circ W \circ Z \circ B$ and let $X_1 = X_2 = \{A, B\}$, then the regular expression complying with the on A, B joined instances will be $A \circ ((X \circ Y) + (W \circ Z)) \circ B$.

Example 5. The XML documents on Fig. 5 conform to the DTD element declarations
Courses1:

```
<!ELEMENT course (Cid,Cname,(Instid,Instn)+)>
```

and
Courses2:

```
<!ELEMENT course (Cid,(Stid,Stn)+)>
```

respectively. The disjunctive join of the two instances results in

JoinedCourses:

```
<!ELEMENT course (Cid,((Cname,(Instid,Instn)+))|((Stid,Stn)+))>
```

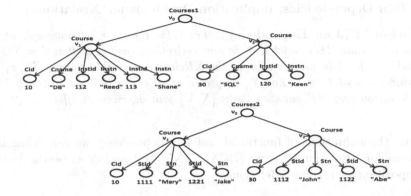

Fig. 5. Example XML documents for natural join

Definition 12 (Concatenative Natural Join). *The concatenative natural join will be defined similarly to its disjunctive counterpart, using concatenation instead of disjunction. That is, we define the regular relational instance* $I = I_1 \bowtie I_2$ *as two (concatenative) joined table instances. If* $t \in I$, *then there exist* $t_1 \in I_1, t_2 \in I_2$ *so that* $t_1[X_1] = t_2[X_2]$ *and* $t[AxyB] = t_1 \bowtie t_2[AxyB] = \{t_1[Ax] \circ t_2[yB] \, | \, t_1[A] = t_2[A], t_1[B] = t_2[B]\}$. *For the special case see Rem. 2.*

Definition 13 (Natural Join of Regular Instances). *The natural (disjunctive or concatenative) join for two regular relational instances will be defined as the set of joined member table instances. That is, if* I_1 *and* I_2 *are regular relational instances, then* $I_1 \bowtie I_2 = \{J | J = J_1 \bowtie J_2; J_1 \in I_1, J_2 \in I_2\}$.

Remark 4. We have defined the join operator for two regular relational instances. We can extend this definition to joining any (finite) number of regular relational instances in a natural way.

Remark 5. If $w_1[X_1] = w_2[X_2] = \bowtie$, then the concatenative join of the two table instances I_1 and I_2 will be $I = I_1 \bowtie I_2 = I_1 \circ I_2$, moreover, *schema* $(I) = (w_1 \circ w_2)$.

Example 6. The concatenative natural join of table instances means in fact concatenation for the background regular expressions. Let $RE_1 = A \circ X \circ Y \circ B$ and $RE_2 = A \circ W \circ Z \circ B$ and let $X_1 = X_2 = \{A, B\}$, then the regular expression complying with the on A, B joined instances will be $A \circ ((X \circ Y) \circ (W \circ Z)) \circ B$.

Example 7. The concatenative join of the two instances realized in the XML documents on Fig. 5 will be

JoinedCourses:

```
<!ELEMENT course (Cid,((Cname,(Instid,Instn)+)),((Stid,Stn)+))>
```

5.3   Join Dependencies, Implication Problems for Xrelations

Definition 14 (Join Dependency). *Let $G(V,E)$ be a schemagraph of the regular expression RE and let X, Y be node-selections over G so that $V = X \cup Y \cup \{IN, OUT\}$. Let I be an instance for the XRelation over RE and $\pi_X(I), \pi_Y(I)$ the projections of I to X and Y, respectively. We say that an instance I for the XRelation over RE satisfies the $\bowtie [X, Y]$ join dependency iff $I = \pi_X(I) \bowtie \pi_Y(I)$.*

Using the Definitions of functional and join dependency we can define normal form for XRelation schemas (BCNF, 4NF etc.) and describe the lossless decomposition for regular XRelations.

Definition 15 (Lossless Decomposition). *Let $G(V,E)$ be a schemagraph of the regular expression RE and let $X_1, \ldots, X_n$ node-selections with $\cup_{i=1}^n X_i \cup \{IN, OUT\} = V$. The set $X_1, \ldots, X_n$ is a lossless decomposition of G if any regular relational instance I for the XRelation over RE satisfies the join dependency $I = \pi_{X_1}(I) \bowtie \pi_{X_2}(I) \bowtie \ldots \bowtie \pi_{X_n}(I)$.*

The logical implication of functional and join dependencies for XRelations is decidable with a special form of the Chase algorithm. We present here an algorithm to decide logical implication of functional dependencies for XRelations.

Definition 16. *Let G be a schemagraph of the regular expression RE. Let Σ be a set of XRFDs and let $X \to Y$ be an XRFD over G, then Σ implies $X \to Y$ (denoted by $\Sigma \models X \to Y$) if for each (finite) regular relational instance I that satisfies Σ $I \models X \to Y$ will also be fulfilled.*

Algorithm 2. *Algorithm for checking implication of XRFDs.*
Input: schemagraph $G = (V, E)$ for an XRelation, a set Σ and $\sigma : X \to Y$ functional dependencies over G
Output: true, if $\Sigma \models \sigma$, false otherwise

1. *Initialization*
Create a counter example from two copies of G (G_1, G_2), the nodes of X colored green on both copies, the nodes of Y colored red on one copy and yellow on the other one.
2. $FDSET := \Sigma$;
3. $greene := X$;
4. *repeat until no more dependency is applicable:*
if $W \to Z \in FDSET$ and $W \subseteq greene$, then
i. $FDSET := FDSET - (W \to Z)$;
ii. $greene := greene \cup Z$;
iii. for all $v \in Z$ set $color(v) := green$ (on both copies)

5. *if the number of yellow nodes and red nodes are both zero, then output is true otherwise output is false.*

Proposition 1 (Functional Dependency Implication). *Let G be a schema-graph of the regular expression RE. Let Σ be a set of XRFDs and let $X \to Y$ be an XRFD over G, then $\Sigma \models X \to Y$ if and only if the Alg. 2 with input G, Σ and $X \to Y$ returns true.*

6 Conclusion and Future Work

This paper presents regular expressions as compact database schemas and defines functional and join dependencies over them, based on the graph representation for the regular expressions. We defined extended relations on the graph representation for regular expressions and determined semantics for the dependencies on instances of extended relations. The logical implication of this kind of functional dependencies is decidable in quadratic time.

Our model offers the tools for a normal form of XRelation. We think that the logical implication for the join dependency, defined here, is decidable similarly to Alg. 2.

We would like to find the connection between our model and data words, that is, to define a register automaton that accepts those data words that satisfy a given functional dependency specified for the corresponding XRelation.

References

1. Abiteboul, S., Hull, R., Vianu, V.: Foundations of Databases. Addison-Wesley (1995)
2. Arenas, M., Libkin, L.: A normal form for XML documents. ACM Transactions on Database Systems 29(1), 195–232 (2004)
3. Berry, G., Sethi, R.: From regular expressions to deterministic automata. Theoretical Computer Science 48(3), 117–126 (1986)
4. Bouyer, P., Petit, A., Thrien, D.: An algebraic approach to data languages and timed languages. Information and Computation 182(2), 137–162 (2003)
5. Brzozowski, J.A.: Derivatives of regular expressions. Journal of the ACM 11(4), 481–494 (1964)
6. Champarnaud, J.-M., Ziadi, D.: Canonical derivatives, partial derivatives and finite automaton constructions. Theoretical Computer Science 289(1), 137–163 (2002)
7. Glushkov, V.M.: The abstract theory of automata. Russian Mathematical Surveys 16, 1–53 (1961)
8. Kaminski, M., Francez, N.: Finite-memory automata. Theoretical Computer Science 134(2), 329–363 (1994)
9. Libkin, L., Vrgoč, D.: Regular expressions for data words. In: Bjørner, N., Voronkov, A. (eds.) LPAR-18. LNCS, vol. 7180, pp. 274–288. Springer, Heidelberg (2012)
10. Murata, M., Lee, D., Mani, M., Kawaguchi, K.: Taxonomy of XML schema languages using formal language theory. ACM Transactions on Internet Technology 5(4), 660–704 (2005)
11. Nicaud, C., Pivoteau, C., Razet, B.: Average Analysis of Glushkov Automata under a BST-Like Model. In: Proc. FSTTCS, pp. 388–399 (2010)
12. Sperberg-McQueen, C.M., Thompson, H.: XML Schema. Technical report, World Wide Web Consortium (2005), http://www.w3.org/XML/Schema

13. Szabó, G. I., Benczúr, A.: Functional Dependencies on Extended Relations Defined by Regular Languages. Annals of Mathematics and Artificial Intelligence (2013), doi: 10.1007/s10472-013-9352-z
14. Vincent, M.W., Liu, J., Liu, C.: Strong functional dependencies and their application to normal forms in XML. ACM Transactions on Database Systems 29(3), 445–462 (2004)
15. Wang, J., Topor, R.W.: Removing XML Data Redundancies Using Functional and Equality-Generating Dependencies. In: Proc. ADC, pp. 65–74 (2005)
16. Watson, B.W.: A taxonomy of finite automata construction algorithms. Computing Science Note 93/43, Eindhoven University of Technology, The Netherlands (1994)

Flexible Relational Data Model – A Common Ground for Schema-Flexible Database Systems

Hannes Voigt and Wolfgang Lehner

Database Technology Group,
Technische Universität Dresden,
01062 Dresden, Germany
{firstname.lastname}@tu-dresden.de
http://wwwdb.inf.tu-dresden.de/

Abstract. An increasing number of application fields represent dynamic and open discourses characterized by high mutability, variety, and pluralism in data. Data in dynamic and open discourses typically exhibits an irregular schema. Such data cannot be directly represented in the traditional relational data model. Mapping strategies allow representation but increase development and maintenance costs. Likewise, NoSQL systems offer the required schema flexibility but introduce new costs by not being directly compatible with relational systems that still dominate enterprise information systems. With the Flexible Relational Data Model (FRDM) we propose a third way. It allows the direct representation of data with irregular schemas. It combines tuple-oriented data representation with relation-oriented data processing. So that, FRDM is still relational, in contrast to other flexible data models currently in vogue. It can directly represent relational data and builds on the powerful, well-known, and proven set of relational operations for data retrieval and manipulation. In addition to FRDM, we present the flexible constraint framework FRDM-C. It explicitly allows restricting the flexibility of FRDM when and where needed. All this makes FRDM backward compatible to traditional relational applications and simplifies the interoperability with existing pure relational databases.

Keywords: data model, flexibility, relational, irregular data.

1 Introduction

Today's databases are deployed in diverse and changing ecosystems. An increasing number of application fields is characterized by high mutability, variety, and pluralism in the data. High mutability is caused by the persistent acceleration of society [16] and technological development [19]. Variety appears in database discourses because information systems extending their scope and strive to cover every aspect of the real world. Pluralism is inevitable with the onging cross linking of information systems and the consolidation of data from different stackholders in a single database. Particular drivers of these developments are end

Y. Manolopoulos et al. (Eds.): ADBIS 2014, LNCS 8716, pp. 25–38, 2014.

user empowerment [20], agile software development methods [6], data integration [14,26], and multi-tenancy [17]. The once stable and closed discourses of databases are rather dynamic and open today.

In such dynamic and open discourses, data often has a irregularly structured schema. Data with an irregular schema exhibits four characteristic traits: (1) multifaceted entities that cannot be clearly assigned to a single entity type, (2) entities with varying sets of attributes regardless of their entity type, (3) attributes occuring completely independent of particular entity types, and (4) attribute-independent technical typing of values. All four traits cannot be directly represented in the relational data model. As a natural reaction, many developers perceive traditional relational data management technologies as cumbersome and dated [18]. Various mapping strategies [15,1,4] allow a presentation but they imply additional costs of implementation and maintenance. Further, the inherent logical schema of the data is not complete visible on the resulting relational data. This schema incompleteness particularly is a problem in the likely case that multiple applications access a database through different channels.

In recent years, the NoSQL movement has introduced a number of new data models, query languages, and system architectures that exhibit more flexibility regarding the schema. Many NoSQL systems allow the direct representation of data with irregular schema as well as the gradual evolution of the schema. Hence, NoSQL systems appear to be a very appealing choice for applications with a dynamic and open discourse. However, the introduction of new data models, query languages, and system architectures is not for free. Particularly in enterprise environments where 90 % of the databases are relational [9] new data models are often a bad fit. They imply additional costs for mapping and transforming data between different data models and require new database management and application development skills. These costs multiply if applications store the different parts of their data in the respectively ideal data model across different database management systems – a scenario often referred to as *polyglot persistency* in the NoSQL context.

With the Flexible Relational Data Model (FRDM) we propose a third way. It allows the direct representation of data with irregular schemas from dynamic and open discourses. This includes multifaceted entities, variable attributes sets, independent attributes, as well as independent technical types. At the same time FRDM remains 100 % backward compatible to the traditional relational data model. Purely relational data with regular schemas can also be represented and relationally processed in FRDM directly. FRDM achieves this centering the data representation around the individual tuples while maintaining the relation as the primary means of data processing.

Additionally, we present the flexible constraint framework FRDM-C that provides explicit restrictions to the flexibility of FRDM. Scope and range of the restriction can be tailored to any requirements ranging from the constraint-free, descriptive nature of pure FRDM to the strictly prescriptive nature of the traditional relational data model. FRDM-C helps to introduce rigidity exactly when and at which parts of data needed. FRDM-C constraints can vary in their effect from simply informing to strictly prohibiting, so that they are not only a tool to maintain data quality but also help achieving data quality.

The remainder of this paper is structured as follows. Section 2 presents the FRDM data model, in particular how it represents data and how data is processed in FRDM. The constraint framework FRDM-C is discussed in Section 3. With both introduced, Section 4 shows how pure relational data can be represented directly in FRDM to demonstrate the backward compatibility of FRDM. This is followed by considerations regarding the implementation of FRDM within the architecture of relational database management systems in Section 5. In Section 6, we compare FRDM with other data models regarding the provided flexibility and backward compatibility. Finally, Section 7 concludes the paper.

2   FRDM

FRDM is a relational data model for structured data. It is free of the relational inflexibilities but remains directly compatible to the relational model. The most prominent feature of FRDM is that it separates the functionality of data representation, data processing, and constraints. Data representation and data processing are realized in separate, dedicated concepts. We detail the data representation of FRDM in Section 2.1 and discuss data processing in Section 2.2. Schema constraints are realized as explicit constraints outside of the core data model in the constraint framework FRDM-C, which is presented in Section 3.

2.1   Data Representation

The data representation of FRDM builds on four concepts. The central concept is the *tuple*:

Tuple. A tuple is the central concept of the flexible relational data model and represents an entity. It consists of values, each belonging to an attribute and is encoded according to a technical type.

The concepts *entity domain*, *attribute*, and *technical type* describe data represented in tuples and provide logical data handles:

Entity Domain. Entity domains are logical data handles allowing to distinguish logical groups of tuples within a database. Tuples belong to at least one entity domain and may belong to multiple entity domains, so that domains can intersect each other.

Attribute. Attributes are logical data handles allowing to distinguish values within a tuple. Each tuple can instantiate each attribute only once.

Technical Type. Technical types determine the physical representation of values. Value operations such as comparisons and arithmetic are defined on the level of technical types.

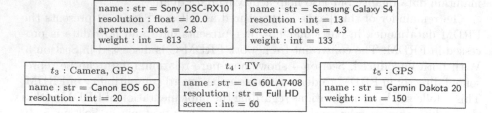

Fig. 1. Example entities representing electronic devices

Tuples:

$$\mathbb{D} = \left\{ \begin{array}{l} t_1 = [\text{Sony DSC-RX10}, 20.0, 2.8, 813], \\ t_2 = [\text{Samsung Galaxy S4}, 13, 4.3, 133], \\ t_3 = [\text{Canon EOS 6D}, 20], \\ t_4 = [\text{LG 60LA7408}, \text{Full HD}, 60], \\ t_5 = [\text{Garmin Dakota 20}, 150] \end{array} \right\}$$

Description elements:

$$\mathbb{A} = \{ aperture, name, resolution, screen, weight \}$$

$$\mathbb{T} = \{ float, int, str \}$$

$$\mathbb{E} = \{ Camera, GPS, Player, Phone, TV \}$$

Schema function:

$$f_s = \left\{ \begin{array}{l} t_1 \rightarrow [name, resolution, aperture, weight], \\ t_2 \rightarrow [name, resolution, screen, weight], \\ t_3 \rightarrow [name, resolution], \\ t_4 \rightarrow [name, resolution, screen], \\ t_5 \rightarrow [name, weight] \end{array} \right\}$$

Membership function:

$$f_m = \left\{ \begin{array}{l} t_1 \rightarrow \{ Camera \}, \\ t_2 \rightarrow \{ Camera, GPS, Phone \}, \\ t_3 \rightarrow \{ Camera, GPS \}, \\ t_4 \rightarrow \{ TV \}, \\ t_5 \rightarrow \{ GPS \} \end{array} \right\}$$

Typing function:

$$f_t = \left\{ \begin{array}{l} (t_1, name) \rightarrow str, (t_1, resolution) \rightarrow float, (t_1, aperture) \rightarrow float, (t_1, weight) \rightarrow int, \\ (t_2, name) \rightarrow str, (t_2, resolution) \rightarrow int, (t_2, screen) \rightarrow double, (t_2, weight) \rightarrow int, \\ (t_3, name) \rightarrow str, (t_3, resolution) \rightarrow int, \\ (t_4, name) \rightarrow str, (t_4, resolution) \rightarrow str, (t_4, screen) \rightarrow int, \\ (t_5, name) \rightarrow str, (t_5, weight) \rightarrow int \end{array} \right\}$$

Fig. 2. Example entities in the flexible relational data model

Formally, a flexible relational database is a septuple $(\mathbb{D}, \mathbb{A}, \mathbb{T}, \mathbb{E}, f_s, f_t, f_m)$. The payload data $\mathbb{D}$ is a set of tuples. A tuple is an ordered set of values $t = [v_1, \ldots, v_m]$. Let $\mathbb{A}$ be the set of all attributes available in the database. Then the tuple schema function $f_s : \mathbb{D} \rightarrow \mathcal{P}(\mathbb{A}) \setminus \emptyset$ denotes the schema of each tuple, i.e., the set of attributes a tuple instantiates. $f_s(t) = [A_1, \ldots, A_m]$ if t instantiates the attributes $A_1, \ldots, A_m$ so that $t \in A_1 \times \cdots \times A_m$. For convenience, we denote with $t[A] = v$ that tuple t instantiates attribute A with value v. $\mathbb{T}$ is the set of all available technical types T. The typing function $f_t : \mathbb{D} \times \mathbb{A} \rightarrow \mathbb{T}$ shows the encoding of values, with $f_t(t, A) = T$ if the value $t[A]$ is encoded according to the technical type T. Finally, $\mathbb{E}$ is the set of all available entity domains E, while the membership function $f_m : \mathbb{D} \rightarrow \mathcal{P}(\mathbb{E}) \setminus \emptyset$ denotes which tuples belong to these domains. $f_m(t) = \{ E_1, \ldots, E_k \}$ if t belongs to the entity domains $E_1, \ldots, E_k$.

As an example, Figure 1 shows six entities in a UML object diagram like notation. The entities represent electronic devices as they could appear in a product catalog. Note that this small example exploits all the flexibilities of FRDM.

Relation *GPS*

name	resolution	screen	weight
Samsung Galaxy S4	13	4.3	133
Canon EOS 6D	20	∄	∄
Garmin Dakota 20	∄	∄	150

Relation *Camera* ∩ *GPS*

name	resolution	screen	weight
Samsung Galaxy S4	13	4.3	133
Canon EOS 6D	20	∄	∄

Fig. 3. Relation by entity domain   **Fig. 4.** Relation from relational operator

All six entities are self-descriptive and have their individual set of attribute. The order of the attributes within an entity differs, too. Entities t_2 and t_3 belong to multiple entity domains. Attributes, such as *name*, appear independently from entity domains. The technical typing of values, for instance of the attribute *resolution*, varies independently from the attribute. In the flexible relational data model these six entities can be represented directly as shown in Figure 2.

2.2 Data Processing

For data processing, FRDM builds on the well-known concept of a *relation*. It allows processing tuples in a relational manner:

Relation. Relations serve as central processing containers for tuples. FRDM queries operate on relations; query operations have relations as input and produce relations as output. The tuples in a relation determine the schema of the relation. Each attribute instantiated by at least a single tuple in the relation is part of the relation's schema.

Let t be a tuple in relation R, then R has the schema $S_R = \bigcup_{t \in R} f_s(t)$ so that $S_R \subseteq \mathbb{A}$. A relation R with schema S_R does not have to instantiate each tuple in every attribute, rather it is $R \subseteq \bigcup_{S_i \in \mathcal{P}(S_R) \setminus \emptyset} \left(\times_{A \in S_i} A \right)$. In other words, tuples may only instantiate a subset of a relation's schema, except the empty set. While $t[A] = v$ denotes that tuple t instantiates attribute A with value v, $t[A] = \not\exists$ indicates that tuple t does not instantiate attribute A.

Mass operations address tuples by means of entity domains. Hence, each entity domain denotes a relation containing all tuples that belong to this domain. Specifically, an entity domain E denotes a relation R so that $E \in f_m(t)$ holds for all $t \in R$. In the following, we refer to a relation representing tuples of domain E simply as E where unambiguously possible. Figure 3 shows the relation denoted by the entity domain *GPS* in the electronic device example.

The well-known relational operators are applicable directly to FRDM relations. However, the descriptive nature of a FRDM relation requires two minor modifications to their semantics. First, the logic of selection predicates and projection expressions has to take into account that attributes may not be instantiated by a tuple. An appropriate evaluation function for such predicates and expressions is described in [28]. In a nutshell, tuples that do not instantiate an attribute used in a selection predicate are not applicable to the predicate and do not qualify. Tuples that do not instantiate an attribute used in a projection expression do not instantiate the attribute newly defined by the expression. Second, all operations have a strictly tuple-oriented semantics, i.e., the schema of the relation resulting from an operation is solely determined by the qualifying tuples. In consequence, the schema resulting from a selection can differ from the schema of the input relation. More specifically, the resulting schema of a selection is equal to or a subset of the input schema depending on which tuples qualify, so that $S_{\sigma_P(A)} \subseteq S_A$. Likewise, the schemas of the operand relations do not matter for set operations. Tuples are equal if they instantiate the same attributes with equal values. For a union, the resulting schema is the union of the schemas of the operands, so that $S_{A \cup B} = S_A \cup S_B$. For set difference, the resulting schema is equal to or a subset of the left operand's schema, again, depending on which tuples qualify, so that $S_{A \backslash B} \subseteq S_A$. Derived operators, such as join or intersection, are affected similarly. As an example, Figure 4 shows the relation resulting from the intersection of the relation *GPS* (cf. Figure 3) and the relation denoted by entity domain *Camera*.

3 FRDM-C

FRDM-C is a flexible constraint framework meant to accompany FRDM. The flexibility of FRDM originates from its lack of implicit constraints. Nevertheless, constraints are a powerful feature if their effect is desired by the user. For the user, constraints are the primary means to obtain and maintain data quality. Each constraint is a proposition about data in the database. Data either complies to or violates this proposition, i.e., every constraint categorizes data into two disjoint subsets. It is up to the user how to utilize this categorization. At least, constraints inform about which data is compliant and which is violating. At most, constraints prohibit data modifications that would result in violating data. Constraints present themselves as additional schema objects, attached to the schema elements of the data model. The user can add and remove constraints any time.

Formally, constraints take the general form of a triple (q, c, o). q is the qualifier; c is the condition compliant data has to fulfill; o is the effect (or the outcome) the constraint will have. The qualifier determines to which tuples the constraint applies. It is either an entity domain $E_q \in \mathbb{E}$, a attribute $A_q \in \mathbb{A}$, or a pair of both (E_q, A_q). Correspondingly, a constraint applies to all tuples t with $E_q \in f_m(t)$, with $A_q \in f_s(t)$, or with $(E_q, A_q) \in f_m(t) \times f_s(t)$, respectively. We denote the set of tuples a constraint C applies to as $\mathbb{D}_C$. Conditions are either tuple conditions or key conditions, depending on whether they affect individual tuples or groups of tuples. The effect determines the result of the operations that lead to violating data and what happens to the violating data itself. In the following, we will detail conditions and effects.

3.1   Conditions

The first group of conditions is tuple conditions. Tuple conditions restrict data on the level of individual tuples, e.g., by mandating to which entity domains a tuple can belong. Formally, a tuple condition is a function $c : \mathbb{D} \to \{\top, \bot\}$. Then, $\mathbb{D}_C^\top = \{t \mid t \in \mathbb{D}_C \wedge c(t)\}$ are the complying tuples and $\mathbb{D}_C^\bot = \{t \mid t \in \mathbb{D}_C \wedge \neg c(t)\}$ are the violating tuples. Tuple conditions are:

Entity Domain Condition. An entity domain condition requires tuples $t \in \mathbb{D}_C$ to belong to an entity domain E_c so that $E_c \in f_m(t)$. We denote a specified entity domain condition as $entity\text{-}domain(E_c)$.

Attribute Condition. A attribute condition requires tuples $t \in \mathbb{D}_C$ to instantiate a attribute A_c so that $A_c \in f_s(t)$. We denote a specified attribute condition as $value\text{-}domain(A_c)$.

Technical Type Condition. A technical type condition limits values of tuples $t \in \mathbb{D}_C$ in attribute A_c to a specified technical type T_c so that $T_c = f_t(t, A_c)$. We denote a technical type condition as $tech\text{-}type(A_c, T_c)$.

Value Condition. A value condition requires values of tuples $t \in \mathbb{D}_C$ in attribute A_c to fulfill a specified predicate p so that $p(t[A_c])$ holds. We denote a value condition as $value(A_c, p)$.

The second group of conditions is key conditions. Key conditions restrict data on the level of tuple groups. Formally, a key condition is a function $c : \mathcal{P}(\mathbb{D}) \to \{\top, \bot\}$. Key conditions are:

Unique Key Condition. A unique key condition requires tuples to instantiate a set of attributes $\mathbb{A}_K \subseteq \mathbb{A}$ uniquely so that $t_i[\mathbb{A}_K] \neq t_j[\mathbb{A}_K]$ holds for all $t_i, t_j \in \mathbb{D}_C$ with $t_i \neq t_j$. As a result, all complying tuples are unambiguously identifiable on $\mathbb{A}_K$. We denote a unique key condition as $unique\text{-}key(\mathbb{A}_K)$.

Foreign Key Condition. A foreign key condition requires tuples to instantiate attributes $\mathbb{A}_F \subseteq \mathbb{A}$ with values referencing at least one tuple on attributes $\mathbb{A}_R \subseteq \mathbb{A}$ so that for every $t_F \in \mathbb{D}_C$ there is one $t_R \in \mathbb{D}_R$ so that $t_F[\mathbb{A}_F] = t_R[\mathbb{A}_R]$. Similarly to $\mathbb{D}_C$, the set of referenceable tuples $\mathbb{D}_R \subseteq \mathbb{D}$ is identified by either an entity domain $E_R \in \mathbb{E}$, a attribute $A_R \in \mathbb{A}$, or a pair of both (E_R, A_R). We denote a foreign key condition as $foreign\text{-}key(\mathbb{A}_F, \mathbb{A}_R, q_R)$ where q_R is the qualifier of $\mathbb{D}_R$.

If a group of tuples does not fulfill a key condition, not all tuples of the group are considered to be violating. We have to distinguish two cases. In the first case, a constraint already exists in the database and a modification of tuples results in a violation. Here, only the modified tuples become violating tuples. In the second case, the constraint is added to the database and the tuples already existing in the database violate this constraint. Here the smallest subset of tuples that violates the condition becomes the set of violating tuples. For a unique key constraint, these are all duplicates. For a foreign key constraint, these are all tuples with a dead reference.

All conditions can be negated in a constraint. Negation swaps the set of violating tuples with the set of complying tuples. For instance, the negated entity domain condition $\neg entity\text{-}domain(E_c)$ prohibits the entity domain E_c instead of requiring it. For two constraints $C = (q, c, o)$ and $C' = (q, \neg c, o)$, it holds that $\mathbb{D}_{C'}^\top = \mathbb{D}_C^\perp$ and $\mathbb{D}_{C'}^\perp = \mathbb{D}_C^\top$. Which tuples violate a constraint is crucial for the effect of the constraint.

3.2   Effects

We distinguish four types of effects constraints can have. They vary in the rigor the constraint will exhibit.

Informing. Allows all operations. The complying tuples and the violating tuples can be queried by using the constraint as a query predicate.

Warning. Allows all operations and issues a warning upon operations that lead to violating tuples. The creation of the constraint results in a warning about already existing violating tuples.

Hiding. Allows all operations and issues a warning upon operations that lead to violating tuples and hides violating tuples from all other operations. The creation of the constraint results in hiding already existing violating tuples except for operations that explicitly request to see violating tuples by using the constraint as predicate.

Prohibiting. Prohibits operations that lead to violating tuples and issues an error. The creation of the constraint is prohibited in case of already existing violating tuples.

4   Presentation of Purely Relational Data

The presented flexible relational data model is a superset of the traditional relational model. Traditional relations can be represented directly in the flexible model. A relational database is a septuple $(\mathbb{D}, \mathbb{A}, \mathbb{T}, \mathbb{R}, f_\sigma, f_\theta, f_\mu)$, where $\mathbb{R}$ is the set of relations, $\mathbb{A}$ is the set of domains, $\mathbb{T}$ is the set of technical types, $\mathbb{D}$ is the set of tuples, f_σ is the schema function $\mathbb{R} \to \mathcal{P}(\mathbb{A})$, f_θ is the typing function $\mathbb{A} \to \mathbb{T}$, and f_μ is the membership function $\mathbb{D} \to \mathbb{R}$. The corresponding flexible relational database is $(\mathbb{D}, \mathbb{A}, \mathbb{T}, \mathbb{E}, f_s, f_t, f_m)$ with

$$\mathbb{E} = \{name\text{-}of(R) \mid R \in \mathbb{R}\},$$
$$f_s = \{t \to f_\sigma(f_\mu(t)) \mid t \in \mathbb{D}\},$$
$$f_t = \{(t, A) \to f_\theta(A) \mid A \in f_\sigma(f_\mu(t)) \wedge t \in \mathbb{D}\}, \text{ and}$$
$$f_m = \{t \to \{name\text{-}of(f_\mu(t))\} \mid t \in \mathbb{D}\}.$$

To emulate the model-inhernt constraints of the relational model the flexible relational database has to be supplemented with explicit constraints. For each relation $R \in \mathbb{R}$ we add the following prohibitive ($P \hat{=} prohibiting$) constraints:

– Entity domains have to mutually exclude each other, so that tuples can be only part of one entity domain. This can be achieved with constraints of the form $(name\text{-}of(R), \neg entity\text{-}domain(E), P)$ where $name\text{-}of(R) \neq E$ and $R \in \mathbb{R}$.

- Entity domains prescribe the attributes of their corresponding relation. This can be achieved with constraints of the form $(name\text{-}of(R), value\text{-}domain(A), P)$ for $A \in f_\sigma(R)$ and $R \in \mathbb{R}$.
- Entity domains forbid all other attributes. This can be achieved with constraints of the form $(name\text{-}of(R), \neg value\text{-}domain(A'), P)$ for $A' \notin f_\sigma(R)$ and $R \in \mathbb{R}$.
- Attributes prescribe the technical type as defined by the corresponding relation. This can be achieved with constraints of the form $(A, tech\text{-}type(A, f_\theta(A)), P)$ for $A \in f_\sigma(R)$ and $R \in \mathbb{R}$.

5   Implementation Consideration

The FRDM data model is positioned as a flexible descendant of the relational model. Therefore it is suitable to be implemented within the existing and established relational database system architecture. In this section, we briefly discuss how this can be done. The characteristics of FRDM require four main changes to existing relational database system code.

First, plan operators and query processing have to be adapted to handling descriptive relations. More specifically, plan operators must reflect the adapted semantics of their logical counterparts. Logically, operators have to remove attributes from the schema of a relation if no tuple instantiates them. With a tuple-at-a-time processing model, this orphaned attribute elimination is a blocking operation, since the system can determine the schema only after all tuples are processed. Implicit duplicate elimination is similarly impractical and thus it was not implemented in relational database systems. Likewise a practical solution for the elimination of orphaned attributes is that plan operators determine the schema of the resulting operation as narrow as they safely can before the actual tuple processing and accept possible orphaned attributes in the result relation. Similar to the DISTINCT clause, SQL can be extended with a, say, TRIM clause that allows the user to explicitly request orphaned attribute elimination.

Second, the physical storage of tuples has to be adapted to the representation of entity domains. For tuple storage, the existing base table functionality can be reused but needs to be extended to handle uninstantiated attributes. Solutions for such an extension are manifold in literature, e.g., interpreted record [7,11], vertical partitioning [1], and pivot tables [3,13]. Another reasonable approach is a bitmap as it is used for instance by PostgreSQL [24] to mark NULL values in records. Tuples can appear in multiple entity domains. However, for storage economy and update efficiency, tuples should only appear in a single physical table. Replication should be left to explicit replication techniques. Consequently, the database system has to assign each tuple to a single physical table and maintain its logical entity domain membership somehow. In principle, there are two ways how this can be done. One is to encode the domain membership in the physical table assignment. Here, the system would create a physical table for each combination of entity domains occurring in a tuple and store tuples in the corresponding table. The mapping is simple and easy to implement.

The downside is that it may lead to a large number of potentially small physical tables (at worst 2^E tables where E is the number of entity domains in a database) and tuples need to be physically moved if their domain membership is changed. The other way is storing the domain membership, e.g., with a bitmap, directly in a tuple itself. This gives liberty regarding the assignment of tuples to physical tables, up to using a single (universal) table for all tuples. With many tuples having the same domain membership, it comes to the price of storage overhead – negligible in most cases, though.

Third, the physical tuple layout has to be extended to also represent the technical type of values directly in the tuple. This is necessary for independent technical types. To reduce storage needs and decrease interpretation overhead, the system can omit the technical type in the tuple where explicit constraints prescribe a technical type. However, creating and dropping such explicit constraints becomes expensive as the physical representation of the affected tuples has to be changed.

Fourth, independent attributes require a modification of the system catalog. In most system catalogs, attributes have a reference to the base table they belong to. This reference has to be removed to make attributes available to all tuples regardless of their entity domain membership.

6   Related Work

Over decades, research and development have created numerous data models and approaches to represent data. Obviously, we can concentrate only on the most prominent ones used for representing structured data. Data models worth considering can be grouped in four main categories: (1) relational models, (2) software models, (3) document models, (4) tabular models, (5) graph data models, and (6) models from the data modeling theory. In the following, we will briefly discuss these categories with regard to the flexibility to directly represent data of dynamic and open discourses.

Relational models are extensions of the traditional relational model [28,7,2,5]. These extensions intend to free the relational model from one or more implicit constraints. Hence, these extended relational models allow additional flexibility compared to the pure relational model. Specifically, reasonable extensions exist to support variable attribute sets. Besides, all these extensions preserve 100 % compatibility with the relational model. To the best of our knowledge, there are no extensions that add support for multifaceted entities, independent attributes, and independent technical types to the relational model.

Software models originate from programming languages and other software development technologies. Generally, software models consist of elements to structure operations and elements to structure data. The elements to structure data resemble a data model. Two popular software models are object orientation and role modeling [27]. Both build on the notion of an object and encompass a dedicated association element to represent relationships. Accordingly, they provide no direct compatibility with the relational model, a fact also well known as

Table 1. Flexible Data Models vs. Requirements

Category	Data Model	Multi-faceted entities	Variable attribute sets	Independent attributes	Independent technical types	Relational representation and processing
Relational	Pure relational					✓
	Extended NULL semantic [28]	(✓)[1]				✓
	Interpreted column [2]	(✓)[2]				✓
	Interpreted record [7]		✓			✓
	Polymorphic table [5]	(✓)[3]	(✓)[2]			✓
	FRDM	✓	✓	✓	✓	✓
Programming	Object orientation	(✓)[4]			(−)[6]	
	Role modeling [27]	(✓)[5]			(−)[6]	
Document	XML, well-formed [31]			✓	✓	(✓)[7]
	XML, valid [32]			(✓)[3]	✓	
	JSON [12]		(✓)[8]	(✓)[8]	(✓)[8]	✓
	OEM [22]		(✓)[8]	(✓)[8]	(✓)[8]	✓
Tabular	Bigtable [10]		✓	✓		✓
Graph	Property graph [25]			✓	✓	(−)[6]
	Neo4J [21]		✓	✓	✓	✓
	Freebase [8]		✓	(✓)[1]		
	RDF [30]		(✓)[8]	(✓)[8]	(✓)[8]	✓
	RDF w/ RDF Schema [29]		✓	✓	✓	✓
Theory	Intensional classification [23]	✓	✓	✓	(−)[6]	

[1] only generalization   [2] only specialization   [3] extensions   [4] inheritance
[5] roles   [6] not specified   [7] no technical types   [8] no entity types

object-relational impedance mismatch. With inheritance and the notion of roles, these two software models offer limited support for multifaceted entities. Particularly the role concept allows the dynamic leaving and joining of entity types. Nevertheless, which combination of entity types an entity can join has to be modeled upfront.

Document models [31,32,12,22] have been developed for representing documents, e.g., web pages. Typically, document models represent data as a hierarchy of entities, where entities nest other entities. Nesting is the only or the primary means of entity referencing. The identity of an entity solely or primarily depends on the position of an entity within the hierarchy. In consequence, document models offer direct relational compatibility. Document models offer more flexibility than most relational systems or software models. However, most of their flexibility originates from completely omitting entity types. Where document models have schema information, such as DTD or XML Schema, they are similarly strict.

A tabular data model also organizes data in tables like the relational data model but in a significantly different way. The data model of Google's Bigtable system [10] defined the category of tabular data models. Because of its success, it has also remained the only model of its kind that draws considerable attention. Bigtable organizes data in large, distributed, sparse tables. The columns of such a table are grouped in column families. Rows can stretch across multiple column families and are free to instantiate any column in a column family, so that the Bigtable data model supports multifaceted entities as well as variable attribute sets. The Bigtable model also supports independent technical types. However, the row identity is restricted to a user-given row key and the processing is limited to put and get operations on row level. Hence, the Bigtable model cannot be considered completely relational compatible.

Graph data models [25,21,8,30,29] build on the mathematical definition of a graph. They represent data as vertices and edges, where vertices represent entities and edges represent relationships, i.e., references to other entities. In practice, graph models differ in how data is represented in a graph. Beside vertices and edges, graphs can have labels and attribute–value pairs attached to the vertices and even to the edges. [25] distinguishes nine types of graphs. Most prominent are the property graph and the RDF graph. All graph models emphasize the representation of data rather than modeling of schema. Graph models have a descriptive nature and allow in most cases the direct representation of data from dynamic and open discourses. In all graph models, however, entities have an object identity and edges are an explicit representation of references. Consequently, graph models are not directly compatible to relational data.

Finally in the theory of data modeling, intensional classification was proposed to allow for more schema flexibility [23]. Here, entity domains are defined intensionally, i.e., by a set of attributes. All entities that instantiate the set of attributes defining an entity domain belong to that domain. Accordingly, the intensional classification builds on independent attributes and allows multi-faceted entities as well as variable attribute sets. Technical types are not considered in the approach. While intensional classification is appling, it is less flexible than extensional classification used in FRDM, since entities are required to instantiate an defined attribute set to belong to a domain. They cannot be explicitly added to a domain regardless their intension. In that sense, intensional classification is a useful complement to extensional classification.

As a summary, Table 1 shows which flexibilities sample data models in the discussed categories do allow. We can see that none of these models fulfills all flexibility requirements. Graph models, particularly as in Neo4j, are free of implicit constraints regarding entity domains, attributes and technical types, while the relational approaches are the only ones to offer value-based identity and value-based references. FRDM integrates the level of flexibility graph models provide with value-based identity and value-based references, as indicated in Table 1, in a super-relational fashion.

7 Conclusion

As an evolutionary approach to meet the need for more flexible database systems and to build on the still existing dominance of relational database systems we proposed the flexible super-relational data model FRDM. FRDM is entity-oriented instead of schema-oriented. It is designed around self-descriptive entities, where schema comes with the data and does not have to be defined up front. Additionally, FRDM allows multi-faceted entities where entities can belong to multiple entity domains. Attributes can exist independently from entity domains in FRDM. Similarly, FRDM allows technically typing values independently from their attributes. FRDM can express irregular data as well as regular relational data. We demonstrated both by examples. For data retrieval, FRDM builds on the powerful, well-known, and proven set of relational operations. Compared to the relational data model, FRDM is free of implicit constraints. Nevertheless, where these constraints are needed and welcome, the presented constraint framework FRDM-C allows formulating explicit restrictions to the flexibility of FRDM. A lot of technological expertise, knowledge, and experience have accumulated in and around relational database management systems over the last three decades. We are convinced FRDM contributes to the use of that also in the more flexibility-demanding areas of data management.

References

1. Abadi, D.J., Marcus, A., Madden, S., Hollenbach, K.J.: Scalable Semantic Web Data Management Using Vertical Partitioning. In: VLDB 2007 (2007)
2. Acharya, S., Carlin, P., Galindo-Legaria, C.A., Kozielczyk, K., Terlecki, P., Zabback, P.: Relational support for flexible schema scenarios. The Proceedings of the VLDB Endowment 1(2) (2008)
3. Agrawal, R., Somani, A., Xu, Y.: Storage and Querying of E-Commerce Data. In: VLDB 2001 (2001)
4. Aulbach, S., Grust, T., Jacobs, D., Kemper, A., Rittinger, J.: Multi-Tenant Databases for Software as a Service: Schema-Mapping Techniques. In: SIGMOD 2008 (2008)
5. Aulbach, S., Seibold, M., Jacobs, D., Kemper, A.: Extensibility and Data Sharing in evolving multi-tenant databases. In: ICDE 2011 (2011)
6. Beck, K., Beedle, M., van Bennekum, A., Cockburn, A., Cunningham, W., Fowler, M., Grenning, J., Highsmith, J., Hunt, A., Jeffries, R., Kern, J., Marick, B., Martin, R.C., Mellor, S., Schwaber, K., Sutherland, J., Thomas, D.: Manifesto for Agile Software Development (2001), http://agilemanifesto.org/
7. Beckmann, J.L., Halverson, A., Krishnamurthy, R., Naughton, J.F.: Extending RDBMSs To Support Sparse Datasets Using An Interpreted Attribute Storage Format. In: ICDE 2006 (2006)
8. Bollacker, K.D., Evans, C., Paritosh, P., Sturge, T., Taylor, J.: Freebase: A Collaboratively Created Graph Database For Structuring Human Knowledge. In: SIGMOD 2008 (2008)
9. Brodie, M.: OTM"10 Keynote. In: Meersman, R., Dillon, T.S., Herrero, P. (eds.) OTM 2010. LNCS, vol. 6426, pp. 2–3. Springer, Heidelberg (2010)

10. Chang, F., Dean, J., Ghemawat, S., Hsieh, W.C., Wallach, D.A., Burrows, M., Chandra, T., Fikes, A., Gruber, R.: Bigtable: A Distributed Storage System for Structured Data. In: OSDI 2006 (2006)
11. Chu, E., Beckmann, J.L., Naughton, J.F.: The Case for a Wide-Table Approach to Manage Sparse Relational Data Sets. In: SIGMOD 2007 (2007)
12. Crockford, D.: The application/json Media Type for JavaScript Object Notation (JSON), RFC 4627 (July 2006), http://tools.ietf.org/html/rfc4627
13. Cunningham, C., Graefe, G., Galindo-Legaria, C.A.: PIVOT and UNPIVOT: Optimization and Execution Strategies in an RDBMS. In: VLDB 2004 (2004)
14. Franklin, M.J., Halevy, A.Y., Maier, D.: From Databases to Dataspaces: A New Abstraction for Information Management. SIGMOD Record 34(4) (2005)
15. Friedman, C., Hripcsak, G., Johnson, S.B., Cimino, J.J., Clayton, P.D.: A Generalized Relational Schema for an Integrated Clinical Patient Database. In: SCAMC 1990 (1990)
16. Gleick, J.: Faster: The Acceleration of Just About Everything. Pantheon Books, New York (1999)
17. Jacobs, D.: Enterprise Software as Service. ACM Queue 3(6) (2005)
18. Kiely, G., Fitzgerald, B.: An Investigation of the Use of Methods within Information Systems Development Projects. The Electronic Journal of Information Systems in Developing Countries 22(4) (2005)
19. Kurzweil, R.: The Law of Accelerating Returns (March 2001), http://www.kurzweilai.net/the-law-of-accelerating-returns
20. Nagarajan, S.: Guest Editor's Introduction: Data Storage Evolution. Computing Now, Special Issue (March 2011)
21. Neo Technology: Neo4j (2013), http://neo4j.org/
22. Papakonstantinou, Y., Garcia-Molina, H., Widom, J.: Object Exchange Across Heterogeneous Information Sources. In: ICDE 1995 (1995)
23. Parsons, J., Wand, Y.: Emancipating Instances from the Tyranny of Classes in Information Modeling. ACM Transactions on Database Systems 25(2) (2000)
24. PostgreSQL Global Development Group: PostgreSQL 9.2.4 Documentation, chap. 56.6: Database Page Layout (2013)
25. Rodriguez, M.A., Neubauer, P.: Constructions from Dots and Lines. Bulletin of the American Society for Information Science and Technology 36(6) (August 2010)
26. Sarma, A.D., Dong, X., Halevy, A.Y.: Bootstrapping Pay-As-You-Go Data Integration Systems. In: SIGMOD 2008 (2008)
27. Steimann, F.: On the representation of roles in object-oriented and conceptual modelling. Data & Knowledge Engineering 35(1) (2000)
28. Vassiliou, Y.: Null Values in Data Base Management: A Denotational Semantics Approach. In: SIGMOD 1979 (1979)
29. W3C: RDF Vocabulary Description Language 1.0: RDF Schema (February 2004), http://www.w3.org/TR/2004/REC-rdf-schema-20040210/
30. W3C: Resource Description Framework (RDF): Concepts and Abstract Syntax (February 2004), http://www.w3.org/TR/2004/REC-rdf-concepts-20040210/
31. W3C: Extensible Markup Language (XML) 1.0 (Fifth Edition). (November 2008), http://www.w3.org/TR/2008/REC-xml-20081126/
32. W3C: XML Schema Definition Language (XSD) 1.1 Part 1: Structures. (July 2011), http://www.w3.org/TR/2011/CR-xmlschema11-1-20110721/

Defining Temporal Operators
for Column Oriented NoSQL Databases

Yong Hu and Stefan Dessloch

Heterogenous Information Systems Group,
University of Kaiserslautern, Kaiserslautern, Germany
{hu,dessloch}@informatik.uni-kl.de

Abstract. Different from traditional database systems (RDBMSs), each
column in Column-oriented NoSQL databases (CoNoSQLDBs) stores
multiple data versions with timestamp information. However, this implicit temporal interval representation can cause wrong or misleading
results during query processing. To solve this problem, we transform the
original CoNoSQLDB tables into two alternative table representations,
i.e. explicit history representation (EHR) and tuple time-stamping representation (TTR) in which each tuple (data version) has an explicit
temporal interval. For processing TTR, the temporal relational algebra
is extended to TTRO operator model with minor modifications. For processing EHR, a novel temporal operator model called CTO is proposed.
Both TTRO and CTO contain seven temporal data processing operators, namely, *Union, Difference, Intersection, Project, Filter, Cartesian
product* and *Theta-Join* with additional table transformation operations.

Keywords: CoNoSQLDBs, temporal data and temporal operators.

1 Introduction

Recently, a new type of data storage system called "Column-oriented NoSQL"
database (CoNoSQLDB) has emerged. A CoNoSQLDB manages data in a structured way and stores the data which belongs to the same "column" continuously
on disk. Tuples in a CoNoSQLDB are delivered based on unique row keys. Different from RDBMSs, each column in a CoNoSQLDB stores multiple data versions
sorted by their corresponding timestamps and each data version has an implicit valid temporal interval (TI) (derived from the data versions). Well known
examples are "BigTable" [9], which was proposed by Google in 2004, and its
open-source counterpart "HBase" [10].

To consume data in CoNoSQLDBs, users can either write low-level programs
such as a *MapReduce* [8] procedure or utilize high-level languages such as *Pig
Latin* [11] or *Hive* [12]. MapReduce is a parallel data processing framework in
which users code the desired data processing tasks in Map and Reduce functions
and the framework takes the charge of parallel task execution and fault tolerance.
Although this approach gives users enough flexibility, it imposes programming
requirements and restricts optimization opportunity. Moreover, it forces manual
coding of query processing logic and reduces program reusability.

Y. Manolopoulos et al. (Eds.): ADBIS 2014, LNCS 8716, pp. 39–55, 2014.

Pig Latin and Hive are two high-level languages built on top of the MapReduce framework, where each includes various predefined operators. To analyze the data in a CoNoSQLDB, clients utilize the default load function and denote queries either by a set of high-level operators (Pig Latin) or SQL-like scripts (Hive). However, the default load function will transform a CoNoSQLDB table into a *first-normal-form* (1NF) relation [1] by purely loading the latest data values (without TSs) and discarding older versions. If users wish to load multiple data versions, a customized load function has to be coded manually. Each column will then have a *"set"* type instead of atomic values. Generally, this type of table is called *non-first-normal-form* (NF^2) [2] or nested relations. To process NF^2 in Pig Latin or Hive, users need to first flatten the nested relation to 1NF, then apply the desired data processing based on the predefined high-level operators and finally nest the 1NF relation to rebuild the nested relation. However, this approach has several pitfalls: 1) as the data volume of CoNoSQLDB is usually massive, the table reconstructing operations can heavily decrease the performance and exhaust the hardware resources; 2) the predefined high-level operators are traditional relational operators which handle only the data values without considering any temporal information.

In this paper, we study the issues of defining temporal operators for CoNoSQLDBs and several significant aspects need to be taken into account:

- What is the meaning of TS in CoNoSQLDBs, i.e. should it be understood as *valid time* or *transaction time*? This issue will be discussed in Section 4.1.
- The original CoNoSQLDB tables maintain the temporal interval for each data version implicitly. This property can cause wrong or misleading results during query processing. How can we avoid this? The suitable solutions will be proposed in Section 4.2.
- The CoNoSQLDB tables must be closed under the temporal operators, namely, the output of each operator must still be a CoNoSQLDB table. For example, a traditional temporal Project operator will merely produce the columns specified in the projection attributes. However, in the CoNoSQLDB context, the row key column is mandatory for each CoNoSQLDB table.

The remainder of paper will address these issues and is organized as follows: In Section 2, we discuss the related work. Section 3 describes the core properties of CoNoSQLDBs. The formalization of the CoNoSQLDBs is given in Section 4. Section 5 depicts the temporal operators and Section 6 makes the conclusions.

2   Related Work

Extensive research in the temporal relational database area was done over the last decades, and finally database products as well as the SQL standard have picked up capabilities for temporal data modeling [7]. To model temporal dimensions in a table, two main alternatives exist. The first approach is called *tuple time-stamping* [3,6,7] (TTS) which appends two auxiliary columns to the 1NF table to indicate start and end time for the tuple's valid TI. The second approach is called *attribute time-stamping* [4,5] (ATS) in which each attribute value consists of an atomic value with a valid TI. Generally, the ATS relation is viewed as NF^2 [4,5], as the domain of its attribute is not atomic anymore. In ATS, a new concept *"temporal atom"* [5] is proposed. A temporal atom is composed as a pair (*Value, TI*) which is treated as an atomic data unit for each column (analogy to float, integer and etc. in RDBMS). Moreover, each column in an ATS table usually stores multiple temporal atoms as it can reduce the data redundancy and keep the entire history of an object in one tuple instead of splitting it into multiple tuples.

For handling the TTS, traditional relational algebra is extended to the temporal relational algebra (TRA) [3,6]. For the ATS, the relational algebra is extended to the historical relational algebra (HRA) [4,5] with two auxiliary table reconstructing operators, i.e. *Nest* and *Unnest*.

CoNoSQLDBs fall into the ATS modeling, as each data version is attached with a corresponding TS. However, the method adopted to process the ATS is not suitable for CoNoSQLDBs. More discussions will be given in Section 4 and Section 5.

In the context of CoNoSQLDBs, to our best knowledge, our temporal operator models are the first proposals which address temporal data processing and are consistent with the data model and data processing model of CoNoSQLDBs.

3 Characteristics of CoNoSQLDBs

As the temporal operators are applied to CoNoSQLDBs, in this section, we indicate some important features of CoNoSQLDBs compared to RDBMSs, focusing on the aspects that affect the design of our operator model:

- **Data Model.** In addition to the concepts of table, row and column, CoNoSQLDBs introduce a new concept called *"column family"*. Columns which belong to the same column family will be stored continuously on disk. Each table in a CoNoSQLDB is partitioned and distributed based on the row keys. For a given tuple, one column can store multiple data versions sorted by the corresponding timestamps (TSs). Moreover, users can indicate the ttl (time-to-live) property for each column family to denote the life time of data items. When data items expire, CoNoSQLDBs will make them invisible and eventually remove them in a cleanup operation. Note that only timestamps (indicating at what time the value changed) and no explicit time intervals (TIs) are stored!
- **Operations.** Different from RDBMSs, a CoNoSQLDB does not distinguish update from insertion. A new data value for a specific column will be generated by the "Put" command without overwriting the existing ones. When issuing a "Put" command, users need to denote the parameters such as row key, column-family name, column name, value and TS (optional). If no TS is specified in the "Put" operation, a system-generated TS is used as the version-TS. Following the data model of CoNoSQLDB, data deletions are

classified into various granularities, namely, *data version*, *column* and *column family*. A delete operation will not delete data right away but insert a "tombstone" marker to mask the data values whose TSs are equal to or less than the TS of the tombstone.

Web-surfing

	Web: Page ttl=10s	Web:Content ttl=10s	Network:Supplier ttl=∞
Tom	Spiegel.de : 4, Yahoo.com : 1	Politics :4, Sport :1	Telecom:1

Fig. 1. Example of CoNoSQLDBs

Figure 1 shows an example defined in "HBase" to illustrate the aforementioned characteristics. The "Web-Surfing" table records the information when a user browses the internet. It contains two column families and each column family includes several columns. For row "Tom", the "Web:Page" column contains two data versions.

4   Formalization of CoNoSQLDBs

As timestamps (TSs) represent the temporal information for the data versions, in this section, we first discuss their semantics and clarify the usages of TS in CoNoSQLDBs. Then, we pursue the formalization of CoNoDSQLDB tables and use an example to illustrate how the implicit TI strategy supported by the original CoNoSQLDB tables can cause wrong or misleading results during query processing. To overcome this problem, we propose two alternative table representations in which each tuple (data version) has an explicit TI.

4.1   Understanding TS in CoNoSQLDBs

CoNoSQLDBs follow an *attribute time-stamping* (ATS) approach by attaching a TS to each data version. However, in contrast to temporal databases, the explicit TS just represents the start of a time interval. The TI is only implicitly represented when we assume the end of the interval to be determined by the start-TS of the subsequent version (or for the most recent version), which is consistent with the semantics of version timestamps. This interpretation constrains the derived TIs belonging to the same column to form a contiguous time interval, e.g. the TIs of data versions "Yahoo.com:1" and "Spiegel.de:4" in Figure 1 are [1s, 4s) and [4s, 14s), respectively. In addition, the time interval for the most recent version is limited by the ttl property. For example, although the "Spiegel.de:4" is the latest data version, as the ttl is set to 10s, its TI is [4s, 14s) instead of [4s, ∞).

Furthermore, another question in terms of the semantics of TS in CoNoSQLDBs arises. In the temporal database literature, there are two orthogonal time dimensions [3,4,6,7]: 1) *valid time*, which indicates the time interval during which a data value reflects the state of the real world; 2) *transaction time*, which denotes when a data item is recorded in the database. The valid time and the transaction time are usually depicted as a time period $[t_1, t_2)$ which denotes a data value holds at time t where $t_1 \leq t<t_2$. The valid time can be assigned and modified by users, whereas the transaction time is generated and maintained automatically by the database system.

CoNoSQLDBs, due to the different usages of the Put and Delete commands, the TS can be either arbitrarily specified by users or automatically generated by the system. If the TS is denoted by users, this implies that data versions can be inserted or discarded at any point in time in the version history of a CoNoSQLDB column. Consequently, the Put and Delete commands with the explicit TS assignments may cause TI modifications of existing data versions. For example, in Figure 1, if a data version "Google.de:3" is inserted between "Yahoo.com:1" and "Spiegel.de:4", the TI for "Yahoo.com:1" is implicitly changed to [1s, 3s). In this situation, the TS in the CoNoSQLDB has *valid time* semantics. However, when TS is generated by the CoNoSQLDB, 1) for the Put command, the TS of the new generated data version will be greater than all the existing data versions; 2) for the Delete command, either the current (latest) data version will be deleted (data version deletion) or the whole column/column family will be discarded (column and column-family deletions). Hence, either only the current data version will be changed (Put command and data version deletion) or all the data versions will be eliminated (column and column-family deletions). For example, if a data version "Google.de:6" (where 6 is generated by the CoNoSQLDB) is inserted into "Web:Page" column, the TI of "Spiegel.de:4" is changed from [4s, 14s) to [4s, 6s) and the TI of "Yahoo.com:1" is still [1s, 4s). In this situation, the TS in the CoNoSQLDB is close to *transaction time*.

Hence, the temporal semantics of TS in CoNoSQLDBs is ambiguous, namely, it can be understood as either valid time or transaction time based on the usages of Put and Delete commands. Consequently, the user or application has to make consistent use of temporal concepts supported in CoNoSQLDBs. The TS for a single column should have the semantics of either valid time or transaction time but not both. Moreover, if bi-temporal data needs to be maintained (i.e., both transaction and valid time data is needed), additional columns need to be added by the application to keep the time information. For this paper, we assume that the application is aware of this. Since our operator model and the additional representations we propose do not depend on the time semantics (valid vs. transaction), our results are not impacted.

4.2   Representations of CoNoSQLDBs

A schema R for a CoNoSQLDB table is a collection of rules of the form R = (*rk*, CF_1:Col_{11},..., CF_n:Col_{nn}), where *rk* is shorthand for row key and the subscript n denotes the number of column families (CFs). Each CF_i is composed by a set of columns (Col_{i1},..., Col_{ij}). The value of a column is a set of data versions in which each data version D_m can be further decomposed as a pair (*Value, TS*). Value denotes the content of D_m and TS has the semantics of either valid time or transaction time. For each column, TS functionally determines the Value, i.e. TS→Value. The TI for each data version is implicitly represented among columns and the deduced TIs which belong to the same column must form a contiguous time interval.

A CoNoSQLDB table r is an instance of a CoNoSQLDB schema R. $Dom()$ is a function which maps an attribute name into its value domain. In the CoNoSQLDB context, $Dom(rk)$ has usually a string type. $Dom(Value)$ can be any set of atomic values, such as integer, float, string and etc. $Dom(TS)$ is assigned as a discrete time domain which consists of a set of long nonnegative integer with an ascending order. $Dom(CF:Col) = Dom(Value) \times Dom(TS)$ where $Dom(R) = Dom(rk) \times Dom(CF_1:Col_{11}) \times ... \times Dom(CF_n:Col_{nn})$, where $\times$ is Cartesian product. Clearly, a CoNoSQLDB table does not satisfy 1NF, as the attribute type of each tuple is not atomic. However, different from the general NF^2 relations, the nesting level of a CoNoSQLDB table is fixed, i.e. 1 (we view the nesting level of 1NF as 0). In the following, we use t[S] to denote the value of navigation path S and $Attr(A)$ to indicate a set of attributes that belong to A, where A can be a table name, a CF name or a Col name. As the TI for each data version is implicitly represented inside the column, we call the original table representation *implicit history representation* (IHR).

Although the IHR is suitable for data storage, it can cause wrong or misleading results during query processing. Suppose we use the "Network-speed" table in Figure 2 as an input and wish to choose the Tom's internet suppliers whose speed has ever been faster than 1000K. The filter operation will discard "1&1:3" in the Supplier column and "920K:3" in "Speed" column. The right-hand side shows the filter results. As the TI for each data version is implicit, directly discarding data versions will cause TI changes of the remaining data versions, e.g. the valid TI of "Telecom:1" is changed from [1s, 3s) to [1s ,4s). Obviously, this produces incorrect results.

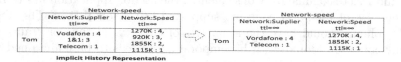

Fig. 2. Select network supplier whose speed is faster than 1000K

The wrong query processing results can be avoided by adopting an explicit TI representation. In contrast to the original data version definition, we model a data version D in the CoNoSQLDBs as a pair (Val, TI) where:

– Val indicates the value of D;
– TI denotes how long D is temporally valid and has a form $[Sta, End)$.

Using the new data version model, Figure 3 shows the equivalent representation of the "Network-speed" table on the left and the correct results of filter processing on the right. We call this new table representation *explicit history representation* (EHR).

As an alternative to grouping multiple data versions with explicit TIs in a single column, we can also adopt the *tuple time-stamping* approach by splitting each IHR tuple into several tuples in which each column contains only a single data version and the row key includes the valid TI to guarantee its uniqueness. We call this table representation *tuple time-stamping representation* (TTR). Figure 4 shows the TTR example derived from "Network-speed" on the left and the correct results for filter processing on the right. For better readability, we specify the row key *rk* in *TTR* as a pair (*srk, TI*). *srk* denotes the original row key value extracted from the corresponding IHR table and *TI* indicates the valid time interval which has the form [Sta, End). Clearly, ERH can also be transformed into TTR.

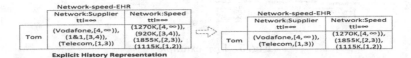

Fig. 3. Select network supplier whose speed is faster than 1000K by using EHR

Network-speed-TTR		
RK/Sta/End	Network:Supplier ttl=∞	Network:Speed ttl=∞
Tom/1/2	Telecom : 1	1115K : 1
Tom/2/3	Telecom : 2	1855K : 2
Tom/3/4	1&1 : 3	920K : 3
Tom/4/∞	Vodafone : 4	1270K : 4

Tuple time-stamping representation

⇢

Network-speed-TTR		
RK/Sta/End	Network:Supplier ttl=∞	Network:Speed ttl=∞
Tom/1/2	Telecom : 1	1115K : 1
Tom/2/3	Telecom : 2	1855K : 2
Tom/4/∞	Vodafone : 4	1270K : 4

Fig. 4. Select network supplier whose speed is faster than 1000K by using TTR

We define 4 table transformation operations which transform IHR to EHR (T_{IE}), IHR to TTR (T_{IT}), EHR to TTR (T_{ET}) and TTR to EHR (T_{TE}), respectively:

- T_{IE} takes an IHR table as an input and outputs its corresponding EHR table. The explicit TI of a data version D_n in an EHR column is derived from its corresponding data version D_i in the IHR column and formed as $[D_i.TS, D_j.TS)$, where D_j is the immediate successor of D_i. When D_i is the current data version, its end point of TI is either denoted by ∞ or calculated by using ttl.
- T_{ET} takes an EHR table as an input and outputs its corresponding TTR table. For every tuple in EHR, T_{ET} will first collect the TIs of all data versions, and then derive TI for each TTL tuple. Finally, the derived TI will be utilized as a selection criterion to select the data versions from EHR columns. We illustrate the T_{ET} operation in Figure 5. The TIs for row "Tom" in EHR are denoted at the top right corner. The corresponding derived TIs are indicated at the bottom right ([1s, 2s) and [2s, 3s)). T_{ET} then exploits each derived TI ([1s, 2s) and [2s, 3s)) as a selection criterion to scan both "Network:Supplier" and "Network:Speed" columns to find the matching data versions. The resulting TTR is shown at the bottom left.
- T_{TE} takes a TTR table as an input and outputs its corresponding EHR table. T_{TE} first groups the TTR tuples which share the same *rk.srk* together. At the same time, the TI in the row key will be attached to each data version.

At last, several data versions that have the same value will be coalesced into a single data version when their TIs are overlapping or adjacent. Figure 6 shows an example of T_{TE}. The two arrows indicate these two data processing tasks, respectively.

– T_{IT} takes an IHR table as an input and outputs its corresponding TTR table. T_{IT} can be represented as a T_{IE} followed by a T_{ET}.

Fig. 5. T_{ET} example

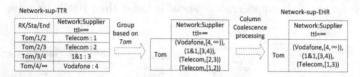

Fig. 6. T_{TE} example

Due to the table transformations, each IHR can be mapped to one EHR and one TTR. Moreover, one EHR can be mapped to one TTR and vice versa. We omit to define the EHR to IHR and TTR to IHR transformations, as not every EHR or TTR can be transformed back to IHR. We can utilize the EHR table at the right-hand side in Figure 3 as a counter-example, as the TIs of column "Network:Supplier" do not form a contiguous time interval. It is impossible to rebuild the corresponding IHR. The same counter-example for TTR can be found at the right-hand side in Figure 4.

The reasons for inapplicable transformations (EHR or TTR to IHR) are the characteristics of IHR, namely, 1) the valid temporal interval for each column can only be $[OSta, \infty)$, where $OSta$ denotes the starting point of the oldest data version in a column; 2) the TI for each data version among the same column has to be contiguous, namely, for any two data versions D_1 and D_2 in an IHR column,

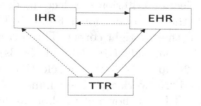

Fig. 7. Transformation between IHR, EHR and TTR

if D_2 is the immediate successor of D_1, it denotes $D_2.TI.Sta = D_1.TI.End$. As the EHR and TTR in Figure 3 and Figure 4 violate the condition 2, they cannot be transformed to IHR.

We represent the table transformations between IHR, TTR and EHR in Figure 7. The dotted lines indicate the transformations may not be possible where the solid lines denote the transformations are always possible.

As IHR is the default table representation supported by CoNoSQLDBs, it is more natural for users to directly issue queries against IHR. However, as we have already seen in Figure 2, the implicit TI representation strategy of IHR can cause wrong or misleading results during query processing. Hence, to guarantee the soundness of query processing, an IHR table has to be translated into either a TTR table or an EHR table. The table transformation tasks could be either automatically inserted by the query processing engine or explicitly specified by users. The former would correspond to a model where users issue the queries against IHR and the EHR and TTR are only used internally for query processing. However, as not every EHR or TTR table can be transformed back to an equivalent IHR, it is possible that users unexpectedly see the internal table representation in the query result. If users are allowed to perform table transformations explicitly, it implies that the users should also have the ability to process the EHR or TTR tables and the corresponding algebra operators need to be defined. As the first approach is not really transparent to the user, the second approach is more preferable. However, both approaches may be worth considering and are supported by the algebra we present in this paper. The temporal operators of EHR and TTR will be introduced in Section 5.

Although both TTR and EHR can guarantee the soundness of query processing, each of them has drawbacks. For storing TTR tables in CoNoSQLDBs, the TI has to be encoded into the row key to guarantee its uniqueness, as composite keys are not supported in CoNoSQLDBs. This strategy can also cause significant data redundancy when splitting IHR or EHR tuples. For example, in Figure 6, the strings "Tom" and "Telecom" appear 4 times and twice, respectively. As CoNoSQLDBs usually manage a tremendous volume of data, the volume of TTR tables may exhaust the disk capacity before any data processing. EHR has an optimal structure for data storage, but its physical representation in CoNoSQLDBs is very complicated. For example, in HBase, we have to "encode" the pair (*Val, TI*) as the data value (e.g. JSON string) and TI.Sta as the TS for each data version. Hence, the data processing tasks for EHR will need more time to "extract" the actual data values compared to TTR. Choosing the table representation for temporal query processing is therefore a trade-off between data capacity (TTR) and data processing complexity (EHR), which we plan to explore further in the future.

5   Temporal Operators for CoNoSQLDBs

As we have already described in Section 4.2, one IHR can be mapped to one EHR and one TTR. Moreover, one EHR can be transformed into one TTR and vice versa. In this section, we first introduce a set of temporal operators (TTRO) for TTR. Then, we define the temporal operators (CTO) for EHR.

5.1 TTRO Operator Model

If temporal relational data is modeled by exploiting tuple time-stamping (TTS), the temporal relational algebra (TRA) [3,6], which is an extension of the relational algebra, can be used for data processing. In the context of CoNoSQLDBs, TTR tables follow the TTS model. Intuitively, we can directly utilize the TRA for processing TTR tables. However, different from the general TTS table, to guarantee the uniqueness of row key in CoNoSQLDBs, each TTR table must integrate the time interval into the row key rather than represents it as two separate columns. We model the row key in TTR as a pair (srk, TI). srk denotes the row key value derived from its corresponding IHR table and TI indicates the valid time interval which has the form [Sta, End). Moreover, as the row key is mandatory for the CoNoSQLDB (TTR) tables, it still must be included in the final results even it is not indicated in the desired attributes, e.g. projection. Hence, to satisfy the characteristics of the TTR tables, we extend and customize TRA to a temporal operator model called TTRO for the TTR relations. Before presenting the details of TTRO operators, let us first adapt some concepts and definitions from [3,6] to the TTR context.

Definition 1 (Value Equivalent). *Let r be any TTR table. Two tuples t_1 and t_2 on r are value equivalent (written $t_1 \cong t_2$) if and only if all families:columns and rk.srk have the same values in both tuples.*

Definition 2 (Coalesce operation). *The functionality of the Coalesce operation (denoted by $\boxplus$) is to combine all the value-equivalent tuples of a TTR table together, when their TIs are overlapping or adjacent.*

To simplify the definition of the TTRO operators, we define a function $overlap()$ which takes two tuples t_1 and t_2 as input and returns $t_1.rk.TI \cap t_2.rk.TI$.

5.1.1 Union Operator $\cup^T$

Let r_1 and r_2 be two TTR tables which share the same schema definitions. The union of these two tables is defined as follows:

$$r_1 \cup^T r_2 = \boxplus(r_1 \cup r_2), \text{where} \cup \text{ is the relational union operator.}$$

In the definition, we first union the tuples from two tables together and then apply coalesce operation to combine multiple tuples which are value equivalent and their TIs are overlapping or adjacent. Figure 8 shows an example of $\cup^T$. Tuples "Tom/1/2" and "Tom/2/3" are value equivalent and hence are coalesced into "Tom/1/3".

Network-speed		
RK/Sta/End	Network:Supplier ttl=∞	Network:Speed ttl=∞
Tom/1/2	Telecom:1	1115K:1

$\cup^T$

Network-speed'		
RK/Sta/End	Network:Supplier ttl=∞	Network:Speed ttl=∞
Tom/2/3	Telecom:2	1115K:2
Tom/4/∞	Vodafone:4	1270K:4

$=$

Network-speed''		
RK/Sta/End	Network:Supplier ttl=∞	Network:Speed ttl=∞
Tom/1/3	Telecom:1	1115K:1
Tom/4/∞	Vodafone:4	1270K:4

Fig. 8. Example of Union operation

5.1.2   Difference Operator $-^T$

Let r_1 and r_2 be two TTR tables which share the same schema definitions. The difference of these two tables is given as follows:

$$r_1 -^T r_2 = \{t | ((t \in r_1) \land (\neg \exists t_2 \in r_2 | (t \cong t_2) \land (overlap(t, t_2) \neq \emptyset))) \lor$$
$$(\exists t_1 \in r_1, \exists t_2 \in r_2 | (t_1 \cong t_2) \land (t \cong t_2) \land (t.rk.TI \in (t_1.rk.TI - overlap(t_1, t_2)))$$
$$\land ((t_1.rk.TI - overlap(t_1, t_2) \neq \emptyset) \land (overlap(t_1, t_2) \neq \emptyset)) \}.$$

In the difference definition, the tuples in r_1 will be directly emitted, when there does not exist any tuples in r_2 in which they are value equivalent and their TIs have overlaps (line 1). Otherwise, the TIs of tuples in r_1 need to be modified. Figure 9 displays 3 possible temporal relationships between r_1 and r_2 to denote the values of (r_1.rk.TI-overlap(r_1,r_2)):

1. $\{[t_1.rk.TI.Sta, t_2.rk.TI.Sta)\}$;
2. $\{[t_2.rk.TI.End, t_1.rk.TI.End)\}$;
3. $\{[t_1.rk.TI.Sta, t_2.rk.TI.Sta), [t_2.rk.TI.End, t_1.rk.TI.End)\}$;

An example of $-^T$ is shown in Figure 10 where the TI of tuple "Tom/1/3" is changed to [2s, 3s).

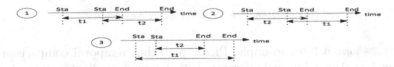

Fig. 9. Various temporal relationships between t_1 and t_2

Network-speed''		
RK/Sta/End	Network:Supplier ttl=∞	Network:Speed ttl=∞
Tom/1/3	Telecom:1	1115K:1
Tom/4/∞	Vodafone:4	1270K:4

$-^T$

Network-speed		
RK/Sta/End	Network:Supplier ttl=∞	Network:Speed ttl=∞
Tom/1/2	Telecom:1	1115K:1

$=$

Network-speed'		
RK/Sta/End	Network:Supplier ttl=∞	Network:Speed ttl=∞
Tom/2/3	Telecom:2	1115K:2
Tom/4/∞	Vodafone:4	1270K:4

Fig. 10. Example of Difference operation

5.1.3   Intersection Operator $\cap^T$

$$r_1 \cap^T r_2 = r_1 -^T (r_1 -^T r_2).$$

Let r_1 and r_2 be two TTR tables which share the same schema definitions and the definition of $\cap^T$ can be derived from $-^T$.

5.1.4   Project Operator π^T

$$\pi_A^T(r_1) = \{t | \exists t_1 \in r_1 | (t.rk = t_1.rk) \wedge (t[A_1] = t_1[A_1]) \wedge ... \wedge (t[A_n] = t_1[A_n])$$

where A=$(A_1,...,A_n)$ and each A_i has a form of CF_i or $CF_i.Col_i\}$

In TRA, a project operation will only keep the columns indicated by the set of desired projection attributes (in our definition, it is denoted by notation "A"). However, in the TTR context, to guarantee that the output of projection is still consistent with the data model of CoNoSQLDBs (TTR tables), the row key must be "implicitly" included in each tuple. We say "implicitly" because the row key may not be specified in A. As the row key already contains the TI, we can also view π^T as temporal projection [6] or slice operator [5].

5.1.5   Filter Operator σ_p^T

Let r_1 be a TTR table. Let p denote a selection condition over the attributes of r_1, where the p is defined as follows:

1. $p = \emptyset$;
2. $p = a\theta b$, where $\theta \in \{<, >, \leq, \geq, \neq, =\}$, a and b can be atomic value constants, $rk.srk$ and $D.Val$;
3. $p = a\theta b$, where $\theta \in \{<, >, \leq, \geq, \neq, =\}$, a and b can be atomic value constants, $rk.TI.Sta$ and $rk.TI.End$;
4. $p = p\theta p$, where $\theta \in \{\wedge, \vee\}$.

For a better explanation, we classify p into four different categories, i.e. 1) no predicates (line 1); 2) atomic value comparisons (line 2); 3) temporal conditions (line 3); 4) predicates with logical connectives (line 4). σ_p^T is defined as follows:

$$\sigma_p^T(r_1) = \{t | \exists t_1 \in r_1 | (t = t_1) \wedge (p(t_1) = true)\}.$$

Figure 4 shows a filter example. Please note that temporal comparison operators such as Allens interval operators [13] or period predicates supported by SQL 2011 [7] can be easily translated by temporal conditions (line 3) with logical connectives (line 4) and therefore could be easily added as syntactic sugar.

5.1.6   Cartesian Product $\bowtie^T$

$$r_1 \bowtie^T r_2 = \{t | \exists t_1 \in r_1, \exists t_2 \in r_2 | (t[Attr(R_1) - rk] = t_1[Attr(R_1) - rk]) \wedge$$
$$(t[Attr(R_2) - rk] = t_2[Attr(R_2) - rk]) \wedge$$
$$(t.rk.srk = concat(t_1.rk.srk, t_2.rk.srk)) \wedge (t.rk.TI = overlap(t_1, t_2)) \wedge$$
$$(t.rk.TI \neq \emptyset)\}.$$

Let r_1 and r_2 be two TTR tables. The Cartesian product of these tables is defined as above. As each TTR (CoNoSQLDB) table can merely contain one row key column, we define a "$concat$" function to concatenate the srk of both tuples (line 2). Moreover, both tuples have to be temporally valid during the same time period (line 2). Figure 11 shows an example of $\bowtie^T$.

Network-speed

RK/Sta/End	Network:Supplier ttl=∞
Tom/1/2	Telecom : 1
Tom/4/∞	Vodafone : 4

$\boxtimes^T$

Manager-infor

RK/Sta/End	Company:Name ttl=∞
Jim/2/3	1&1 : 2
Green/7/9	Vodafone : 7

=

Network-infor

RK/Sta/End	Network:Supplier ttl=∞	Company:Name ttl=∞
Tom/ Green/7/9	Vodafone : 7	Vodafone : 7

Fig. 11. Example of Cartesian product

5.1.7   Theta-Join $\bowtie^T$

The definition of $\bowtie^T$ can be defined from $\boxtimes^T$ and σ_p^T.

5.2   CTO Operator Model

We have already seen that each TTR table can be transformed to its corresponding EHR table (the transformation task is defined by T_{TE}). EHR falls into the attribute time-stamping (ATS) model, as each EHR column maintains multiple data versions attached with the explicit TIs. To process ATS relations in the temporal database context, traditional relational algebra is extended to the historical relational algebra (HRA) with two table restructuring operators (*Nest* and *Unnest*) [4,5]. However, this strategy is not suitable for EHR table processing, as 1) the quantity of EHR tables is always massive. Hence, the table restructuring operators can become very expensive; 2) after an EHR table is processed using the Unnest operator, its corresponding 1NF representation is not closed under HRA. For example, the HRA projection will discard the row key column if it is not specified in the projection attributes.

Hence, to process the temporal data in the EHR context, we propose a novel temporal operator model called *CTO*. The CTO model is defined under the following considerations: 1) each CTO operator can be directly applied to EHR tables without first changing the table structure; 2) the class of EHR tables is closed under the CTO model, namely, the output of each operator must still be an EHR table. In the following, we utilize the TTRO operators together with the table transformation operations (T_{ET} and T_{TE}) to define the operational semantics of CTO operators. Note that this is only for definitional purposes. The CTO operator implementations do not perform transformations to TTR and back.

5.2.1   Union Operator $\cup^C$

Let r_1 and r_2 be two EHR tables which share the same schema definitions. $r_1 \cup^C r_2 = T_{TE}(T_{ET}(r_1) \cup^T T_{ET}(r_2))$. Figure 12 shows this example.

5.2.2   Difference Operator $-^C$

Let r_1 and r_2 be two EHR tables which share the same schema definitions. $r_1 -^C r_2 = T_{TE}(T_{ET}(r_1) -^T T_{ET}(r_2))$. Figure 13 shows this example.

Fig. 12. $\cup^C$ Example

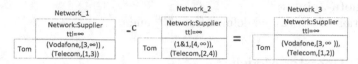

Fig. 13. $-^C$ Example

5.2.3   Intersect Operator $\cap^C$

Let r_1 and r_2 be two EHR tables which share the same schema definitions. The definition of $\cap^C$ is derived from $-^C$: $r_1 \cap^C r_2 = (r_1) -^C (r_1 -^C r_2)$.

5.2.4   Project Operator π_A^C

Let r_1 be an EHR table. $\pi_A^C(r_1) = T_{TE}(\pi_A^T(T_{ET}(r_1)))$.

5.2.5   Filter Operator σ_p^C

Let r_1 be an EHR table. $\sigma_p^C(r_1) = T_{TE}(\sigma_{p'}^T(T_{ET}(r_1)))$. The corresponding example is shown in Figure 3. Please note that, the data version temporal comparisons in p have to be translated to row key temporal comparisons in p'.

5.2.6   Cartesian Product $\bowtie^C$

As the value of each column in EHR is non-atomic (multiple data versions), the EHR tables satisfy NF^2. This property implies that it is possible to do a Cartesian product at various nested levels. However, different than the general NF^2, the nested depth of any EHR table is fixed. This characteristic prohibits doing Cartesian product at the arbitrary nested level. Figure 14 shows this situation. Suppose we wish to do the product operation at the level of CF_1:Col_{11} in R_1 and the level of table R_2. The desired schema is denoted at the right-hand side (R_3) which cannot be represented in CoNoSQLDBs.

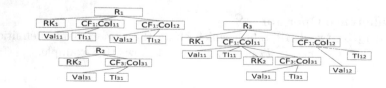

Fig. 14. A desired product results which cannot be represented in EHR

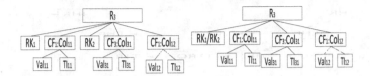

Fig. 15. Solution of Figure 14

The only solution of this problem is to reduce the nested depth of R_3. The left-hand side in Figure 15 shows the new structure. Clearly, its schema representation looks the same as the outermost Cartesian product (table level). As each CoNoSQLDB table can only has one row key column, we need to concatenate rk_1 and rk_2 (shown at the right-hand side in Figure 15). Hence, we define the Cartesian product for EHR as: $r_1 \boxtimes^C r_2 = T_{TE}(T_{ET}(r_1) \boxtimes^T T_{ET}(r_2))$, where the group key for T_{TE} is composed by $r_1.rk$ and $r_2.rk$.

5.2.7 Theta-Join $\bowtie^C$
The definition of $\bowtie^C$ can be defined from $\boxtimes^C$ and σ_p^C.

5.3 Query Examples

In this section, we show the query examples by using CTO and TTRO models. The input table is "Network-speed" (NS) shown at the left in Figure 2.

Query: What is the name of internet suppliers whose speed has at any time been faster than 1000K?

- CTO expression: $\pi_{Network.Supplier}^C (\sigma_{Network.Speed.Val \geq 1000K}^C (T_{IE}(NS)))$.
- TTRO expression: $\pi_{Network.Supplier}^T (\sigma_{Network.Speed.Val \geq 1000K}^T (T_{IT}(NS)))$.
- CTO$\Rightarrow$TTRO: $T_{ET}(\pi_{Network.Supplier}^C (\sigma_{Network.Speed.Val \geq 1000K}^C (T_{IE}(NS))))$.
- TTRO$\Rightarrow$CTO: $T_{TE}(\pi_{Network.Supplier}^T (\sigma_{Network.Speed.Val \geq 1000K}^T (T_I(NS))))$.

	Network-speed-EHR
	Network:Supplier ttl=∞
Tom	(Vodafone,[4,∞)), (Telecom,[1,3))

	Network-speed-TTR	
RK/Sta/End	Network:Supplier ttl=∞	
Tom/1/2	Telecom : 1	
Tom/2/3	Telecom : 2	
Tom/4/∞	Vordafone : 4	

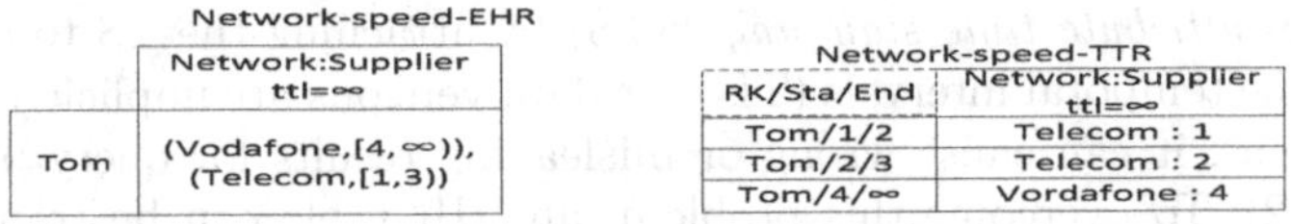

Fig. 16. Query results

As the IHR can cause misleading results during the filter processing, users first issue the table transformation operation T_{IT} or T_{IE}. Then, either TTRO operators or CTO operators can be exploited due to the table representations. Figure 16 depicts the results of query processing. Please note that, although the row key is not specified in the projection attributes, it is still included in the final results.

Fig. 17. Data processing and table transformation stack

5.4   Summary

As we have seen the example in Section 4.2, IHR can cause wrong or misleading results during query processing because of its implicit TI strategy. To overcome this problem, we can either translate IHR to EHR or to TTR. TTR utilizes the *tuple time-stamping model* and we extended the temporal relational algebra to the TTRO model. As can be seen from the definitions, the class of *TTR* tables is closed under the *TTRO* operators. EHR follows the *attribute time-stamping model*. Simply using historical relational operators with two table restructuring operations is not appropriate in the EHR context (See the discussion in 5.2). We hence proposed a novel temporal operator model (CTO). Obviously the class of *EHR* tables is closed under *CTO* operators.

Figure 17 shows an overview of the temporal data processing and table transformation stack which is a refinement of Figure 7. To process the temporal data in CoNoSQLDBs, users can either write a script using CTO or TTRO operators as well as representation transformation operators (the same strategy as Pig Latin) or a SQL-like language can be built on top of CTO or TTRO.

6   Conclusions and Further Work

To our knowledge, our work is the first proposal for defining temporal operators based on the characteristics of CoNoSQLDBs. We first clarify the meaning of TS and describe various table representations, such as *implicit history representation* (IHR), *explicit history representation* (EHR) and *tuple time-stamping representation* (TTR). IHR is the original CoNoSQLDB table representation which utilizes *attribute time-stamping* (ATS) by attaching the TS to each data version. As the temporal intervals (TIs) for data versions are implicitly included among columns, it can cause wrong or misleading results for query processing (See Figure 2). To overcome this problem, an IHR table can be translated to either EHR or TTR table format with explicit TI representations. EHR uses the ATS model where TTR falls into the *tuple time-stamping model*. For processing TTR tables, we introduce the TTRO operator model as a minor extension of the temporal relational algebra. For processing EHR tables directly, we propose a novel temporal operator model called CTO which can be applied to EHR without additional table restructuring. We showed that not every resulting EHR or TTR table can be transformed back into IHR and pointed out in which situations this is in fact possible. Both TTRO and CTO include seven temporal operators, such as *Union, Difference, Intersection, Project, Filter, Cartesian product* and *Theta-Join* with auxiliary table transformation operators. Moreover, the

TTR and EHR tables are closed under the TTRO model and the CTO model, respectively.

In further work, we plan to give more deliberate classifications of EHR and TTR to denote when the transformations of EHR to IHR or TTR to IHR are possible. Moreover, we consider extending the *TTRO* and *CTO* operators with temporal aggregation functions and the ways to efficiently implement the TTRO and CTO operators.

References

1. Codd, F.: A Relational Model of Data for Large Shared Data Banks. Community, 377–387
2. Makinouchi, A.: A Consideration on Normal Form of Not-necessarily-normalized Relation in the Relational Data Model. In: VLDB 1977, pp. 447–453 (1977)
3. Richard, S.: The TSQL2 Temporal Query Language. Kluwer (1995) ISBN 0-7923-9614-6
4. Clifford, J., et al.: On completeness of historical relational query languages. PACM Transactions on Database Systems, 64–116 (March 1994)
5. Tansel, A.: Temporal Relational Data Model. IEEE Transactions on Knowledge and Data Engineering, 464–479 (May 1997)
6. Dey, D., et al.: A complete temporal relational algebra. Journal the VLDB Journal 5(3) (May 1997)
7. Kulkarni, K., et al.: Temporal features in SQL: 2011. ACM SIGMOD, 34–43 (September 2012)
8. Dean, J., Ghemawat, S.: MapReduce: Simplified Data Processing on Large Clusters. In: OSDI, pp. 137–150 (2004)
9. Change, F., et al.: Bigtable: A Distributed Storage System for Structured Data. In: OSDI, pp. 205–218 (2006)
10. Apache HBase, http://hbase.apache.org/
11. http://pig.apache.org/
12. http://hive.apache.org/
13. Allen, J.: Maintaining knowledge about temporal intervals. Communications of ACM 26 (November 1983)
14. Hu, Y., Dessloch, S.: Extracting Deltas from Column Oriented NoSQL Databases for Different Incremental Applications and Diverse Data Targets. In: Catania, B., Guerrini, G., Pokorný, J. (eds.) ADBIS 2013. LNCS, vol. 8133, pp. 372–387. Springer, Heidelberg (2013)

Analyzing Sequential Data
in Standard OLAP Architectures

Christian Koncilia[1], Johann Eder[1], and Tadeusz Morzy[2]

[1] Alpen-Adria-Universität Klagenfurt
Dep. of Informatics-Systems
{eder,koncilia}@isys.uni-klu.ac.at
[2] Poznan University of Technology
Institute of Computing Science
morzy@put.poznan.pl

Abstract. Although nearly all data warehouses store sequential data, i.e. data with a logical or temporal ordering, traditional data warehouse or OLAP approaches fail when it comes to analyze those sequences. In this paper we will present a novel approach which generates query-specific subcubes, i.e. subcubes that consist only of data which fulfill a given sequential query pattern. These subcubes may then be analyzed using standard OLAP tools. Our approach consists of two functions which both return such subcubes. Hence, the user can still use all the well-known OLAP operations like drill-down, roll-up, slice, etc. to analyze the cube. Furthermore, this approach may be applied to all data warehousing architectures.

1 Introduction

Business Intelligence (BI), Data Warehousing (DWH), and On-Line Analyical Processing (OLAP) enable users to perfomantly analyze mass data by storing data in No-SQL database systems, e.g. multidimensional database systems, or by applying DWH specific logical schemas to relational database systems, e.g. the Star Schema, Snowflake Schema, etc. [6].

Traditional business intelligence tools analyze facts along dimensions. Facts describe what a user wants to analyze whereas dimensions describe how the user analyses his data [6]. Typical examples for facts are Turnover, Profit, the Stock of Inventory, etc. These facts may then be analyzed along a set of dimensions like Time, Products or Geography.

This approach succeeded to proof its feasibility in innumerable implementations in many industrial sectors. However, this approach fails when it comes to efficiently analyze sequential data, i.e. data with a logical or temporal ordering [9].

Why does the traditional DWH approach fail when it comes to sequential data analysis? Assume that we store data about treatment costs and diagnoses for patients in a DWH. Traditional data warehouses are built to answer questions like "what are the total costs for patients in 2010" or "what are the average costs

Y. Manolopoulos et al. (Eds.): ADBIS 2014, LNCS 8716, pp. 56–69, 2014.

for all patients diagnosed cerebral infarction". However, they are not prepared to answer queries like "what are the follow-up costs of patients diagnosed cerebral infarction within 12 months after the diagnose". This even gets more complicated, when analyzing data along several events, e.g. when analyzing follow-up costs for patients with a certain diagnose who received a certain treatment within a given time period after the diagnose.

Although sequential data representation is not a new research area, the fact that most data sets in an OLAP system are sequential by nature has been ignored until recently, e.g. in [1,9,8]. These approaches focus on developing novel data warehouse / OLAP architectures. This allows to develop new operators, query languages, indexing and caching strategies, etc. However, in our opinion there is also an evident need to analyze sequential data in existing OLAP infrastructures.

Contribution: In this paper we will present a sophisticated approach which enables the user to analyze simple atomic events and complex sequences of events. In contrast to other approaches (which will be discussed in section 7), our approach smoothly integrates into a standard OLAP architecture. Basically, our approach consists of the following steps:

1. The user defines the sequence she / he wants to analyze, e.g. all patients who had a specific diagnose A after a diagnose B.
2. A subcube is generated which contains all relevant data, e.g. all patients records for all patients who had a diagnose A after a diagnose B.
3. An additional dimension Relative Time Axis is created enabling the user to analyze data in a very flexible way.

The result of a sequential query in our approach is itself a standard OLAP (sub-)cube. Hence, the user can still use all the well-known OLAP operations like *drill-down, roll-up, slice, dice* and so on to analyze this cube.

This paper is organized as follows: In section 2 we will briefly describe a motivating example which we will use throughout the rest of this paper to depict the application of our approach. Section 3 will provide a formal model of a data warehouse which we will extend in section 4 with our sequential OLAP approach. We will present the prototypical implementation of our approach in section 5. In section 6, we are going to briefly discuss some application areas for a sequential OLAP approach. Related work will be discussed in section 7. Finally, we will conclude this paper in chapter 8.

2   Motivating Example

In this section, we will present our motivating example which we will use as running example throughout the rest of the paper. Consider a database with the following table storing information about patients, diagnoses and treatment costs:

Patient	Diag	Date	Costs
Tim	I26	1-1	50
Tim	C11	1-2	70
Walter	I26	1-8	45
Tim	I27	1-8	110
John	B32	1-2	80
Walter	C11	1-2	60

In this example the patient Tim went to a doctor on 1/1/10 and was diagnosed with ICD (International Classification of Diseases) code I26 (the code for the disease *pulmonary embolism*). The next day, he wanted to get a second opinion and went to a different doctor who diagnosed a different disease encoded C11. Then, a few days later, he went to a third doctor who diagnosed I27.

The star schema for a data warehouse to analyze this information consists of a fact table storing the costs, and three dimension tables (*Patient, Diagnose, Date*). Easily one can use this data warehouse to answer queries like "what are the total costs for patient Tim in 2010" or "what are the average costs for all patients diagnosed I26". However, such a data warehouse structure would not be suitable to answer queries where a dimension member depends on another dimension member, i.e. where we have sequences.

As we will discuss in section 4.1, such a sequential OLAP query may be based on *Atomic Sequences* or on *Complex Sequences*. Queries based on atomic sequences are queries that make use of only one single event, e.g. "What are the follow-up costs of patients during three month after she/he has been diagnosed I26". In this example, the single event would be the diagnose I26.

In contrast to such a query, a query based on complex sequences consists of two or more events. An example for such a query would be: "How many patients have been diagnosed H35 (Retinopathie, a disease often caused by diabetes which can lead to blindness) within 12 months after they have been diagnosed E10 (diabetes)." This query would consist of two events, namely the diagnose E10 and the diagnose H35.

Of course, such a complex sequence query is not restricted to two events nor is it restricted to events that stem from one dimension in the data warehouse. For instance, if we would store prescriptions in our data warehouse we could also state queries like: "What are the average follow-up costs of a diagnose I26 for patients that have been prescribed Heparin within 6 months after the diagnose". This query would consist of two events stemming from two different dimensions.

3 Formal OLAP Model

In this section we will give a formal definition of a data warehouse based on the model presented in [7]. Later on we will extend this data warehouse model such that the user is able to state all kinds of sequential queries. Please note that our approach for sequential OLAP simply extends the standard OLAP approach.

The result of a sequential query is itself a standard (sub-)cube, extended with a set of relative time axes. Hence, the user can still use all the well-known OLAP operations like drill-down, roll-up, slice, dice and so on to analyze her or his cube.

Intuitively, we define the schema of a data warehouse as a set of cubes which again are defined as a set of dimensions. The schema of each dimension is defined by a set of categories, e.g., the dimension $Date$ might consist of the categories $Year$, $Month$ and Day organized in a hierarchical relation $Year \rightarrow Month \rightarrow Day$, where for example $Year \rightarrow Month$ means that a month rolls-up to a year.

Each category consists of a set of dimension members. Dimension members define the instances of a data warehouse schema. For instance, $January$, $February$ and $March$ are dimension members assigned to the category $Month$.

Formally, the schema of a data warehouse is defined by:

i.) A number of dimensions J.
ii.) A set of dimensions $\mathbb{D} = \{D_1, ..., D_J\}$, where $D_i = < ID, D_{Key} >$. ID is a unique identifier of the dimension. D_{Key} is a user defined key (e. g. , the name of the dimension), which is unique within the data warehouse.
iii.) A number of categories K.
iv.) A set of categories $\mathbb{C} = \{C_1, ..., C_K\}$ where $C_i = < ID, C_{Key} >$. ID is a unique identifier of the category. C_{Key} is a user defined key (e. g. , the name of the category) which is unique within the data warehouse.
v.) A set of assignments between dimensions and categories $\mathbb{A}_{DC} = \{A^1_{DC}, ..., A^N_{DC}\}$, where $A^i_{DC} = < D.ID, C.ID >$. $D.ID$ represents the identifier of the corresponding dimension. $C.ID$ represents the identifier of the corresponding category.
vi.) A number of hierarchical category assignments O.
vii.) A set of hierarchical category assignments $\mathbb{HC} = \{HC_1, ..., HC_O\}$ where $HC_i = < ID, C.ID_C, C.ID_P >$. ID is a unique identifier of the hierarchical category assignment. $C.ID_C$ is the identifier of a category, $C.ID_P$ is the category identifier of the parent of $C.ID_C$ or $\emptyset$ if the category is a top-level category.
viii.) A number of cubes I.
ix.) A set of cubes $\mathbb{B} = \{B_1, ..., B_I\}$ where $B_i = <ID, B_{Key}, S >$. ID is a unique identifier of the cube (similar to $O_{id's}$ in object-oriented database systems). B_{Key} is a user defined key (e. g. , the name of the cube), which is unique within the data warehouse.
   S represents the schema of the cube. The tuple S consists of all dimensions and hierarchical category assignments that are a part of this cube. Therefore, S is defined as $S = (\mathbb{D}, \mathbb{A})$ where $\mathbb{D} = \{D_1.ID, ..., D_N.ID\}$ $(N \leq J)$ and $\mathbb{A} = \{HC_1.ID, ..., HC_M.ID\}$ $(M \leq O)$.

The instances of a data warehouse are defined by:

i.) A number of dimension members P.
ii.) A set of dimension members $\mathbb{M} = \{M_1, ..., M_P\}$ where $M_i = < ID, M_{Key}, \mathbb{CA}, >$. ID is a unique identifier of the dimension member. M_{Key} is a user

defined key (e. g., the name of the dimension member), which is unique within the data warehouse. The set $\mathbb{CA}$ represents the set of categories, to which the corresponding dimension member is assigned.

iii.) A set of hierarchical member assignments $\mathbb{HM} = \{ HM_1, ..., HM_O \}$ where $HM_i = <ID, M.ID_C, M.ID_P, f>$. ID is a unique identifier of the hierarchical member assignment. $M.ID_C$ is the identifier of a dimension member, $M.ID_P$ is the dimension member identifier of the parent of $M.ID_C$ or $\emptyset$ if the dimension member is at the top-level. f represents the consolidation function between $M.ID_C$ and $M.ID_P$, e. g. $+$ for addition, $-$ for subtraction, etc.

iv.) A function $cval : (M_{D_1}, ..., M_{D_N}) \rightarrow measure$, which uniquely assigns a measure to each vector $(M_{D_1}, ..., M_{D_N})$ where $(M_{D_1}, ..., M_{D_N}) \in \mathcal{M}_{D_1} \times ... \times \mathcal{M}_{D_N}$. The domain of this function is the set of all cell references. The range of this function are all measures of a cube.

4   Sequential OLAP Model

In this chapter, we will extend the OLAP model presented in section 3. The extension basically consists of two items: 1) we will introduce the concept of sequential OLAP functions and 2) we will enrich this model with the concept and definition of a relative time axis.

Intuitively, a sequential OLAP function can be considered as an extended slice operation and a relative time axis represents the time difference between a given event and any other event.

4.1   Sequential OLAP Function and Events

Basically, a sequential OLAP function takes a cube, a grouping dimension, an ordering dimension and a sequence of events as input and returns a subcube as output. The terms *grouping dimension, ordering dimension* and *sequence of events* will be defined in section 4.2. The query "fetch all patient records for patients which have been diagnosed retinopathie after they have been diagnosed diabetes" could be an example for a sequential OLAP function. This query would result in a subcube that consists of all dimensions of the corresponding cube and all dimension members and measures which belong to patients that have been diagnosed retinopathie after a diagnose diabetes. This subcube may then serve as basis for analysis which for instance easily enable the user to compute follow-up costs.

The fundamental basis for our sequential OLAP function are sequences. We distinguish between two different kinds of sequences:

1.) **Complex Sequence:** A complex sequence forms a path through a set of events, e.g. a sequence $\mathcal{E}_1 \rightarrow \mathcal{E}_2 \rightarrow ... \rightarrow \mathcal{E}_n$ where $\mathcal{E}_i$ is an event.

2.) **Atomic Sequence:** An atomic sequence is a subset of complex sequence. It represents a one stepped path, i.e. $\mathcal{E}_1$. For instance, $\mathcal{E}_1$ may be the event "diagnosed diabetes".

An event $\mathcal{E}$ is an appearance of an incident at a given point of time. In our context, we can define an event $\mathcal{E}$ as the existence of a function $cval$ with $cval(M_t, M_e, \dots) \neq null$ with a given dimension member M_e that defines the incident and a dimension member M_t that defines the point of time.

$\mathcal{E}_j \rightarrow \mathcal{E}_k$ means that the event $\mathcal{E}_k$ occurred directly after event $\mathcal{E}_j$, i.e. there exists no event $\mathcal{E}_l$ between $\mathcal{E}_j$ and $\mathcal{E}_k$ along an ordering dimension defined by the user. Usually, this ordering dimension will be the time dimension.

4.2 Sequential OLAP Function for Atomic Sequences

As a complex sequence can be decomposed to a set of atomic sequences, we will start by defining the sequential OLAP function for atomic events.

Pre-Conditions: The following pre-conditions for a sequential OLAP function on atomic sequences have to be fulfilled:

1.) A cube B_i has to be defined as in section 3. This cube serves as input, i.e. it defines the base for the sequential OLAP function.
2.) B_i has to contain at least one ordering dimension D_o. An ordering dimension is a dimension on which an ordering function f_{order} has been defined. $f_{order}(M_j, M_k)$ takes any two dimension members M_j and M_k and returns -1 if $M_j < M_k$, 0 if $M_j = M_k$ or $+1$ if $M_j > M_k$.
3.) The user has to define a grouping dimension D_g. This grouping dimension defines the subject of the analysis. Hence, D_g defines which dimension the event refers to, i.e. which dimension the order of the ordering dimension refers to. D_g may be any dimension of B_i.
4.) Furthermore, the user has to define a single (atomic) event $\mathcal{E}$ with $\mathcal{E} = M_E$ where M_E is a dimension member of D_E and D_E is a dimension in B_i.

Definition: Now, the function $solap$ can be defined as follows: Given the input B_i, D_o, D_g and $\mathcal{E}$ the function $solap(B_i, D_o, D_g, \mathcal{E}, \mathcal{E}_p)$ returns a subcube B_o which consists of all dimensions $D_i \in B_i$ and all dimension members M_i with $M_g \in D_g \wedge \exists cval(M_{D_1}, \dots, M_i, M_g, M_E, \dots, M_{D_N}) \neq null \wedge M_E = \mathcal{E}$.

Please note that $\mathcal{E}_p$ is not used in atomic sequences and will be discussed later on in section 4.3.

Intuitively we can say that a $solap(B_i, D_o, D_g, \mathcal{E}, \mathcal{E}_p)$ returns a subcube which consists of all the data of all dimension members in the grouping dimension for which there exists at least one entry in the fact table that represents the given event.

Example: Assume that B_i is the cube as defined in our running example in section 2. The ordering dimension D_o is the dimension $Date$. The grouping dimension D_g, i.e. the subject of our analysis, is the dimension $Patient$. The event $E = I26$.

Taking these input parameters, the function $solap(B_i, Date, Patient, I26)$ would return a subcube which consists of all the data of all patients who had a diagnose I26, i.e. it would return a subcube which consists of the data represented in the following table:

Patient	Diag	Date	Costs
Tim	I26	1-1	50
Tim	C11	1-2	70
Walter	I26	1-8	45
Tim	I27	1-8	110
Walter	C11	1-2	60

4.3   Sequential OLAP Function for Complex Sequences

In the previous section we defined the function *solap* for atomic events. We will now extend this function to work on complex sequences.

Pre-Conditions: The pre-conditions are the same as defined in section 4.2 except the fact that the user may define any sequence of events $\mathbb{E} =< \mathcal{E}_1, \ldots, \mathcal{E}_n >$ with $\mathcal{E}_i = M_{E_i}$ where M_{E_i} is a dimension member of D_{E_i} and D_{E_i} is a dimension in B_i.

Furthermore, as complex sequences have to consider the ordering of several events, we have to extend the *solap* function with an additional parameter, namely $\mathbb{E}_p$. In an atomic sequence, $\mathbb{E}_p$ is always *null*. In a complex sequence, $\mathbb{E}_p$ is the previous event in the sequence of events or *null*, if no previous event has been defined, i.e. if applying *solap* to the first event in a sequence of events.

Definition: First, extending the definition given in 4.2 with the parameter $\mathcal{E}_p$, the function *solap* can be defined as follows: Given the input $B_i, D_o, D_g, \mathcal{E}$ and $\mathcal{E}_p$ the function $solap(B_i, D_o, D_g, \mathcal{E}, \mathcal{E}_p)$ returns a subcube B_o which consists of all dimensions $D_i \in B_i$ and all dimension members M_i with $M_g \in D_g \wedge$ $\exists cval(M_{D_1}, \ldots, M_i, M_g, M_E, \ldots, M_{D_N}) \neq null \wedge M_E = \mathcal{E} \wedge$ $\exists cval(M_{D_1}, \ldots, M_i, M_g, M_{E_p}, \ldots, M_{D_N}) \neq null \wedge M_{E_p} = \mathcal{E}_p \wedge f_{order}(\mathcal{E}, \mathcal{E}_p) > 0$.

Secondly, with the extended definition of the *solap* function, we can define *solap* for complex sequences: Given the input B_i, D_o, D_g and $\mathbb{E}$ the function $solap(B_i, D_o, D_g, \mathbb{E}, \mathbb{E}_p)$ can now be defined as a composition of *solap* functions on atomic sequences:

$$solap(B_i, D_o, D_g, \mathbb{E}) =$$
$$solap(solap(\ldots solap(B_i, D_o, D_g, \mathcal{E}_1, null)\ldots, \tag{1}$$
$$D_o, D_g, \mathcal{E}_{n-1}, \mathcal{E}_{n-2}), D_o, D_g, \mathcal{E}_n, \mathcal{E}_{n-1})$$

Example: Again, let B_i be the cube as defined in our running example in section 2, *Date* be the ordering dimension D_o and *Patient* be the grouping dimension D_g. Now, the user would like to analyze all patient records about patients who had a diagnose I26 and afterwards a diagnose I27. Hence, $\mathbb{E} =< I26, I27 >$.

Taking these input parameters, the function $solap(B_i, Date, Patient, < I26, I27 >)$ would result in a function $B_{o_1} = solap(B_i, Date, Patient, I26)$ whose result B_{o_1} would serve as input parameter for $B_{o_2} = solap(B_{o_1}, Date,$

Patient, I27). Therefore, the resulting cube would consist of the data represented in the following table:

Patient	Diag	Date	Costs
Tim	I26	1-1	50
Tim	C11	1-2	70
Tim	I27	1-8	110

4.4 Relative Time Axis

The relative time axis function generates a new dimension in the cube which stores the difference between a given event and any other event. We will use the term relative time axis, although the concept of a relative time axis may be applied to any ordering dimension which doesn't necessarily have to be a time or date dimension.

In contrast to other time dimensions in the cube, the relative time axis is not a set of absolute timestamps like *12-30-2010* or *8-15-2010 10:42*, but a set of time intervals which are relative to the ordering dimension D_o (as described above, this ordering dimension is usually a time dimension). Thus, the relative time axis could for instance be a dimension with a set of dimension members $\{-n \ days, \ldots, -1 \ day, 0, +1 \ day, \ldots, +m \ days\}$.

Pre-Conditions: In order to compute a relative time axis, the following pre-conditions have to be fulfilled:

1.) A cube B_i has to be defined. Usually, this cube will be the result of a *solap()* function as defined in sections 4.2 and 4.3.
2.) As defined in section 4.2 this cube B_i has to contain at least one ordering dimension D_o. Furthermore, the user has to define a grouping dimension D_g (the subject of the analysis) with $D_g \in B_i$.
3.) The user has to define a single event $\mathcal{E}$ with $\mathcal{E} = M_E$ where M_E is a dimenson member of D_E and D_E is a dimension in B_i.
4.) As there might exist several cell values in the cube referred to by a function $cval(M_1, \ldots, M_g, \mathcal{E}, \ldots, M_n)$ with M_g being a dimension member assigned to D_g, the user has to define which occurrence of $\mathcal{E}$ should serve as base. Currently, this can be done by applying a *first()* or *last()* function, which sets the first or last occurrence $\mathcal{E}$ as base. Other functions could be implemented.

Definition: we define a function $rta()$ (relative time axis) whish uses a function $diff()$ to compute the difference between any two event occurrences. $diff()$ takes two records, i.e. two $cval()$ functions as defined in section 3, and the ordering dimension D_o as input and computes the differences between the two entries. The granularity of $diff()$ is equal to the granularity of D_o, e.g. if the granularity of D_o is a day, then $diff()$ will return the difference in days.

The function $diff()$ may be defined by the user. Usually, it simply computes the difference between two dates:

$$diff(cval(M_{o_1}, M_g, M_E, \ldots), cval(M_{o_2}, M_g, \ldots)) =$$
$$M_{o_1} - M_{o_2}$$
$$with M_{o_1}, M_{o_2} \in D_o \wedge M_g \in D_g \wedge M_E \in \mathcal{E}. \tag{2}$$

Using the defined function $diff()$ we can formally define the $rta()$ function. $rta(B_i, D_o, \mathcal{E})$ returns a cube B_o where B_o consists of the same schema $\mathcal{S}$ as B_i and all dimension members $\mathbb{M}$, hierarchical member assignments $\mathbb{HM}$ and all measures assigned to B_i. Furthermore, B_o consists of an additional dimension D_{RTA} with a set of dimension members $\mathbb{M}_{RTA} = \{M_1, \ldots, M_n\}$ assigned to D_{RTA} (via $\mathbb{CA}$, $\mathbb{C}$ and $\mathbb{A}_{DC}$ as defined in section 3). For each $M_i \in \mathbb{M}_{RTA}$ we can define that $M_i.M_{Key} = diff(x, y)$ where $x = cval(M_{O_1}, \mathcal{E}, \ldots)$ and $y = cval(M_{O_2}, \ldots)$ and $x \neq y$.

Example: Assume that B_o is the resulting cube of the function $solap(B_i, Date, Patient, I26)$ as described in section 4.2. Again, the ordering dimension D_o is the dimension $Date$. The grouping dimension D_g, i.e. the subject of our analysis, is the dimension $Patient$. The event $E = I26$.

Taking these input parameters, the function $rta(B_o, Date, Patient, I26)$ would return a subcube which consists of all the data of all patients who had a diagnose I26. Furthermore, this subcube would consist of an additional dimension named RTA which stores the difference between the occurrence of a diagnose I26 and any other event. The following table depicts the resulting cube:

Patient	Diag	Date	Costs	RTA
Tim	I26	1-1	50	0
Tim	C11	1-2	70	+1
Tim	I27	1-8	110	+7
Walter	C11	1-2	60	-6
Walter	I26	1-8	45	0

4.5 Workflow Example

In this section we will discuss how a user may use SOLAP() and RTA() to state sequential OLAP queries and how she may analyse the resulting cube.

Assume that a user would like to state a query like "what are the follow-up costs for patients diagnosed I26 within 12 month after they have been diagnosed I26"? To answer this query the user would select the $Date$ dimension as ordering dimension and $Patient$ as grouping dimension. Furthermore, he defines an atomic sequence with one event "Diagnose = I26". Now, the application would use the functions $SOLAP()$ and $RTA()$ (with the corresponding parameters) to generate a cube as depicted in Table 4.4.

This cube would enable the user to easily analyze the follow-up costs that occurred within 12 month after the diagnose I26. This could be done by applying standard OLAP functions to the cube. In this example, the user could simply

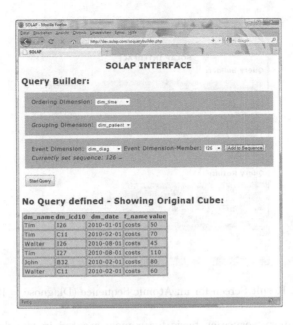

Fig. 1. Start Screen of our Prototype

select the dimension members $0 \dots 12$ of the dimension RTA (which would correspond to a slice and dice operation) and calculate the sum of the fact *cost*. The same method could be applied to analyze which diagnoses occurred within 3 months before a diganose I26.

5   Proof of Concept

We implemented a prototype of our approach as proof of concept. This prototype has been implemented as a web-client using a PostgreSQL 9.0.0 database, PHP 5.3.2 and jQuery 1.4.2. Technically, the data warehouse itself has been built using the traditional Star Schema approach. Hence, we have one fact table and several tables representing the dimensions of the cube. For our running example, this results in a fact table that consists of the costs and foreign keys to the three dimensions: Patient, Diagnose and Date.

Figure 1 shows a screenshot of the start screen of the prototype. For this paper, we imported the data from our running example.

Using the prototype depicted in Fig. 2 the user may select an ordering dimension and a grouping dimension. Furthermore, she or he may define a sequence of events, i.e. an atomic sequence or a complex sequence. Currently, the prototype does not support using wildcards in sequences. In this example, the user selected a single event, i.e. Diagnose I26.

Basically, the application takes the user inputs, extracts the sequence defined by the user, and dynamically generates an SQL query for the first step in this

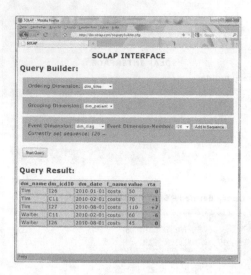

Fig. 2. Result Screen for an Atomic Sequence (Diagnose = I26)

sequence. This query serves as basis for a view created in the database. This view represents the subcube returned by the function $solap()$ as presented in section 4.2. For all subsequently defined sequence steps we repeat this process as defined in section 4.3. In contrast to the first step, all further steps work on the view defined in the previous step. Finally, the implementation calls the $rta()$ function as defined in section 4.4 to compute the relative time axis.

The result of this query is being depicted in figure 2. As can be seen, a new dimension "rta" has been created, representing the relative time axis.

6   Application Examples

In section 2 we discussed an application example originated in the health care sector. Basically, such a sequential OLAP approach would enrich each data warehouse that stores any kind of events, e.g. diagnoses, prescriptions, workflow tasks, sensor values and so on.

In this section we would like to briefly discuss some application examples for sequential OLAP are:

1.) **Workflow Systems:** Usually, a workflow system consists of several tasks. These tasks are linked with control structures like conditional branches, loops, joins and so on [13]. Analyzing worklow instances with OLAP or data warehouse techniques is tedious and sometimes impossible because of these control structures [5]. However, applying our sequential OLAP technique would enable us to reduce the complexity of an unlimited amount of possible instance structures to a limited amount of instance structures which follow a specific pattern, e.g. $A \rightarrow B \rightarrow * \rightarrow D$ would select all instances which used the task A followed by task B followed by any other set of tasks followed by task D.

2.) **Detecting Pharmacological Interactions:** Another application example would be a medical system to support doctors in avoiding dangerous pharmacological interaction. For instance, if a patient has already been prescribed Ciclosporin (an immunosuppressant drug usually used after organ transplants) and now gets a prescription from a different doctor for a barbiturate (drugs that act as central nervous system depressants). Taking both medicins at the same time may have dangerous interactions. To be more precise, a barbiturate negatively influences the effective level of Ciclosporin which may lead to organ repulsion. A sequential analysis would allow doctors to avoid prescribing such combinations of drugs.

3.) **Ticketing systems** for light rail traffic, skiing resorts or multi-storey car parks would be another application example. Here, a user could want to analyze different sets of customers which for instance took a specific route $A \rightarrow * \rightarrow X$, which means that they entered the subway at station A, changed trains at any station, and left the subway at station X.

4.) **Sensor data warehouses** would also be an interesting application area for a sequential data warehousing approach. Consider a data warehouse that stores information which stems from dozens sensors mounted at a power turbine. Analyzing sequences in this data warehouse could provide very useful information, e.g. to reduce down-times. For instance, we could want to analyze the allocation of heat of certain parts of the turbine within 30 seconds after a specific sensor reported a defined temperature.

7   Related Work

While the support of sequential data in traditional database management systems in general and specifically on time-sequences isn't a new topic (see [11], [12], [10], [2]), the term of Sequence OLAP or S-OLAP has been coined recently in [9]. In [9] the authors present an approach where a user defines a query based on pattern templates to analyze sequence data. A pattern template consists of a sequence of symbols where each symbol corresponds to a domain of values. In contrast to a pattern template, e.g. (A, B, A) a pattern is an instantiation of cell values corresponding to a pattern template. A prototypical implementation of such an S-OLAP system has been presented in [3].

The approach presented in [9] has been extended by the same research group in [4]. In [4] the authors focus on the efficient evaluation of ranking pattern based aggregate queries. As in [9] the number of dimensions of the defined cube is equal to the number of distinct values of the selected attribute in the source table.

In order to avoid an overwhelming amount of data to be presented to the user, [4] introduces support for top-k queries.

Another interesting approach has been presented in [8]. The authors combine two existing technologies, namely OLAP (Online Analytical Processing) and CEP (Complex Event Processing) to analyze real-time event data streams. They introduce patterns and pattern hierarchies. If a pattern A contains a subset of event types compared to a pattern B, then A is at a coarser level then B in the resulting pattern hierarchy. Based on these hierarchical relationships, the authors present different strategies how to exploit these hierarchies for query optimization.

The approach presented in [1] discusses a model to analyze time-point-based sequential data. The authors introduce a formal model and define several operators to create and analyze sequences. Furthermore, it formaly defines and discusses the notion of facts, measures and dimensions in the context of sequential OLAP.

Our approach differs from the approaches discussed in this section as follows: our approach is not a redefinition of the well know OLAP approach and architecture as for instance presented in [6], but an extension. To the best of our knowledge, it is the first sequential OLAP approach that smoothly integrates into existing OLAP systems.

8   Conclusion

Traditional data warehouse and OLAP approaches still fail when it comes to efficiently analyze sequential data, i.e. data with a logical or temporal ordering [9]. For instance, a query like "what are the follow-up costs of patients diagnosed cerebral infarction within 12 months after the diagnose" cannot be answered without a relative time axis defined in the data warehouse for the event defined in the query (here: diagnose cerebral infarction). A naive approach to solve this problem would be to create a relative time axis in advance for all combinations of events. However, such a naive approach will fail as the number of possible combinations will quickly blast the capacity of the cube.

In this paper we presented a novel and sophisticated approach that enables the user to analyze sequential data in a standard OLAP environment. The user may state simple queries that require only an atomic event or complex queries with a defined sequence of events. The result of our approach is itself a standard OLAP cube, extended with a new dimension representing the relative time axis. Thus, it is easy to implement our approach into an existing OLAP solution. Furthermore, the user may use her or his OLAP solution to analyze the resulting data.

We implemented this approach as a proof of concept. Basically, this implementation enables the user to define a sequence of events and automatically apply the defined functions $solap()$ and $rta()$ to a given data warehouse.

Future work will focus on wildcard support in sequence definitions. A wildcard may be a question mark "?", represeting any single event, an asterisk "$*$", representing any sequence of events or a plus "+", representing any sequence of events which consists of at least one event.

References

1. Bębel, B., Morzy, M., Morzy, T., Królikowski, Z., Wrembel, R.: Olap-like analysis of time point-based sequential data. In: Castano, S., Vassiliadis, P., Lakshmanan, L.V., Lee, M.L. (eds.) ER 2012 Workshops 2012. LNCS, vol. 7518, pp. 153–161. Springer, Heidelberg (2012)
2. Chandra, R., Segev, A.: Managing Temporal Financial Data in an Extensible Database. In: VLDB (1992)
3. Chui, C., Kao, B., Lo, E., Cheung, D.: S-OLAP: an OLAP System for Analyzing Sequence Data. In: SIGMOD (June 2010)
4. Chui, C., Lo, E., Kao, B., Ho, W.: Supporting Ranking Pattern-Based Aggregate Queries in Sequence Data Cubes. In: CIKM (2009)
5. Eder, J., Olivotto, G.E., Gruber, W.: A Data Warehouse for Workflow Logs. In: Han, Y., Tai, S., Wikarski, D. (eds.) EDCIS 2002. LNCS, vol. 2480, pp. 1–15. Springer, Heidelberg (2002)
6. Kimball, R.: The Data Warehouse Toolkit, 2nd edn. John Wiley & Sons (1996)
7. Koncilia, C.: The COMET Temporal Data Warehouse (PhD). In: UMI (2002)
8. Liu, M., Rundensteiner, E., Greenfield, K., Gupta, C., Wang, S., Ari, I., Mehta, A.: E-cube: Multi-dimensional event sequences processing using concept and pattern hierarchies. In: ICDE (2010)
9. Lo, E., Kao, B., Ho, W., Lee, S., Chui, C., Cheung, D.: OLAP on Sequence Data. In: SIGMOD (June 2008)
10. Segev, A., Shoshani, A.: Logical Modeling of Temporal Data. In: SIGMOD (1987)
11. Seshadri, P., Livny, M., Ramakrishnan, R.: Sequence query processing. In: SIGMOD (1994)
12. Seshadri, P., Livny, M., Ramakrishnan, R.: The Design and Implementation of a Sequence Database System. In: VLDB (1996)
13. van der Aalst, W., ter Hofstede, A., Kiepuszewski, B., Barros, A.: Workflow patterns. In: Distributed and Parallel Databases (2003)

Hybrid Fragmentation of XML Data Warehouse Using K-Means Algorithm

Mohamed Kechar and Safia Nait Bahloul

University of Oran, LITIO Laboratory, BP 1524, El-M'Naouer, 31000 Oran, Algeria
{mkechar,nait1}@yahoo.fr

Abstract. The efficiency of the decision-making process in an XML data warehouse environment, is in a narrow relation with the performances of decision-support queries. Optimize these performances, automatically contribute in improving decision making. One of the important performances optimization techniques in XML data warehouse is fragmentation with its different variants (horizontal fragmentation and vertical fragmentation). In this paper, we develop a hybrid fragmentation algorithm combining a vertical fragmentation based on XPath expressions and a horizontal fragmentation based on selection predicates. To control the number of fragments, we use the K-Means algorithm. Finally, we validate our approach under Oracle Berkeley DB XML by several experiments done on XML data, derived from the XWB benchmark.

Keywords: XML Data Warehouse, Hybrid Fragmentation, XPath Expressions, Selection Predicates.

1 Introduction

With the emergence of XML, a large amount of heterogeneous XML data is manipulated by enterprises. Various works [11], [16], [26], and [27] have proposed to integrate and store the XML data to exploit them in decision-making (the birth of XML data warehouses). However, in a decision-making system, time is considered as a major constraint. The managers of the company should take appropriate decisions timely. Unfortunately, their decisions are based on analyzing done on the results of several quite complex queries, called decision-support queries. Characterized by join operations, selection operations and aggregation operations, the response times of these queries is generally quite high. Optimize the performances of such queries, contributes significantly to the improvement of decision-making. In this context, several performance optimization techniques have been proposed in the field of data warehouses, such as indexes, materialized views and data fragmentation. Among these techniques, fragmentation has received much interest by the researcher's community. Its efficiency has been proven in the relational databases [1], [13], [25], the object-oriented databases [5,6] and the relational data warehouses [3], [4], and [14]. However, few works on fragmentation have been proposed in the XML data warehouses. To fragment an XML data warehouse modeled by star schema [10], the authors in [22]

Y. Manolopoulos et al. (Eds.): ADBIS 2014, LNCS 8716, pp. 70–82, 2014.

use the primary horizontal fragmentation and derived horizontal fragmentation. They use the K-Means algorithm to group the selection predicates into disjoint classes defining the horizontal XML fragments. In [23], the authors propose two horizontal fragmentation techniques of an XML data warehouse. The first is based on the concept of minterms [25] and the second is based on predicates affinities[30]. The authors in [28], propose different models of partitioning of a multi-version XML data warehouses. They propose the partitioning model of XML documents, the partitioning model based on the XML schema of the XML data warehouse and the mixed model that combines the first two models. The approach proposed in [9], vertically fragment the XML data warehouse based on all frequently paths used by queries. The authors use the association rules to find the set of paths from which they derive the vertical fragmentation schema. To the best of our knowledge, no hybrid fragmentation approach, combining the vertical fragmentation and the horizontal fragmentation has been proposed to date in the context of XML data warehouse. Although its efficiency has been already proven in the relational databases [24], the Object Oriented databases [2], and the relational data warehouses [15]. For this fact, we present in this paper a hybrid fragmentation of an XML data warehouse. We partition vertically the structure of the data warehouse into vertical fragments by a classification of XPath expressions. Then we fragment horizontally the XML data of each vertical fragment by a classification of selection predicates. We use in our classification the K-Means algorithm[18] with the euclidean distance. In addition to its simplicity and its rapidity, it allows us to control the number of fragments.

The remainder of this paper is organized as follows. In Sect.2, we survey the different multidimensional models and we focus on the flat model that we use as a reference model. In Sect.3 we detail our hybrid fragmentation. Finally, we present some experimental results of our evaluations in Sect.4.

2   Multidimensional Modeling of XML Data

In the literature, different XML data warehouse models have been proposed. In [11] the XML data Warehouse is represented by a collection of homogeneous XML documents. Each XML document represents a fact with its measures and its dimensions. In [8], the authors propose the hierarchical model in which they use a single XML document containing all facts and all dimensions. Each fact is represented by an XML element containing its measures and the references to the XML elements containing its dimensions. In addition to the hierarchical model, they define the flat model represented by a single XML document. Each fact in this document is represented by a single XML element containing its measures and its dimensions in the form of XML sub-elements. The XCube model proposed by [17], uses an XML document named FaitsXCube to represent facts and another XML document named DimensionsXCube to represent dimensions. By analogy to the relational star model [19], the authors in [10] and [27], model the XML data warehouse by a central XML document containing all facts with their measures surrounded by several XML documents representing the dimensions. These XML documents are linked by primary keys and foreign keys.

Performance evaluations of these different models of XML data warehouses have been conducted in several works. For example in [8], the authors have conducted evaluations and comparisons of performances between the hierarchical model, the flat model, and XCube model. They noticed that the flat model provides better performance compared with the other two models, except that it introduces redundancy of the dimensions. A performance comparison between the star model, the flat model, and the model proposed in [11] has been carried out in [10]. The authors have shown that the star model provides improved performance for queries that use two joins. However, from three joins, the performances decrease in favour of the flat model. In order to improve the response time of XQuery queries, a join index has been proposed in [20]. By carefully inspecting this index, we found that his representation is in compliance with a flat model (a single XML document containing all the facts with their measures and dimensions). Based on these performance evaluations, we use the flat model depicted in Fig.1) as a reference model to represent the XML data warehouse.

In the following sections, we describe our hybrid fragmentation approach.

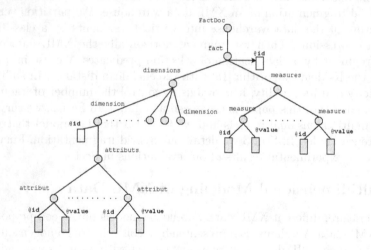

Fig. 1. Reference Model of the XML Data Warehouse

3   Hybrid Fragmentation of the XML Data Warehouse

In this section, we detail the three main phases of our hybrid fragmentation approach. For the remaining sections, letters E, T and D refer respectively to, the set of the names of distinct XML elements, the set of names of distinct XML attributes and the set of distinct data values. Δ represents the XML data warehouse modeled by the flat model and W is the workload executed on Δ. We use in our approach the two concepts of the XPath expression (Definition 1) and the selection predicate (Definition 2).

Definition 1. *A path expression EC is a sequence $root/e_1/.../\{e_n|@ak\}$, with $\{e_1, ..., e_n\} \in E$ and $@a_k \in T$. The expression EC may contain the symbol '*' which indicates an arbitrary element of E, the symbol '//' indicating a sequence of elements $e_i/.../e_j$ such as $i < j$ and the symbol '[i]' which indicates the position of the element e_i in the XML tree [7].*

Definition 2. *A selection predicate is defined by the expression $Pred_j := P$ θ value $| \phi_v(P) \theta$ value $| \phi_b(P) | Q$, with P a terminal XPath expression, $\theta \in \{=, <, >, \leq, \geq, \neq\}$, value $\in D$, ϕ_v is an XPath function, that returns values in D, ϕ_b is a Boolean function and Q denotes an arbitrary XPath expression [7].*

3.1 Vertical Fragmentation Based on XPath Expressions

We define the vertical fragmentation of XML data warehouse Δ, by partitioning its structure into K vertical fragments $VF_1, ..., VF_K$. Each fragment is a projection of a set of XPath expressions frequently accessed by the workload. In this phase we proceed by:

Extraction of XPath Expressions. Each XQuery query belonging to W is in conformity with the basic syntax of the $FLWOR$ expression *(For, Let, Where, Order by, Return)* [29]. For each query, we perform a syntactic analysis by clause and we extract all its XPath expressions. Thus we identify the overall set of XPath expressions EC used by the workload W.

XPath Expressions-Queries Usage Matrix (XPQUM). Defines the use of each XPath expression by the set of queries. We create in $XPQUM$, a line i for each XPath expression $EC_i \in EC$ and a column j for each query $Qj \in W$. If the query Q_j use EC_i, then $XPQUM(i, j) = 1$, else $XPQUM(i, j) = 0$.

Vertical Fragmentation. In this step, we use the K-Means classification algorithm [18] (the choice of K-Means is justified by its simplicity and rapidity) to partition the set of XPath expressions into subsets (classes) that present a usage similarity by queries. With the $XPQUM$ matrix as classification context and an integer K indicating the number of vertical fragments, the K-Means algorithm generates K disjoint classes of XPath expressions. The XPath expressions of the class C_i describe the structure of the vertical fragment VF_i and the set of fragments VF_i ($i = 1...K$) defines our vertical fragmentation schema noted VFS. After this partitioning (fragmentation), we assign every query to the vertical fragments needed to its processing. We formalize this assignment as following: Let:

- $C_1, C_2, ..., C_k$ the sets of XPath expressions defining respectively the vertical fragments $VF_1, VF_2, ..., VF_k$,
- SQ_i is the set of query assigned to VF_i,
- d is the number of queries requiring join operations in VFS schema,
- A the set of XPath expressions used by the query Q_j,

Then

1. If $A \subseteq C_i$ then $SQ_i \leftarrow SQ_i \cup \{Q_j\}$.
2. If $A \subseteq (C_x \cup ... \cup C_y)$ then $SQ_x \leftarrow SQ_x \cup \{Q_j\},...,SQ_y \leftarrow SQ_y \cup \{Q_j\}$ and $d \leftarrow d + 1$.

In case (2), the processing of the query Q_j, requires a join operations between the vertical fragments $VF_x, ..., VF_y$. These join operations are among the causes of performance deterioration. For this fact, we minimize the number of join queries (the d number) appearing in the vertical fragmentation schema VFS. We vary N times the value of the number K of vertical fragments (N is random integer) and for each value, we generate a vertical fragmentation schema. Among these N schemas, we select the optimal according to the following rule:

Rule.1. A vertical fragmentation schema is optimal if and only if it contains a minimum of queries requiring join operations between the vertical fragments. Formally:

$$VFS_i \; is \; optimal \equiv \forall j \in [1..N], \exists i \in [1..N] \; / \; (d_i < d_j) \; with \; i \neq j \; . \quad (1)$$

d_i is the number of join queries in the fragmentation schema VFS_i.

Then for each vertical fragment $VF_i \in VFS_i$, we create a vertical script VS_i represented by a XQuery query. The execution context (the clause *for*) of this query is the XML data warehouse Δ and its clause *return* represents the projection of all XPath expressions belonging to C_i. The selected vertical fragmentation schema is the final result of this first phase as represented by the Fig.2.

In the next section we detail the horizontal fragmentation of each vertical fragment belonging to this schema.

3.2 Horizontal Fragmentation Based on Selection Predicates

In the second phase of our hybrid fragmentation, we fragment horizontally the XML data of each vertical fragment VF_i into L horizontal fragments $FH_{i1},...,FH_{iL}$. The following steps are executed for each vertical fragment as represented by the Fig.3.

Extraction of Selection Predicates. We perform a syntactical parsing of the *where* clause of each query belonging to the set SQ_i (the set of queries assigned to the vertical fragment VF_i). This parsing allows us to extract the set of selection predicates noted PS_i.

Selection Predicates-Queries Usage Matrix (SPQUM). It defines the use of selection predicates of PS_i by the queries of SQ_i. The SPQUM lines correspond to the selection predicates and its columns represent queries. if the predicate p_x exists in the *where* clause of the query Q_y then $SPQUM(x,y) = 1$, else $SPQUM(x,y) = 0$.

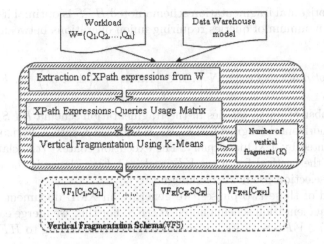

Fig. 2. Vertical fragmentation of the XML data warehouse

Horizontal Fragmentation. Using the K-Means algorithm, we group into classes the selection predicates that present a usage similarity by queries. Specifying the number L of the horizontal fragments, the algorithm partitions the set of the selection predicates of the $MUPSR$ matrix in L disjoint classes representing the horizontal fragmentation schema noted HFS_i. The selection predicates of each class C_{ij} (i the index of the vertical fragment and $j = 1..L$) define the XML data of the horizontal fragment FH_{ij}. According to this partitioning, we assign each query belonging to SQ_i to the horizontal fragments needed to its processing as follows:

Let:

- $Q_h \in SQ_i$,
- $C_{i1}, C_{i2}, ..., C_{iL}$ are the sets of selection predicates corresponding to the horizontal fragments $FH_{i1}, FH_{i2}, ..., FH_{iL}$.
- PSQ_h the set of the selection predicates used by the query Q_h,
- d' the number of queries requiring union operations between FH_{ij},
- SQ_{ij} the set of queries assigned to the fragment FH_{ij}

Then

1. If $PSQ_h \subseteq C_{ij}$ then $SQ_{ij} \leftarrow SQ_{ij} \cup Q_h$.
2. if $PSQ_h \subseteq (C_{ix} \cup ... \cup C_{iy})$ then $SQ_{ix} \leftarrow SQ_{ij} \cup \{Q_h\}, ..., SQ_{iy} \leftarrow SQ_{iy} \cup \{Q_h\}$ and $d' \leftarrow d' + 1$.

In the case (2), the processing of the query Q_h, requires the union of the horizontal fragments $FH_{ix}, ..., FH_{iy}$. In order to reduce these union operations, we vary N' times the value of the number of horizontal fragments L and we generate a horizontal fragmentation schema for each value. Among these N' fragmentation schemas, we select the best according to the following rule:

Rule.2. An horizontal fragmentation schema noted HFS is optimal if and only if it contains a minimum of queries requiring union operations between horizontal fragments

$$HFS_i \ \ is \ \ optimal \equiv \forall j \in \left[1..N'\right], \exists i \in \left[1..N'\right] \ / \ \left(d'_i < d'_j\right) \ \ with \ \ i \neq j. \quad (2)$$

d'_i is the number of union queries in the fragmentation schema HFS_i.

For each horizontal fragment $HF_{ij} \in HFS_i$, we create a horizontal script HS_{ij} represented by a XQuery query. The execution context (the clause *for*) of this query is the vertical fragment VF_i and its *where* clause is the disjunction between the selection predicates belonging to C_{ij}.

At the end of these two phases, we generate an XML document containing the hybrid fragmentation schema noted $HDFS$. For this, we merge each vertical fragment $VF_i \in VFS$ with its horizontal fragments belonging to HFS_i.

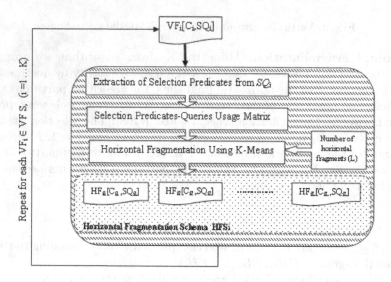

Fig. 3. Horizontal fragmentation of each vertical fragment

3.3 Query Processing on the Fragmented Data Warehouse

The access to the XML data, after fragmentation, should be transparent to the users of the warehouse. To ensure transparency, query processing must be performed on fragmented XML data warehouse. For this, we rewrite the queries according to their assignments carried out during the previous two phases. For each query of the workload:

1. We run through the hybrid fragmentation schema (the XML document) and we identify all fragments needed to its processing.
2. In its execution context, we replace the unfragmented data warehouse by the already identified fragments.

3. If it requires join operations between fragments, we adjust its *where* clause by adding a join qualifications.
4. If it requires union operations between hybrid fragments, we add to its clause *for* the XQuery function *distinct-deep* which removes the duplicate XML data from its result.

In order to prove the effectiveness of the hybrid fragmentation detailed in the previous sections, we have conducted various evaluations that we present in the following section.

4 Experimental Studies

4.1 Experimental Conditions

We have conducted our evaluations under Oracle Berkeley DB XML[12] (an XML native database allowing the storage of voluminous XML documents and implements the XQuery1.0 queries execution engine). We have used the XML dataset from the XML Data Warehouse Benchmark (XWB) proposed in [21]. Modeled with a star schema, the XML data warehouse of the XWB contains the sales facts characterized by the measures: quantity of purchased product and amount of purchased product. These facts are analyzed by the dimensions: products, consumers, suppliers and time. While respecting the definition of flat model (Sect.2), we have merged the facts and dimensions into a single XML document representing our data warehouse. As a programming language, we have used the Java language to implement our hybrid fragmentation algorithm in which we have used the *K-Means*[1] library. The machine used for our experiments is equipped with a Intel Pentium processor and 02 GB of main memory.

4.2 Experimental Assessment and Analysis

In order to prove the effectiveness of our hybrid fragmentation algorithm, we have performed various experiments. In the first, we have used a XML data warehouse composed of 2000 facts and we have (i) calculated the global response time of 19 queries executed on the original XML data warehouse, (ii) fragmented this data warehouse into 02 vertical fragments VF_1 and VF_2, (iii) calculated the global response time of the same queries on the vertically fragmented data warehouse. In the second experiment, we have (i) fragmented respectively VF_1 and VF_2 into 04 and 06 horizontal fragments (ii) calculated the global response time of the 19 queries on the new hybrid fragments. Figure 4, summarizes the results of this two experiments, and the Fig.5 shows the details of the queries response time before fragmentation, after the vertical fragmentation, and after the hybrid fragmentation.

According to the results shown in Fig.4, and compared to the unfragmented XML data warehouse, we observe that the vertical fragmentation improves the

---

[1] https://www.hepforge.org/downloads/jminhep/

global response time of the workload to 30%. As against, the global response time of the same workload is improved to 82% after applying the hybrid fragmentation on the XML data warehouse. The detailed results shown by the Fig.5, allows us to see clearly the effect of the hybrid fragmentation on queries response times. Indeed, after a vertical fragmentation of the data warehouse, the processing of the queries Q_3, Q_5, Q_7, Q_9, Q_{11}, Q_{14}, and Q_{15}, requires a join operation between the two vertical fragments VF_1 and VF_2. Their response time have not been improved, on the contrary we notice a significant deterioration in the performances of the queries Q_7, Q_{14}, and Q_{15}. However, only the response times of the queries requiring a single vertical fragment VF_1 or VF_2, have benefited from some improvement. But after the hybrid fragmentation, we observe a meaningfully enhancement in the response time of each query, in particularly join queries, that which proves the effectiveness of our hybrid fragmentation algorithm.

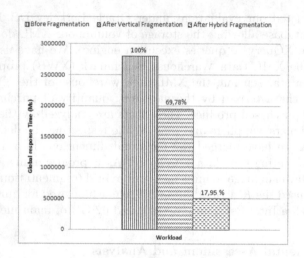

Fig. 4. Global response time of the workload on 2000 facts

In the Third experiment, we have applied our hybrid fragmentation algorithm on three XML data warehouses of different sizes: 2000, 4000, and 8000 facts. We have fragmented each data warehouse according to the same previous hybrid fragmentation schema and we have calculated the global response time of 19 queries before and after fragmentation on each data warehouse. The obtained results shown by the Fig.6, confirm that our hybrid fragmentation always guarantee an improvement of the performances even after the increase of the size of the XML data warehouse.

Indeed, fragmenting XML data warehouse by our algorithm allows us to:

1. Group in hybrid fragments (XML documents) the XPath expressions (vertical fragmentation) and the XML data (horizontal fragmentation) needed in processing queries.
2. Generate fragments of small sizes compared to the size of the unfragmented data warehouse.

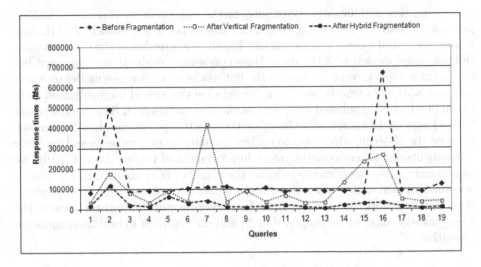

Fig. 5. Response time by query

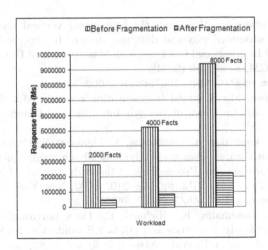

Fig. 6. Response time of the workload on different sizes of data warehouses

The first point, allows us to improve the search time of the XML data to satisfy a query. On the other side, the second point, allows us to improve the time needed to browse the XML structure of the unfragmented data warehouse to search data. According to these two points, we justify the performances improvement provided by our hybrid fragmentation algorithm.

5   Conclusion

The processing time of the decision-support queries on an XML data warehouse is quite high especially on a large volume of XML data. However, minimizing this processing time significantly contributes to the improvement of decision-making process. In this context, we proposed a new fragmentation approach of XML data warehouse called hybrid fragmentation. Firstly, we introduced the different multidimensional models of XML data. Based on several evaluations conducted between these models, we have chosen the flat model as a reference model to represent the XML data warehouse. Then, we detailed our hybrid fragmentation algorithm in which we combined a vertical fragmentation based on XPath expressions with a horizontal fragmentation based on the selection predicates. In our approach we used the K-Means algorithm to control the number of fragments and generate a fragmentation schema offering more improvement of performance. Finally, we conducted various experiments to prove the validity of our algorithm. The results obtained allowed us to confirm the effectiveness of our proposed hybrid fragmentation. In future work, we plan to conduct an experimental comparison between the fragmentation algorithms proposed in [9] and [22], and our hybrid fragmentation algorithm.

References

1. Agrawal, S., Narasayya, V., Yang, B.: Integrating vertical and horizontal partitioning into automated physical database design. In: Proceedings of the 2004 ACM SIGMOD International Conference on Management of Data, SIGMOD 2004, pp. 359–370. ACM, New York (2004),
   http://doi.acm.org/10.1145/1007568.1007609
2. Baio, F., Mattoso, M.: A mixed fragmentation algorithm for distributed object oriented databases. In: Proc. of the 9th Int. Conf. on Computing Information, pp. 141–148 (1998)
3. Bellatreche, L., Bouchakri, R., Cuzzocrea, A., Maabout, S.: Horizontal partitioning of very-large data warehouses under dynamically-changing query workloads via incremental algorithms. In: Proceedings of the 28th Annual ACM Symposium on Applied Computing, SAC 2013, pp. 208–210. ACM, New York (2013),
   http://doi.acm.org/10.1145/2480362.2480406
4. Bellatreche, L., Boukhalfa, K., Richard, P.: Data partitioning in data warehouses: Hardness study, heuristics and ORACLE validation. In: Song, I.-Y., Eder, J., Nguyen, T.M. (eds.) DaWaK 2008. LNCS, vol. 5182, pp. 87–96. Springer, Heidelberg (2008)
5. Bellatreche, L., Karlapalem, K., Simonet, A.: Horizontal class partitioning in object-oriented databases. In: Tjoa, A.M. (ed.) DEXA 1997. LNCS, vol. 1308, pp. 58–67. Springer, Heidelberg (1997),
   http://dl.acm.org/citation.cfm?id=648310.754717
6. Bellatreche, L., Karlapalem, K., Simonet, A.: Algorithms and support for horizontal class partitioning in object-oriented databases. Distrib. Parallel Databases 8(2), 155–179 (2000), http://dx.doi.org/10.1023/A:1008745624048
7. Berglund, A., Boag, S., Chamberlin, D.: andez, M.F.F.: Xml path language (xpath) 2.0, 2nd edn. (December 2010)
8. Boucher, S., Verhaegen, B., Zimányi, E.: XML Multidimensional Modelling and Querying. CoRR abs/0912.1110 (2009)

9. Boukraâ, D., Boussaïd, O., Bentayeb, F.: Vertical fragmentation of XML data warehouses using frequent path sets. In: Cuzzocrea, A., Dayal, U. (eds.) DaWaK 2011. LNCS, vol. 6862, pp. 196–207. Springer, Heidelberg (2011), http://dblp.uni-trier.de/db/conf/dawak/dawak2011.html#BoukraaBB11

10. Boukraa, D., Riadh Ben, M., Omar, B.: Proposition d'un modèle physique pour les entrepôts XML. In: Premier Atelier des Systèmes Décisionnels (ASD 2006), Agadir, Maroc (2006)

11. Boussaid, O., BenMessaoud, R., Choquet, R., Anthoard, S.: Conception et construction d'entrepôts XML. In: 2ème journée francophone sur les Entrepôts de Données et l'Analyse en ligne (EDA 2006), Versailles. RNTI, vol. B-2, pp. 3–22. Cépaduès, Toulouse (Juin 2006)

12. Brian, D.: The Definitive Guide to Berkeley DB XML (Definitive Guide). Apress, Berkly (2006)

13. Ceri, S., Negri, M., Pelagatti, G.: Horizontal data partitioning in database design. In: Proceedings of the 1982 ACM SIGMOD International Conference on Management of Data, SIGMOD 1982, pp. 128–136. ACM, New York (1982), http://doi.acm.org/10.1145/582353.582376

14. Dimovski, A., Velinov, G., Sahpaski, D.: Horizontal partitioning by predicate abstraction and its application to data warehouse design. In: Catania, B., Ivanović, M., Thalheim, B. (eds.) ADBIS 2010. LNCS, vol. 6295, pp. 164–175. Springer, Heidelberg (2010), http://dl.acm.org/citation.cfm?id=1885872.1885888

15. Elhoussaine, Z., Aboutajdine, D., Abderrahim, E.Q.: Algorithms for data warehouse design to enhance decision-making. WSEAS Trans. Comp. Res. 3(3), 111–120 (2008), http://dl.acm.org/citation.cfm?id=1466884.1466885

16. Golfarelli, M., Rizzi, S., Vrdoljak, B.: Data warehouse design from XML sources. In: Proceedings of the 4th ACM international workshop on Data warehousing and OLAP, DOLAP 2001, pp. 40–47. ACM, New York (2001), http://doi.acm.org/10.1145/512236.512242

17. Hümmer, W., 0004, A.B., Harde, G.: XCube: XML for Data Warehouses. In: DOLAP, pp. 33–40 (2003)

18. MacQueen, J.: Some Methods for Classifcation and Analysis of Multivariate Observations. In: Proceeding of Fifth Berkley Symposium on Mathematical Statistics and Probability, vol. 1, pp. 281–296 (1967)

19. Kimball, R.: A dimensional modeling manifesto. DBMS 10, 58–70 (1997), http://portal.acm.org/citation.cfm?id=261018.261025

20. Mahboubi, H., Aouiche, K., Darmont, J.: Un index de jointure pour les entrepôts de données xml. In: 6émes Journées Francophones Extraction et Gestion des Connaissances (EGC 2006), Lille. Revue des Nouvelles Technologies de l'Information, vol. E-6, pp. 89–94. Cépadués, Toulouse (2006)

21. Mahboubi, H., Darmont, J.: Benchmarking xml data warehouses. In: Atelier Systemes Décisionnels (ASD 2006), 9th Maghrebian Conference on Information Technologies (MCSEAI 2006), Agadir, Maroc (December 2006)

22. Mahboubi, H., Darmont, J.: Data mining-based fragmentation of xml data warehouses. In: DOLAP, pp. 9–16 (2008)

23. Mahboubi, H., Darmont, J.: Enhancing xml data warehouse query performance by fragmentation. In: Proceedings of the 2009 ACM Symposium on Applied Computing, SAC 2009, pp. 1555–1562. ACM, New York (2009), http://doi.acm.org/10.1145/1529282.1529630

24. Navathe, S.B., Karlapalem, K., Ra, M.: A mixed fragmentation methodology for initial distributed database design. Journal of Computer and Software Engineering 3(4), 395–426 (1995)

25. Ozsu, M.T.: Principles of Distributed Database Systems, 3rd edn. Prentice Hall Press, Upper Saddle River (2007)
26. Pokorný, J.: XML Data Warehouse: Modelling and Querying. In: Proceedings of the Baltic Conference, BalticDB&IS 2002, vol. 1, pp. 267–280. Institute of Cybernetics at Tallin Technical University (2002),
http://portal.acm.org/citation.cfm?id=648170.750672
27. Rusu, L.I., Rahayu, J.W., Taniar, D.: A methodology for building xml data warehouses. IJDWM 1(2), 23–48 (2005)
28. Rusu, L.I., Rahayu, W., Taniar, D.: Partitioning methods for multi-version xml data warehouses. Distrib. Parallel Databases 25(1-2), 47–69 (2009),
http://dx.doi.org/10.1007/s10619-009-7034-y
29. Walmsley, P.: XQuery. O'Reilly Media, Inc. (2007)
30. Zhang, Y., Orlowska, M.E.: On fragmentation approaches for distributed database design. Information Sciences - Applications 1(3), 117–132 (1994),
http://www.sciencedirect.com/science/article/pii/1069011594900051

Do Rule-Based Approaches Still Make Sense in Logical Data Warehouse Design?

Selma Bouarar, Ladjel Bellatreche, Stéphane Jean, and Mickaël Baron

LIAS/ISAE-ENSMA, Poitiers University, France
{selma.bouarar,bellatreche,jean,baron}@ensma.fr

Abstract. As any product design, data warehouse applications follow a well-known life-cycle. Historically, it included only the physical phase, and had been gradually extended to include the conceptual and the logical phases. The management of phases either internally or intranally is dominated by rule-based approaches. More recently, a cost-based approach has been proposed to substitute rule-based approaches in the physical design phase in order to optimize queries. Unlike the traditional rule-based approach, it explores a huge search space of solutions (e.g., query execution plans), and then based on a cost-model, it selects the most suitable one(s). On the other hand, the logical design phase is still managed by rule-based approaches applied on the conceptual schema. In this paper, we propose to propagate the cost-based vision on the logical phase. As a consequence, the selection of a logical design of a given data warehouse schema becomes an optimization problem with a huge space search generated thanks to correlations (e.g. hierarchies) between data warehouse concepts. By the means of a cost model estimating the overall query processing cost, the best logical schema is selected. Finally, a case study using the *Star Schema Benchmark* is presented to show the effectiveness of our proposal.

1 Introduction

Over the last four decades, databases ($\mathcal{DB}$) technology has evolved constantly to satisfy the growing needs of applications built around it, whether in terms of data volume or technology trends. Once the $\mathcal{DB}$ technology became mature, a design life-cycle of $\mathcal{DB}$-based applications, has emerged. The definition of this life-cycle has undergone *several evolutionary* stages before being accepted as it stands actually. In fact, the first generations of $\mathcal{DB}$ systems can be summarized in one phase: *the physical design*. Physical data independence has become thereafter a necessity because a $\mathcal{DB}$-based application is never written in stone from the first draft, but requires several updates. To do so, the need of a much more thorough analysis arises, which leads to insulate the analysis task from the physical design, so that it becomes a step of its own: *the conceptual design*. It consists of a model-based data representation while ensuring what we call the *data abstraction*. This latter evolution resulted in the *three-tier architecture ANSI/SPARC* [19] that clearly distinguishes the conceptual schema from the internal (physical) one.

Y. Manolopoulos et al. (Eds.): ADBIS 2014, LNCS 8716, pp. 83–96, 2014.

This schema-insulation has implied a *mapping phase* between the two abstraction levels, named the logical design. Pioneer in this Field, Codd [8] has proposed the relational model, a mathematical abstraction of $\mathcal{DB}$ content in the 70s. Since then several models have been introduced namely Object-oriented, multi-dimensional, XML etc.

By examining the current $\mathcal{DB}$ design life-cycle, we found out that either the inter- or intra- phases tasks is managed by means of *rules*. At the conceptual level, for instance, business rules have been applied to generate the conceptual schema [10,16]. In the logical phase, some fixed rules like the type of *applied normal forms*, *grouping or not* dimensional hierarchies in a single dimensional table (star schema), etc. are applied. Rule-based optimization has been largely used in the physical design to optimize queries. It has been supported by most of commercial database systems [11,4]. This optimization applies a set of rules on a query tree in order to optimize it. Pushing down selections and projections is one of the most popular used rules. Rule-based approaches are also applied to pass from one phase to another. For instance a logical model is obtained by translating a conceptual model using child and parent relationships.

The rule-based approach has shown its limitations in the physical phase since it *ignores* the parameters of database tables (size, length of instances, etc.), selectivity factors of selection and join operations, the size of intermediate results, etc. These parameters have a great impact on the query evaluation cost. As a consequence, it has been substituted by a cost-based approach. At first, the cost approach considers a wide search space of solutions (e.g. query plans), then based on a cost-model, the most suitable one(s) is/are selected using advanced algorithms (e.g. dynamic programming). Driven by the success of cost-based physical design, and the modest attention paid to the logical [13], we propose to transpose the cost-based aspect into the logical modelling and to change the *one-logical model vision*. To achieve that, we propose to exploit the correlations between life-cycle objects (entities, attributes, instances, etc.). In this vein, several recent research efforts have focused on exploiting these correlation to improve performance, to name but a few, Agrawal et al. [2] have exploited the similarity interaction between materialized views ($\mathcal{MV}$) and indexes to improve the physical phase. Kimura et al. [12] have implemented the project $CORADD$, where they exploited the correlations linking the attributes to define $\mathcal{MV}$ and indexes. This latter project has motivated us to exploit the correlations in favor of $\mathcal{DB}$ logical design.

In this paper, we focus on how to exploit the correlations in the definition of a cost-based logical model in the context of data-warehouses ($\mathcal{DW}$). A cost-based approach has a sense if the logical phase is associated to research space representing a large number of logical model schemes. To do so, we fix three main objectives: **(i)** identification the concepts and properties sensitive to correlations. To satisfy this objective, we propose to use ontologies due to their strong similarities with conceptual models and their capability of representing the correlations (availability of formal languages such as description logic) and their ability of reasoning on them. **(ii)** The definition of a cost model that corresponds to a predefined metric to select the best logical model schema. We consider the query processing cost as a metric. **(iii)** The development of a query rewriting process to support the change of the logical schema.

The paper is organized as follows: Section 2 shows a thorough analysis of the correlations. In Section 3, we focus on how to choose (theoretically and empirically) the appropriate $\mathcal{DW}$ logical schema. As for Section 4, a case study validating our proposal is detailed, to finally conclude in Section 5.

2   Exploration of Correlations

The purpose of this section is to highlight what we believe to be the key concepts in the design process of any information system: *correlations* (A.K.A: integrity constraints, dependencies, relationships) linking classes, properties.

2.1   Types of Correlations

Fig.1 provides an overview of an ontology covering the domain of the Star Schema Benchmark $\mathcal{SSB}$, which is used further down for our experiments, and right below to illustrate the different types of existing correlations that we have identified and classified that way:

– *Definition/equivalence relations or Generalization (DEF)*: when concepts/roles are defined in terms of other concepts/roles. E.g. a *Supplier* is a *TradeAgent* that *Supplies* a *LineItem*.
– *Inclusion dependencies or Specialization (ID)* Also called is-a relation or subsumption: it occurs when a concept/role is subsumed by another concept/role. E.g. *Customer* subsumes *TradeAgent*. When it concerns attributes, there is another application of this type: the notion of foreign keys, which states that the domain of one attribute must be a subset of the other correlated attribute.

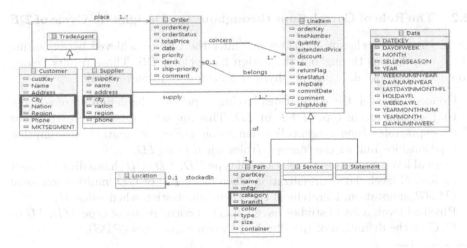

Fig. 1. $\mathcal{SSB}$ ontology

- *Functional dependencies (CD/FD)*: CD stands for functional dependencies between concepts and FD between attributes. They figure when a set of concepts/roles (or their instances) *determine* an other set of the same type. E.g. *custKey* determines the *name* of *Customer*.
- *Multi-valued dependencies (MD)* or *soft dependencies*: specific to attributes, it is a generalization of the functional dependencies. Formally, the difference between the two is the abolition of the determination criterion, in other words, to a value set, we can associate more than one value set of the same type. Examples are given in §. 3.3.
- *Conditional Integrity constraints (CIC)*: specific to attributes, they denote the DB integrity constraints (algebraic or semantic)[1] involving more than one attribute [17] and holding on instances of the relations. Note that definitions and dependencies are considered as simple integrity constraints which are valid on entire relations, contrary to conditional ones where the correlation is accompanied with condition(s) that must be respected. This latter aspect moves the application level from attribute range level to attribute values level. In other words, only a subset of the member attributes domain is involved (reduced range). E.g. $Customer.City{=}Paris{\rightarrow}LineItem.discount{\succ}$ 20%. We distinguish two main categories: (i) conditional functional dependencies (CFD) [6] whereby the (FD) has to hold on a subset of the relation (a set of tuples) that satisfies a specific attribute pattern $([Customer.city{=}London, Customer. name] \rightarrow [Customer.phone])$, rather than on the entire relation *Customer*, and (ii) more specifically, association rules that apply for particular values of some attributes [1] $(Part.color{=}\text{`red'} \rightarrow Part.size{=}50)$.
- *Hierarchies (H)*: specific to attributes, and more present in DWs, where a set of attributes makes up a dimension hierarchy (e.g. Industry, category, product). They can be assimilated to the *part-whole* relationships. The particularity of this type, is that we could plan the construction of a new class for each hierarchy level.

2.2   The Role of Correlations throughout the Design Life-Cycle of DB

From the former classification, we can infer the results achieved by exploiting these correlations throughout the design life-cycle of DB. Those results belong to either conceptual, logical or physical SDB levels. In more detail:

- Conceptual level: the correlations having impact on the definition of conceptual schema are of type: DEF or ID. This impact consists of creating new concepts/roles (non canonical) when using DEF, or creating subsumption relationships linking the concepts/roles when using ID.
- Logical level: exploiting correlations of type CD, FD or H, has a direct impact on logical level: data normalization when using FD or CD, multidimensional $OLAP$ annotation, hierarchical $OLAP$ normalization when using H.
- Physical level: a lot of studies have exploited correlations of type MD, ID or CIC in the definition of the Physical Design Structures (PDS).

---

[1] IC specify conditions/propositions that must be maintained as true ($Part.size{>}0$).

Table 1. Related work on correlations exploitation over the design life-cycle of $\mathcal{DW}$

Studies \ Phases	MC	ML	MP	OLAP	Other
Anderlik & al. [3]				*DEF/ID* Roll-up	
Stohr & al. [18]		*H* Fragmentation			
Kimura & al. [12]			*FD/MD* $\mathcal{MV}$/indexes		
Brown & al. [7]			*CIC* Query optimizer		
Agrawal & al. [1]					*CIC* Data-mining
Petit & al. [15]	*ID/FD* ER schema	*ID/FD* Relational schema			*ID/FD* Reverse engineering

In the light of the foregoing, we believe that any evolution/transition throughout the design life-cycle of $\mathcal{DB}$ can be controlled by correlations. Table 1 shows different studies in this field. In fact, thanks to the formal power of ontologies, and their strong similarity with conceptual models, we can store correlations (identified by the $\mathcal{DB}$ users notably the designer) right from the conceptual phase. Afterwards, the transition to the logical level is henceforth based on correlations: namely the dependencies (CD, FD) for $\mathcal{DB}$, and hierarchies for $\mathcal{DW}$, as for the transition to the physical, it becomes controlled by either MD, ID or CIC. Indeed, several studies have shown that $\mathcal{DB}$ performance can be vastly improved by using $\mathcal{PDS}$ defined upon correlations, and even more when exploiting the interaction - generated upon correlations - between these $\mathcal{PDS}$, as is the case concerning $\mathcal{MV}$ and indexes in CORADD[12](see Table 1).

3 Proposed $\mathcal{DW}$ Design Methodology

Readers are reminded that our objective is to take advantage of correlations so as to set up a cost-based transition from the $\mathcal{DW}$ conceptual phase to the logical one. Actually, the big interest of $\mathcal{DW}$ community is given to the physical design, yet while most problems can be solved by fine-tuning the physical schema, some performance problems are caused by a non-optimized logical schema [9]. In this vein, design process for $\mathcal{DW}$ is based on the multidimensional annotation of the conceptual model. Currently, the designer selects one logical schema among a wide variety (star or the snowflake). To ensure a more efficient selection task, we suppose that we have the $\mathcal{DW}$ semantic multidimensional model in its simple form, that definitely includes the hierarchies correlations in the form of stored axioms. These latter will be exploited in defining the different possible logical schemas (one star and various snowflakes), on which we will apply a cost model, to choose the most suitable one i.e. the best possible compromise between normalization to ensure space savings and efficient updates, and de-normalization to improve performance by minimizing and simplifying query joins.

It is important to highlight the fact that, contrary to what is usually thought, sometimes a pure star schema might suffer serious performance problems. This can occur when a de-normalized dimension table becomes very large and penalizes the star join operation. Conversely, sometimes a small outer-level dimension table does not incur a significant join cost because it can be permanently stored in a memory buffer. Furthermore, because a star structure exists at the center of a snowflake, an efficient star join can be used to satisfy part of a query. Finally, some queries will not access data from outer-level dimension tables. These queries effectively execute against a star schema that contains smaller dimension tables. Not to mention the normalization benefits, and the space gain. Therefore, under some circumstances, a snowflake schema is more efficient than a star schema[14].

3.1   Explanation of our $\mathcal{DW}$ Design Methodology

Our approach can be described through Algorithm 1. It is processed while transiting to the logical phase, hence, real deployment information not available. Instead, we exploit the conceptual knowledge: (i) semantics associated with data (correlations, in our context hierarchies), (ii) semantics associated with the $\mathcal{DB}$ content (table sizes, attribute domains, etc), (iii) semantics associated with future queries workload. The latter two information are the input of our cost-model, and they can be deduced while analyzing users' requirements. Indeed, we can always have an idea about the load of frequent queries (considering Pareto principle), and estimate sizes. These information are generally useful to the assessment of $\mathcal{DB}$ processes.

3.2   Generating the Different Possible Logical Schemas

We assume that the input semantic multi-dimensional schema: $\mathcal{IS} = \{F, D_1, D_2,$ ..., $D_n\}$, such as F for "fact table", and D_i, for dimension tables, and which definitely include hierarchies between D_i attributes (thanks to the semantic aspect), then:

- each dimension D_i having h hierarchical levels, can be decomposed 2^{h-1} times. e.g. Dim(*location*) = {*Country, Region, Department*} $\Rightarrow 2^{3-1} = 4$ possibilities of normalization: (i) each hierarchical attribute in a separate sub-dimension, (ii) both *Country* and *Region* in a separate sub-dimension, and *Department* in another, and so on...
- For the whole set of dimensions $(D_1, D_2, ..., D_n)$, there will be $\prod_{\{d=1\}}^{n} 2^{h_d-1}$ possible schemas. For example, considering 4 dimensions with 3 levels to each one, there will be 256 alternatives.

This process is accomplished by Algorithm 2. Note that evaluating all possible combinations seems to be naive, but (i) this evaluation is done once, before deploying the $\mathcal{DW}$, and optionally with evolution occurrences. Both cases are long-term tasks, (ii) our algorithm gives results in reasonable time since it would make non-sense if the granularity of hierarchies is too long (4 max).

Algorithm 1. The general algorithm of our theoretical approach

Input: $\mathcal{DW}$ semantic multidimensional model (de-normalized form):
$\qquad \mathcal{IS} = \{F, D_1, D_2, ..., D_n\}$; the query workloads $\mathcal{Q} = \{Q_1, Q_2, ..., Q_m\}$
Output: $\mathcal{DW}$ most suitable logical schema
Generate the different possible logical schemas;
for *each generated schema* **do**
   Calculate the new sizes of the pre-existing dimensions and the new
   sub-dimensions (hierarchical levels);
   for *each query in the workload* **do**
      Rewrite the query conforming to the target schema;
      Calculate the query cost using a cost model;

Choose the most suitable schemas i.e. the ones having minimum costs (top-k
set);
Load this latter top-k set into a $\mathcal{DW}$ (empirical evaluation), and select the most
adequate schema, i.e. the one having minimum execution times;

Algorithm 2. Algorithm for generating the different possible $\mathcal{DW}$ logical
schemas

Input: $\mathcal{DW}$ semantic multidimensional model
Output: A set of the different $\mathcal{DW}$ logical schemas
for *each dimension table in the input schema* **do**
   Generate the tree of the different combinations of hierarchy levels;
   for *each combination* **do**
      Create the new sub-dimension table;
      Update the attributes;
      Add the current combination into the list corresponding to the current
      dimension;

Calculate the Cartesian product of the different hierarchical combinations of
each dimension;
Model-building of the corresponding $\mathcal{DW}$ for each element belonging to the
resulting Cartesian product;

3.3 Calculating the New Table Sizes

After each creation of new dimension/sub-dimension tables during the normalization process, the sizes change systematically. This new information is crucial
to calculate the cost model.

In our approach we deal with two table types: (i) the original dimension
table, and (ii) the newly created sub-dimension tables, knowing that the size
of any table is calculated according to two parameters as follows: $Size(T_i) = rowSize * nbRows$. Below, we will explore how these parameters will change for
the two table types within every generated logical schema. We distinguish the
following two scenarios:

The Tables Are at Least in the Second Normal Form (2NF): For the original dimension table: the number of rows does not change, because we suppose that the original table is in 2NF, unlike the size of the row, which does indeed, because there are less attributes than before. So the new size of the concerned dimension is:

$Size(D\prime_i) = Size(D_i) - \sum(Size(H_k)) + Size(a_{FK}) = nbRows * (rowSize - \sum(Size(H_k)) + size(a_{FK}))$, such that (H_k) denotes the attributes (levels) of the hierarchy, and a_{FK} denotes the foreign key attribute, which will relate the dimension table to its sub-dimension tables.

For the sub-dimension tables newly created: firstly, the size of the row is the sum of the attributes' sizes in the concerned sub-dimension in addition to the foreign key size (if present):

$rowSize = \sum(Size(levelAttribute_i)) + Size(a_{FK})$. The computation of rows is more complicated because it depends on the type of correlation between the attributes forming the hierarchy. We distinguish three main scenarios:

(i) Apart from the *hierarchy* correlation (H), There is no dependency between these attributes. In other words, for each value of the higher level attribute (the *whole*), there exists uniformly the same values' domain to the attribute from the lesser level (the *part*). A concrete example of such category, is the simple date hierarchy *{year, month, day of week}*, where whatever the *year* is (respectively, the *month*), the cardinal of the *month* values domain is always equal to 12 (respectively, the cardinal of the *day* values domain is always equal to 7). In this case, the number of rows becomes the Cartesian product of the domain cardinality of every hierarchical attribute of the current level: $nbRows = (\prod |domain(levelAttribute_i)|)$. Considering just two years, the numerical application gives $(2 * 12 * 7)$ possible combinations.

(ii) Besides the *hierarchy* correlation (H), there is a functional dependency $(FD)^2$ between attributes. i.e, to one value of the higher level attribute, we can match values from only a determined set of values of the lesser level attribute. In line with the previous example, we take the more sophisticated date hierarchy *{year, season, month, day}*. We notice that the *season* attribute share a *FD* correlation with the *month* attribute. Indeed, every *season* value can be linked to only three specific *month*'s values. Let us consider a more concrete example, where the dependency correlation is fully applied: the location hierarchy *{continent, country, city}* example, in which, to each higher level attribute value, only a specific set of lesser level attribute values can be associated (e.g. the *continent* value: *Europe*, has it specific set of values of *country* such as France, Spain, Germany, etc. different from the values set which can be associated for instance to the *continent* value: *America*. In this case, the number of rows is: $nbRows = \max(|domain(levelAttribute_i)|)$. Considering 2 values of *continent*, 4 of *country* and 8 of *city*, the number of possible rows of this level is equal to $nbRows = \max(|domain(levelAttribute_i)|) = \max(2, 4, 8) = 8$.

---

[2] The sets of values of the lesser level attribute, which can be associated to the values of the higher level attribute, are pairwise disjoint.

It should be noted that numerical application of the sophisticated date hierarchy $\{year, season, month, day\}$, gives considering just 2 years as above: $(2*\max(|domain(levelAttribute_i)|)*7 = 2*\max(|domain(season, month)|)*7 = 2*12*7)$ possible combinations instead of $(2*4*12*7)$.

(iii) As for the last scenario, there exists in addition to the H correlation, a MD correlation[3] between those attributes. For example, the following simple nationality hierarchy: $\{nationality, spoken\ languages\}$, in which two different nationalities can share the same languages. Thus, the number of rows is approximatively equal to:

$nbRows \simeq |domain(levelAttribute_{dominant})| * \delta(MD)$. If we consider a MD correlation between two attributes A and B, such that A is the dominant attribute and B is the dependent one, then $\delta(MD)$ is the average number of values that B can have for a value of $A (\delta(MD) < |domain(levelAttribute_{dependant})|)$. E.g. we know that: (i) there is a MD correlation between the attributes $nationality$ (A) and $spoken\ language$ (B), (ii) and we know that generally each nationality can have on average three possible spoken languages so $\delta(MD) = 3$.

It is worth mentioning that this case can be reduced to the first scenario in extreme cases, where $\delta(MD)$ is as large as the cardinal of the concerned attribute domain, in other words, the lesser level attribute sets are completely non-disjoint (the same).

The Tables are at Least in the First Normal Form (1NF): As compared to the previous case, the original dimension tables, and more precisely, the number of rows will be affected. Indeed, a given instance of a relation (object) might need to appear more than once (an attribute is dependent of only a part of the candidate key) causing a redundancy, which may extend to the cardinal of the responsible attribute. We can get inspired from the previous scenario, and define $\delta(Attribute)$ as the maximal number of values belonging to this attribute domain, and can be associated to one instance (individual) of the concerned relation. Or simply suggest that $\delta(Attribute)$ is equal to the average (half) of its domain cardinal. In both cases, it is about an overestimated approximation, where the number of rows becomes: $nbRows\prime = \lceil nbRows/\delta(Attribute)\rceil$

It emerges from the foregoing that the designer knowledge of the domain is fundamental in the determination of table sizes. However, this knowledge is easy to acquire since it concerns the following parameters: Attribute size (implies tuple size), attribute domain size, number of tuples of the original tables, correlation type linking the hierarchical attributes and possibly the delta measure. When conceiving any information system, the designer must have at least an approximate idea about these values.

3.4   Rewriting the Query Conforming to the Target Schema

The query workload $\mathcal{Q} = \{Q_1, Q_2, ..., Q_m\}$ is the set of the more recurrent/interesting queries (any form) to-be/being submitted about the domain in question

---

[3] Where the sets of values of the lesser level attribute are non-disjoint.

represented by the $\mathcal{DW}$ semantic multidimensional model. These queries will obviously change according to the underlying schema, for the sheer fact that the attributes being referenced by the query, can move to another table while breaking down the dimensions into hierarchical levels (sub-dimensions). Algorithm 3 describes the query rewriting process. It is important not to confuse this

---

Algorithm 3. Algorithm for rewriting queries according to the logical schema

---

Input: The $\mathcal{DW}$ logical schema and $\mathcal{Q} = \{Q_1, Q_2, ..., Q_m\}$
Output: $\mathcal{Q} = \{Q\prime_1, Q\prime_2, ..., Q\prime_m\}$
for *each query in the workload* **do**

> Extract the attributes from the "Select" clause;
> Find in which new tables of the new $\mathcal{DW}$ logical schema (the input), those attributes are;
> Set the new query ($Q\prime$) "Select" and "from" clauses;
> Extract the attributes from the "Where" clause;
> Find in which new tables of the new $\mathcal{DW}$ logical schema (the input), those attributes are;
> Set the new query ($Q\prime$) "Where" clause, and update its "from" clause;
> Manage the extra-joins resulting from the addition of hierarchies;

---

rewriting process with query translation problem. In fact, based on a query-model, we rewrite input queries into the same language, while considering the same (local) environment.

3.5 Calculating the Query Cost Using a Cost Model

After the rewriting process of queries, we will obtain a set of logical schemas, and to each schema, a corresponding set of queries. The next step consists in generating the best execution plan for each query, i.e. the best order of the relational algebra operators in the query tree, to finally apply the cost model. This latter is a tool designed to quantify the efficiency of a solution. Such a tool is useful to evaluate the performance of a solution without having to deploy it on a DBMS (Simulation), and then to compare different solutions. Our cost model is based upon the one described in [5], adapted to the context of snowflake join queries. It estimates the number of inputs/outputs between disk and main memory while executing each query. It is worth noting that it is about a logical cost model which may differ from physical ones ($\mathcal{DBMS}$ cost optimizers) which, moreover, are not usually available.

4   Case Study

In order to instantiate the design approach described above, some experiments are conducted on Oracle $\mathcal{DBMS}$ with 8192 as block size, hosted on a server machine with 32GB of RAM. We have used the SSB Benchmark with a scale factor of 100. We have at first proposed its ontology to move up to ontological level (Fig.1), where correlation will be saved. There are: $H(Customer) * H(Part) * H(Supplier) * H(Date) = 2^{3-1} * 2^{2-1} * 2^{3-1} * 2^{4-1} = 256$ possible logical schemas. Every schema would have 13 characteristics: a size, and 12 costs values (one for each submitted query). The users' requirements can intervene at this stage by possibly indicating which queries are more important than others. Then according to this information and the sets of results, schemas having the lower size/execution costs ratio (better compromise) will be chosen.

4.1   Theoretical Evaluation

The SSB load of queries is composed of 4 query flights, each one has on average 3 queries. In order to overcome the large number of studied parameters (what is generally done by *skyline* algorithms) that yielded to a pretty complicated analysis (multi-parameters solutions: 13 parameters for each one), we begin by analyzing the costs of every query, according to which we pick up the best schema, then we compare this latter with the star schema and schemas having the minimum size.

1. *Query Flight 1.* It has selections on merely one dimension involving the date hierarchy (*year* attribute), therefore, minimum costs belong to schemas whose date dimension is not normalized including the initial one. Otherwise, the cost is maximum, reporting an average increase of 40% with reference to the star schema. Note that we mean by *average increase ($\nearrow$)* of cost values: the ratio between the cost of queries execution in the current schema, and their cost execution in the initial star schema.

2. *Query Flight 2.* This query type has selections on two dimensions: *Supplier* and *Part*, involving their hierarchies (respectively Location, and category). Contrary to expectations, we notice that the minimum cost does not belong to the Star schema, it rather does to every schema whose Part dimension is normalized using the category hierarchy, such that $i\prime = i+5$ and $i \notin 0, 1, 2, 3$). This can be explained by the large size gained through this normalization (the new loaded table is about only 0.01% of the original size, compared to 0.5% concerning the date normalization), and we can also relate this to the fact that the hierarchy granularity is small.

3. *Query Flight 3.* The selections are placed on three dimensions: *Date, Customer* and *Supplier*, involving their hierarchies as well (respectively *Customer* Location, *Supplier* Location and *category*). The minimum costs belong to the original schema, and to the 4th one, where only the Part dimension is normalized. The next nearest cost with an average increase of only 2% belong to schemas normalized on *Customer* dimension. Same as previous explanation, this is due to the size gained by this latter normalization (The new loaded *Customer* Location hierarchy tables are about only 0.006% of the original size), compared with the size gained through the *Supplier* normalization (0.12% of the initial size), or the date one (0.5%).

4. All dimensions are involved. The costs are significantly different from a query
   to another as detailed below:
   - The first query (Q4.1) uses mainly *Date*, *Customer* and *Supplier* hierar-
     chies. The minimum costs belong to schemas where either *Part* or *Date* or
     both, are normalized, and the next nearest cost having an average increase
     of 15% comes with the normalization of *Customer* dimension. Unlike the
     previous case (where *Date* and *Customer* are requested by the two main
     operations: join, and selection) *Date* gets ahead of *Customer* because here,
     it is not used in the selections (used only in the join) contrary to *Customer*
     which is used in both operations.
   - The second query (Q4.2) uses mainly *Date*, *Customer* and *Supplier* hierar-
     chies in both operations. The minimum costs belong to the original schema,
     and to the 4th one, where only the *Part* dimension is normalized. As ex-
     pected, the next nearest cost having an average increase of 9% is when
     the *Customer* is normalized, and then comes the *Supplier* normalization
     with an increase of 39%, and finally the date normalization with entirely a
     double cost value.
   - The third query (Q4.3) uses all possible hierarchies in both operations,
     which explains the fact that the star schema is the unique schema which
     owns the minimum cost. Then, logically, comes the schema where the *Part*
     dimension is normalized with an increase of 0.4%, and right after the
     schema where *Customer* dimension is normalized (1% ↗), and then when
     the *Supplier* is normalized too (49% ↗), and finally the Date (92% ↗).

We have noticed that the 4th schema (where only *Part* dimension is normalized)
is providing almost the best costs to all queries as illustrated in Fig. 2.

4.2   Empirical Evaluation and Results Analysis

After the theoretical pass, we pick up the schemas selected as the best ones during the later step, and we deploy them on Oracle $\mathcal{DBMS}$ (Empirical pass). The
deployment consists in distributing the original real data over the tables of the
current normalized schema while loading them into the $\mathcal{DW}$. Fig. 3 illustrates
a comparison of the execution times of $\mathcal{SSB}$ queries submitted to the original
schema with those submitted to the 4h schema. The results depicted in Fig. 3
are expected to be similar to those depicted in Fig. 2. Although the results are
not broadly identical, they still are coherent. Coherent in that they keep which
schema is better than the other in execution of queries: the queries $Q2, Q4$ are
executed faster when submitted to the 4th schema, the same goes for $Q3$ which is
executed slightly faster. As for $Q1$, it is quickly executed when submitted to the
original schema. These facts are true to some extent, no matter what evaluation
type we have used (theoretical based on the cost model, or empirical using the
real $DBMS$). However, the results are not identical due to the overstatement of
some queries costs $Q2, Q3, Q4$. That difference may be explained by the number

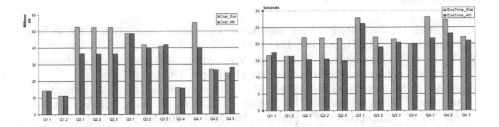

Fig. 2. Queries costs in the 4thschema   **Fig. 3.** Execution Time on Oracle11G (4th schema)

of joins contained in the concerned queries. Indeed, $Q2, Q3, Q4$ queries have several joins, which must be ordered. We justify this costs overstatement by our chosen scheduling technique which is based on dimensions' share value [5].

The schemas generation and their cost-based evaluation has been executed in less than 4 seconds. The reader may consider the results obtained by the SSB case study not convincing enough, because of the modest gain, obtained from hierarchies exploitation when generating the DW logical schema, in terms of size and execution times of SSB queries. We remind that this methodology is **context-dependent**. In fact, the small sizes of the dimension tables of SSB were not in favor of our methodology. On the other hand, snowflake queries can be optimized. However, there are definitely a lot of cases where the gain could be much more important.

5   Conclusion

In this paper, we showed how to exploit the different relationships between ontological concepts and properties such as functional dependencies, hierarchy relationships between properties, etc along DB life-cycle. To do so, we proposed to integrate these correlations in **domain ontologies** that recently contribute in designing semantic data warehouses. Through this paper, we want to start a debate to give more importance to the logical phase as we did for the physical phase. To perform a good logical phase, we proposed a mathematical cost model that evaluates the execution cost of queries workload. Extensive experiments have been conducted using this cost model and the obtained results are implemented on Oracle11G to show the efficiency and effectiveness of our proposal.

Currently, we are studying the process of selecting physical optimization structures by varying the logical models. Another important direction consists in proposing a generic approach which starts from fully denormalized schema (one flat table) to fully normalized one, even splitting wide fact tables.

References

1. Agrawal, R., Srikant, R.: Fast algorithms for mining association rules in large databases. In: VLDB, pp. 487–499 (1994)
2. Agrawal, S., Chaudhuri, S., Narasayya, V.R.: Automated selection of materialized views and indexes in sql databases. In: VLDB, pp. 496–505 (2000)
3. Anderlik, S., Neumayr, B., Schrefl, M.: Using domain ontologies as semantic dimensions in data warehouses. In: Atzeni, P., Cheung, D., Ram, S. (eds.) ER 2012. LNCS, vol. 7532, pp. 88–101. Springer, Heidelberg (2012)
4. Becker, L., Güting, R.H.: Rule-based optimization and query processing in an extensible geometric database system. ACM Trans. Database Syst. 17(2), 247–303 (1992)
5. Bellatreche, L., Boukhalfa, K., Richard, P., Woameno, K.Y.: Referential horizontal partitioning selection problem in data warehouses: Hardness study and selection algorithms. IJDWM 5(4), 1–23 (2009)
6. Bohannon, P., Fan, W., Geerts, F., Jia, X., Kementsietsidis, A.: Conditional functional dependencies for data cleaning. In: ICDE, pp. 746–755 (2007)
7. Brown, P.G., Hass, P.J.: Bhunt: Automatic discovery of fuzzy algebraic constraints in relational data. In: VLDB, pp. 668–679 (2003)
8. Codd, E.F.: A relational model of data for large shared data banks. Commun. ACM 13(6), 377–387 (1970)
9. Golfarelli, M., Rizzi, S.: Data warehouse testing: A prototype-based methodology. Information and Software Technology 53(11), 1183–1198 (2011)
10. Herbst, H.: Business Rule-Oriented Conceptual Modeling. Contributions to Management Science. Physica-Verlag HD (1997)
11. Hong, M., Riedewald, M., Koch, C., Gehrke, J., Demers, A.: Rule-based multiquery optimization. In: EDBT, pp. 120–131. ACM, New York (2009)
12. Kimura, H., Huo, G., Rasin, A., Madden, S., Zdonik, S.: Coradd: Correlation aware database designer for materialized views and indexes. PVLDB 3(1), 1103–1113 (2010)
13. Marchi, F.D., Hacid, M.-S., Petit, J.-M.: Some remarks on self-tuning logical database design. In: ICDE Workshops, p. 1219 (2005)
14. Martyn, T.: Reconsidering multi-dimensional schemas. SIGMOD Rec. 33(1), 83–88 (2004)
15. Petit, J.-M., Toumani, F., Boulicaut, J.-F., Kouloumdjian, J.: Towards the reverse engineering of denormalized relational databases. In: ICDE, pp. 218–227 (1996)
16. Ram, S., Khatri, V.: A comprehensive framework for modeling set-based business rules during conceptual database design. Inf. Syst. 30(2), 89–118 (2005)
17. Rasdorf, W., Ulberg, K., Baugh Jr., J.: A structure-based model of semantic integrity constraints for relational data bases. In: Proc. of Engineering with Computers, vol. 2, pp. 31–39 (1987)
18. Stöhr, T., Märtens, H., Rahm, E.: Multi-dimensional database allocation for parallel data warehouses. In: VLDB, pp. 273–284 (2000)
19. Tsichritzis, D., Klug, A.C.: The ansi/x3/sparc dbms framework report of the study group on dabatase management systems. Inf. Syst. 3(3), 173–191 (1978)

High Parallel Skyline Computation
over Low-Cardinality Domains

Markus Endres and Werner Kießling

Department of Computer Science, University of Augsburg,
86135 Augsburg, Germany
{endres,kiessling}@informatik.uni-augsburg.de
http://www.informatik.uni-augsburg.de/dbis

Abstract. A Skyline query retrieves all objects in a dataset that are not
dominated by other objects according to some given criteria. Although
there are a few parallel Skyline algorithms on multicore processors, it is
still a challenging task to fully exploit the advantages of such modern
hardware architectures for efficient Skyline computation. In this paper
we present high-performance parallel Skyline algorithms based on the
lattice structure generated by a Skyline query. We compare our methods with the state-of-the-art algorithms for multicore Skyline processing.
Experimental results on synthetic and real datasets show that our new
algorithms outperform state-of-the-art multicore Skyline techniques for
low-cardinality domains. Our algorithms have linear runtime complexity
and fully play on modern hardware architectures.

Keywords: Skyline, Parallelization, Multicore.

1   Introduction

The Skyline operator [1] has emerged as an important and very popular summa-
rization technique for multi-dimensional datasets. A Skyline query selects those
objects from a dataset D that are not dominated by any others. An object p
having d attributes (dimensions) dominates an object q, if p is better than q in
at least one dimension and not worse than q in all other dimensions, for a defined
comparison function. This dominance criteria defines a partial order and therefore transitivity holds. The *Skyline* is the set of points which are not dominated
by any other point of D. Without loss of generality, we consider the Skyline with
the *min* function for all attributes.

Most of the previous work on Skyline computation has focused on the development of efficient sequential algorithms [2]. However, the datasets to be processed
in real-world applications are of considerable size, i.e., there is the need for improved query performance, and parallel computing is a natural choice to achieve
this performance improvement, since multicore processors are going mainstream
[3]. This is due to the fact that Moore's law of doubling the density of transistors
on a CPU every two years – and hence also doubling algorithm's performance –
may come to an end in the next decade due to thermal problems. Thus, the chip

Y. Manolopoulos et al. (Eds.): ADBIS 2014, LNCS 8716, pp. 97–111, 2014.

manufactures tend to integrate multiple cores into a single processor instead of increasing the clock frequency. In upcoming years, we will see processors with more than 100 cores, but not with much higher clock rates. However, since most applications are build on using sequential algorithms, software developers must rethink their algorithms to take full advantage of modern multicore CPUs [3]. The potential of parallel computing is best described by Amdahl's law [4]: the speedup of any algorithm using multiple processors is strictly limited by the time needed to run its sequential fraction. Thus, only high parallel algorithms can benefit from modern multicore processors.

Typically an efficient Skyline computation depends heavily on the number of comparisons between tuples, called *dominance tests*. Since a large number of dominance tests can often be performed independently, Skyline computation has a good potential to exploit multicore architectures as described in [5–7]. In this paper we present algorithms for high-performance parallel Skyline computation which do *not* depend on tuple comparisons, but on the *lattice structure* constructed by a Skyline query over *low-cardinality domains*. Following [8, 2] many Skyline applications involve domains with small cardinalities – these cardinalities are either inherently small (such as star ratings for hotels), or can naturally be mapped to low-cardinality domains (such as price ranges on hotels).

The remainder of this paper is organized as follows: In Section 2 we discuss some related work. In Section 3 we revisit the Hexagon algorithm [9], since it is the basic idea behind our parallel algorithms. Based on this background we will present our parallel Skyline algorithms in Section 4. We conduct an extensive performance evaluation on synthetic and real datasets in Section 5. Section 6 contains our concluding remarks.

2   Related Work

Algorithms of the *block-nested-loop* class (BNL) [1] are the most prominent algorithms for computing Skylines. In fact the basic operation of collecting maxima during a single scan of the input data can be found at the core of several Skyline algorithms, cp. [10, 2]. Another class of Skyline algorithms is based on a straightforward *divide-and-conquer* (D&C) strategy. D&C uses a recursive split-and-merge scheme, which is definitely applicable in parallel scenarios [11].

There is also a growing interest in distributed Skyline computation, e.g., [12–16], where data is partitioned and distributed over net databases. Also there are several approaches based on the MapReduce framework, e.g., [17]. All approaches have in common that they share the idea of partitioning the input data for parallel shared-nothing architectures communicating only by exchanging messages. The nodes locally process the partitions in parallel, and finally merge the local Skylines. The main difference of such a parallel Skyline computation resides in the partitioning schemes of the data. The most used partitioning scheme is grid-based partitioning [14]. Recent work [18] focus on an angle-based space partitioning scheme using hyperspherical coordinates of the data points. In [19], the authors partition the space using hyperplane projections to obtain useful partitions of the dataset for parallel processing.

Im et al. [6] focuses on exploiting properties specific to multicore architectures in which participating cores inside a processor share everything and communicate simply by updating the main memory. They propose a parallel Skyline algorithm called *pSkyline*. pSkyline divides the dataset linearly into N equal sized partitions. The local Skyline is then computed for each partition in parallel using sSkyline [6]. Afterwards the local Skyline results have to be merged. Liknes et al. [7] present the *APSkyline* algorithm for efficient multicore computation of Skyline sets. They focus on the partitioning of the data and use the angle-based partitioning from [18] to reduce the number of candidate points that need to be checked in the final merging phase. The authors of [5] modified the well-known BNL algorithm to develop parallel variants based on a shared linked list for the Skyline window. In their evaluation, the lazy locking scheme [20] is shown to be most efficient in comparison to continuous locking or lock-free synchronization. There is also recent work on computing Skylines using specialized parallel hardware, e.g., GPU [21] and FPGA [22]. In contrast to previous works, our approach is based on the parallel traversal of the lattice structure of a Skyline query.

3 Skyline Computation Using the Lattice Revisited

Our parallel algorithms are based on the algorithms *Hexagon* [9] and *LS-B* [8], which follow the same idea: the partial order imposed by a Skyline query over a low-cardinality domain constitutes a *lattice*. This means if $a, b \in D$, the set $\{a, b\}$ has a least upper bound and a greatest lower bound in D. Visualization of such lattices is often done using *Better-Than-Graphs* (*BTG*) (Hasse diagrams), graphs in which edges state dominance. The nodes in the BTG represent *equivalence classes*. Each equivalence class contains the objects mapped to the same feature vector. All values in the same class are considered substitutable.

An example of a BTG over a 2-dimensional space is shown in Figure 1a. We write [2, 4] to describe a two-dimensional domain where the first attribute A_1 is an element of $\{0,1,2\}$ and attribute A_2 an element of $\{0,1,2,3,4\}$. The arrows show the dominance relationship between elements of the lattice.

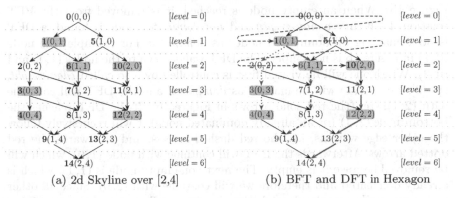

(a) 2d Skyline over [2,4] (b) BFT and DFT in Hexagon

Fig. 1. The Hexagon algorithm revisited [9]

The node $(0,0)$ presents the *best node*, i.e., the least upper bound for two arbitrary nodes a and b in the lattice. The node $(2,4)$ is the *worst node* and serves as the greatest lower bound. The bold numbers next to each node are *unique identifiers* (ID) for each node in the lattice, cp. [9]. Nodes having the same level are incomparable. That means for example, that neither the objects in the node $(0,4)$ are better than the objects in $(2,2)$ nor vice versa. They have the same overall *level 4*. A dataset D does not necessarily contain representatives for each lattice node. In Figure 1a the gray nodes are occupied (*non-empty*) with real elements from the dataset whereas the white nodes have no element (*empty*).

The method to obtain the Skyline can be visualized using the BTG. The elements of the dataset D that compose the Skyline are those in the BTG that have *no path leading to them from another non-empty node in D*. In Figure 1a these are the nodes $(0,1)$ and $(2,0)$. All other nodes have direct or transitive edges from these both nodes, and therefore are *dominated*. The algorithms in [9, 8] exploit these observations and in general consist of three phases:

1) **Phase 1:** The *Construction Phase* initializes the data structures. The lattice is represented by an *array* in main memory with the size of the lattice, i.e., the number of nodes. Each position in the array stands for one node ID in the lattice. Initially, all nodes of the lattice are marked as *empty*.
2) **Phase 2:** In the *Adding Phase* the algorithm iterates through each element t of the dataset D. For each element t the unique ID and the node of the lattice that corresponds to t is determined. This node is marked *non-empty*.
3) **Phase 3:** After all tuples have been processed, the nodes of the lattice that are marked as *non-empty* and which are not reachable by the transitive dominance relationship from any other *non-empty* node of the lattice represent the Skyline values. Nodes that are *non-empty* but are reachable by the dominance relationship, and hence are not Skyline values, are marked *dominated* to distinguish them from present Skyline values.

From an algorithmic point of view this is done by a combination of *breadth-first traversal* (BFT) and *depth-first traversal* (DFT). The nodes of the lattice are visited level-by-level in a breadth-first order (the blue dashed line in Figure 1b). When an empty node is reached, it is removed from the BFT relation. Each time a *non-empty* and *not dominated* node is found, a DFT is done marking all dominated nodes as *dominated*. For example, the node $(0,1)$ in Figure 1b is not empty. The DFT walks down to the nodes $(1,1)$ and $(0,2)$. Which one will be visited first is controlled by a so called *edge weight*, cp. [9]. Here, $(1,1)$ will be marked as *dominated* and the DFT will continue with $(2,1)$, etc. (the red solid arrows in Figure 1b). If the DFT reaches the bottom node $(2,4)$ (or an already dominated node) it will recursively follow the other edge weights, i.e. the red dashed arrows, and afterwards the red dotted arrows. Afterwards the BFT will continue with node $(1,0)$, which will be removed because it is empty. The next non-empty node is $(1,1)$, which is already dominated and therefore we will continue with $(2,0)$. Since all other nodes are marked as dominated, the algorithm will stop and the remaining nodes $(0,1)$ and $(2,0)$ present the Skyline.

4   Parallel Skyline Algorithms

In this section we describe our parallel algorithms, the used data structures, discuss some implementation issues, and have a look at the complexity and memory requirements of our algorithms.

4.1   Parallel Skyline Computation

For the development of our parallel Skyline algorithms we combine a *split* approach of the input dataset with a *shared data structure* supporting fine grained locking and apply them to the Hexagon algorithm described in Section 3.

The general idea of parallelizing the Hexagon algorithm is to parallelize the adding phase (Phase 2) and the removal phase (Phase 3). Phase 1 is not worth to parallelize because of its simple structure and minor time and effort for the initialization. Parallelizing Phase 2 can be done using a simple partitioning approach of the input dataset, whereas for Phase 3 two different approaches can be used: In the first variant the parallel Phase 3 starts *after* all elements were added to the BTG. We call this algorithm **ARL-Skyline** (Adding-Removal-Lattice-Skyline). The second approach runs the adding and removal simultaneously. This algorithm is called **HPL-Skyline** (High-Parallel-Lattice-Skyline).

The ARL-Skyline Algorithm (ARL-S) is designed as follows:

- **Phase 1:** Initialize all data structures.
- **Phase 2:** Split the input dataset into c partitions, where c is the number of used threads. For each partition a worker thread iterates through the partition, determines the IDs for the elements and marks the corresponding entries in the BTG as *non-empty*.
- **Phase 3:** After adding *all* elements to the BTG a breadth-first walk beginning at the top starts (blue line in Figure 2a). For each *non-empty* and *not dominated* node run *tasks*[1] for the depth-first walk with the dominance test. In parallel continue with the breadth-first walk.

  For example, if the node $(0,1)$ is reached in Figure 2a, two further tasks can be started in parallel to run a DFT down to $(0,2)$ and $(1,1)$ (red solid arrows). Continuing with the BFT we reach the already dominated node $(1,1)$ and afterwards $(2,0)$. A new DFT task follows the red dashed arrows to mark nodes as dominated. Note that the BFT task might be slower or faster than the DFT from node $(0,1)$ and therefore the DFT could follow different paths in the depth-first dominance search. The pseudocode for ARL-S reduced to its essence is decpicted in Figure 2b; the fork/join task for the DFT can be found in Figure 3b.

---

[1] We use the `ForkJoinPool` from Java 7 to manage the recursive DFT tasks.

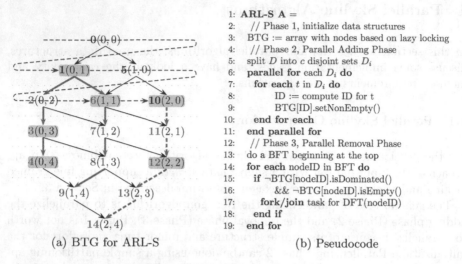

(a) BTG for ARL-S

```
1:  ARL-S A =
2:    // Phase 1, initialize data structures
3:    BTG := array with nodes based on lazy locking
4:    // Phase 2, Parallel Adding Phase
5:    split D into c disjoint sets D_i
6:    parallel for each D_i do
7:      for each t in D_i do
8:          ID := compute ID for t
9:          BTG[ID].setNonEmpty()
10:     end for each
11:   end parallel for
12:   // Phase 3, Parallel Removal Phase
13:   do a BFT beginning at the top
14:   for each nodeID in BFT do
15:     if ¬BTG[nodeID].isDominated()
16:        && ¬BTG[nodeID].isEmpty()
17:       fork/join task for DFT(nodeID)
18:     end if
19:   end for
```

(b) Pseudocode

Fig. 2. The ARL-Skyline algorithm

The HPL-S algorithm combines Phase 2 and 3 of ARL-S to *one* phase.

The HPL-Skyline Algorithm (HPL-S) is designed as:

- **Phase 1:** Initialize all data structures.
- **Phase 2+3:** Similar to Phase 2 in ARL-S we split the dataset into c partitions, for each partition a worker thread c_i. If one of the worker threads marks a node in the BTG as *non-empty*, it immediately starts a task for the DFT dominance test (if not done yet) and continues with adding elements, cp. Figure 3a. The simplified pseudocode is shown in Figure 3b.

   For example, thread c_1 adds an element to $(0,3)$ and immediately starts additional tasks for the DFT (red arrows). Simultaneously another thread c_2 adds an element to the node $(0,1)$ and starts tasks for the DFT and dominance tests (red dotted arrows). After thread c_1 has finished, it wants to add an element to $(1,1)$. However, since it is already marked as dominated, thread c_1 can continue with adding elements to other nodes in the BTG without performing a DFT dominance test.

   After all threads have finished, a breadth-first traversal is done on the remaining nodes (blue line in Figure 3a). Again, the non-empty and not dominated nodes present the Skyline.

The advantage of the HPL-S in comparison to ARL-S is that the DFT search will mark dominated nodes as *dominated* and other parallel running threads do not have to add possible elements to these already dominated nodes. This saves memory and runtime.

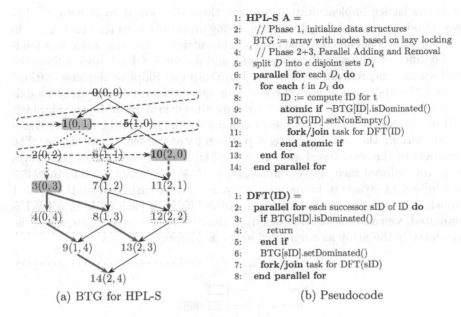

```
1:  HPL-S A =
2:    // Phase 1, initialize data structures
3:    BTG := array with nodes based on lazy locking
4:    // Phase 2+3, Parallel Adding and Removal
5:    split D into c disjoint sets D_i
6:    parallel for each D_i do
7:      for each t in D_i do
8:        ID := compute ID for t
9:        atomic if ¬BTG[ID].isDominated()
10:         BTG[ID].setNonEmpty()
11:         fork/join task for DFT(ID)
12:       end atomic if
13:     end for
14:   end parallel for


1:  DFT(ID) =
2:    parallel for each successor sID of ID do
3:      if BTG[sID].isDominated()
4:        return
5:      end if
6:      BTG[sID].setDominated()
7:      fork/join task for DFT(sID)
8:    end parallel for
```

(a) BTG for HPL-S   (b) Pseudocode

Fig. 3. The HPL-Skyline algorithm

4.2 Data Partitioning and Choosing the Right Data Structure

Data Partitioning. The performance of known parallel and distributed BNL and D&C style algorithms (and many variants) are heavily influenced by the underlying partitioning of the input dataset. [13] suggests a grid-based partitioning, [18, 7] uses an angle-based partitioning, and [19] uses hyperplane projections to divide the dataset into disjoint sets. The *lattice algorithms* are independent from the partitioning, because the dominance tests are done on the lattice structure instead of relying on a tuple-to-tuple comparison. This is also the reason why the underlying data distribution (i.e., whether the dataset attributes are correlated, independent, or anti-correlated) does not influence performance.

Choosing the Right Data Structure. In general concurrency on a shared data structure requires a fine grained locking amongst all running threads to avoid unnecessary locks. In addition, one has to ensure that no data is read or written which has just been accessed by another thread (dirty reads or writes) in order to avoid data inconsistency. When considering for example the parallel removal phase in HPL-S (Figure 3b), a critical situation may occur if two threads try to append (line 10 in HPL-S) or delete an element (line 6 in DFT) on the same node simultaneously. This problem can be tackled by synchronization and locking protocols, cp. [5]. The *lazy locking* approach uses as few locks as possible. Locks are only acquired when they are really needed, i.e., when modifying nodes. Reading can be done in parallel without inconsistency problems. From a performance point of view lazy locking is definitely superior to all other locking protocols like continuous locking, full, or lock-free synchronization.

For the lattice implementation we used three different data structures: **Arrays, HashMaps,** and **SkipLists** [23]. Using an array means that each index in the array represents an ID in the lattice. The entries of the array are *nodes* holding the different states *empty, non-empty,* and *dominated.* Each *node* follows the lazy locking synchronization[2]. For the HashMap and SkipList implementation[3] we used the approach of a level-based storage, cp. Figure 4. An array models the *levels* of the BTG. Then the *nodes* are stored in a HashMap or SkipList. Adding an element to the BTG means computing the ID and the level it belongs to and marking the node at the right position as *non-empty* or *dominated.* The advantage of the level-based storage using SkipLists in contrast to HashMaps lies in the reduced memory requirements, because we do not have to initialize the whole data structure in main memory. A node is initialized on-the-fly if it is marked as *non-empty* or *dominated.* Additionally, if each node in a level is dominated, we can remove all nodes from the corresponding SkipList, mark the level-entry in the array as *dominated* and free memory.

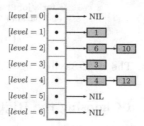

Fig. 4. Level-based storage of the BTG using SkipLists

In [20, 5] a LazyList with some advantages against the concurrent SkipList implementation was proposed to use for concurrent programming. Nevertheless, we decided to use SkipLists instead of LazyLists, because the traversal of a SkipList is faster than that of a LazyList due to the additional pointers which skip some irrelevant elements. Since not all nodes in the lattice are present and we have to find some nodes in the lattice during the DFT search quickly, the concurrent SkipList is the better choice.

4.3   Complexity Analysis and Memory Requirements

Complexity Analysis. The original lattice based algorithms [9, 8] have linear runtime complexity. More precisely, the complexity is $\mathcal{O}(dV + dn)$, where d is the dimensionality, n is the number of input tuples, and V is the product of the cardinalities of the d low-cardinality domains from which the attributes are drawn. Since there are V total entries in the lattice, each compared with at most d entries, this step is $\mathcal{O}(dV)$, cp. [8]. In the original version of Hexagon all entries in the lattice are positioned in an array. Since array accesses are $\mathcal{O}(1)$, the pass through the data to mark an entry as *non-empty* is $\mathcal{O}(dn)$.

---

[2] Implemented with `ReentrantReadWriteLock` in Java 7.
[3] We use `ConcurrentHashMap` and `ConcurrentSkipListMap` from Java 7.

The ARL-S and HPL-S algorithms with an *array* as BTG representation follow the original implementation of [9, 8] and therefore have a complexity of $\mathcal{O}(dV + dn)$. Using a level-based representation of the BTG with a HashMap for each level, we have a constant access for each level and $\mathcal{O}(1)$ for the look-up in the HashMap, since we can use a perfect hash function due the known width of the BTG in each level, cp. [24]. In summary this leads to $\mathcal{O}(dV + dn)$, too. For the SkipList based BTG implementation we have $\mathcal{O}(dV + dn \log w)$, since operations on SkipLists are $\mathcal{O}(\log w)$ [23], where w is the number of elements in the SkipList, i.e., the width of the BTG in the worst case.

Memory Requirements. Given a discrete low-cardinality domain $dom(A_1) \times \ldots \times dom(A_m)$ on attributes A_i, the number of nodes in the BTG is given by $\prod_{i=1}^{m}(\max(A_i) + 1)$ [9]. Each node of the BTG has one of three different states: *empty*, *non-empty*, and *dominated*. The easiest way to encode these three states is by using two bits with 0x00 standing for *empty*, 0x01 for *non-empty* and 0x10 for *dominated*. This enables us to use the extremely fast bit functions to check and change node states. Since one byte can hold four nodes using two bits each, we have in summary that the BTG for a Skyline query may require the following maximal amount of memory, i.e, it is linear w.r.t. the size of the BTG.

$$mem(BTG) := \left\lceil \frac{1}{4} \prod_{i=1}^{m}(\max(A_i) + 1) \right\rceil$$

4.4   Remarks

Concurrent programming usually increases performance when the number of used threads is equal or less than the number of available processor cores and idling of threads can be prevented. Otherwise it can decrease performance due to waiting or mutual locking program codes. Our algorithms use high parallelism to complete the running tasks. This might be a performance problem for very small BTGs, if many threads work on a lattice where the size is much smaller than the number of threads. In this case there could be a lot of synchronization necessary. However, in practical Skyline problems this should not occur.

Another question concerns the speedup of concurrently programmed algorithms with larger input data. A good description of potentially benefits is given by Gustafson's Law [4], which says that computations involving arbitrarily large datasets can be efficiently parallelized. Our algorithms depend on the lattice size and the dominance tests on the lattice nodes, but not on a tuple-to-tuple comparison. Therefore, for larger datasets only the adding phase influences the performance, but not the removal phase, because the BTG size is independent from the input size. In addition, the HPL-S algorithms have the advantage of an "premature domination" of nodes, i.e., we filter out unnecessary elements early.

Due to Amdahl's law the sequential part of concurrent programs must be reduced to a minimum. In our algorithms only the initialization of the data structure (Phase 1), and the last tuple scanning is sequential, because initializing an array, a SkipList or HashMap is just instantiating these objects.

The reader will notice that the lattice based algorithms require two scans of the dataset to output the Skyline, the first to mark positions in the lattice structure and a second to output Skyline elements from values derived from the lattice. Another approach used in [9] is to mark the nodes in the lattice as *non-empty* and additionally hold pointers to the elements in the dataset. Obviously, this requires more memory, but avoids the second linear scan of the dataset.

In summary that means that we have high parallel algorithms with a minimal sequential part and therefore expect an enormous speed-up in Skyline evaluation.

5   Experiments

This section provides our comprehensive benchmarks on synthetic and real data to reveal the performance of the outlined algorithms. Due to the restricted space of the paper we only present some selected characteristic tests. However, all results show the same trends as those presented here.

5.1   Benchmark Framework

For our synthetic datasets we used the data generator commonly used in Skyline research [1]. We generated *anti-correlated* (anti), *correlated* (corr), and *independent* (ind) distributions and varied three parameters: (1) the data cardinality n, (2) the data dimensionality d, and (3) the number of distinct values for each attribute domain. For real data we used the entries from *www.zillow.com*. This dataset contains more than 2M entries about real estate in the United States. Each entry includes number of *bedrooms* and *bathrooms*, *living area* in sqm, and *age* of the building. The Zillow dataset also serves as a real-world application which requires finding the Skyline on data with a low-cardinality domain.

Our algorithms have been implemented using Java 7 using only built-in techniques for locking, compare-and-swap operations, and thread management. All experiments are performed on a single node running Debian Linux 7.1. The machine is equipped with two Intel Xeon 2.53 GHz quad-core processors using Hyper-Threading, that means a total of 16 cores.

5.2   Experimental Results

Comparison of ARL-S and HPL-S. For our algorithms **ARL-S** and **HPL-S** we used different data structures, i.e. **A** (Array), **HM** (HashMap), **SL** (SkipList) as described in Section 4. For comparison we used synthetic datasets, because they allow us to carefully explore the effect of various data characteristics.

Figure 5a presents the runtime performance of our algorithms on different data cardinality n. We used $n = 1 \cdot 10^6$ to $10 \cdot 10^6$ tuples and 5 dimensions, since this is realistic in practical cases. We fixed the number of threads to $c = 8$ and used a domain derived from $[1, 2, 5, 100, 100]$. In this case the lattice has 367236 nodes. The array based implementations ARL-S A and HPL-S A perform best, whereas the level-based versions with non-linear time complexity are worser, cp. Section

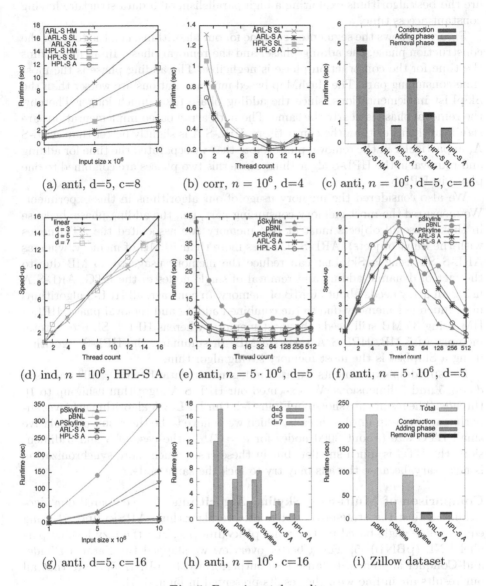

Fig. 5. Experimental results

4.3. Interestingly, the HashMap implementation of the BTG is not as good as the SkipList based version. Maybe this is due to the additional computation of the hash function. The Skyline size ranges from 392 ($n = 10^6$) to 3791 objects ($n = 10 \cdot 10^6$).

In Figure 5b we compared the SkipList and Array variants of ARL-S and HPL-S. We used correlated data on 4 dimensions and varied the number of threads up to 16 ('one thread per core'). As expected, ARL-S A and HPL-S A are the best algorithms exploiting a high parallelism on a data structure having constant access time.

Figure 5c shows the segmented runtime for our algorithms, i.e., the time for the construction phase, the adding phase, and the removal phase. In all algorithms the time for the construction phase is negligible. The adding phase is the most time consuming part. The HashMap based implementations are worser than the SkipList implementations, since the adding phase takes much longer Thereby, the removal phase is nearly the same. The both array based implementations are significantly faster than the competitors. HPL-S A is slightly better than ARL-S A, in particular in the removal phase. Note that we separated the time for adding and removal in the HPL-S algorithms, since this two phases are combined to one phase in HPL-S.

We also considered the memory usage of our algorithms in this experiment. We measured the most memory consuming part, i.e., the adding phase, because in that phase all objects must fit into memory (we associated the BTG nodes with the input objects). ARL-S HM uses more than 70 MB of memory, whereas ARL-S SL using a SkipList can reduce the memory usage to 45 MB due to the fact of dynamic adding and removal of single nodes of the BTG. ARL-S A using an array needs about 50 MB of memory. In contrast, all HPL-S algorithms need much less memory due to the combined adding and removal phase. HPL-S HM using 35 MB still needs the most memory, whereas HPL-S SL uses a total memory of 5 MB. HPL-S A takes 25 MB. In summary, the HPL-S algorithm using a SkipList is the most memory saving algorithm.

For speed-up experiments we used independent data, fixed $n = 10^6$ and used $d = 3, 5$ and 7 dimensions. We executed our HPL-S A algorithm using up to 16 threads. The results are shown in Figure 5d. Our HPL-S A algorithm achieves superlinear speed-up until 6 threads, which we believe to be the result of a relative small BTG size (about 3000 nodes for $d = 7$). In the case of 3 and 5 dimensions the BTG is much smaller, but in these cases much more synchronization is necessary, because threads may try to lock the same node.

Comparison of Multi-core Skyline Algorithms. We compared our algorithms against the state-of-the-art multicore algorithms **APSkyline** [7] using equi-volume angle-based partitioning, **pSkyline** [11], and the lazy locking parallel BNL (**pBNL**) [5]. For a better overview we skipped the Parallel Divide-and-Conquer approach, because it is outperformed by pBNL [5]. Note that all our results are in line with the results presented in [5] and [7].

In Figure 5e we show the results in the case of an anti-correlated dataset as we increase the number of threads from 1 to 512. We expected to reach peak performance at 16 threads, which is the maximum number of hardware threads (two quad-core processors using Hyper-Threading). As the number of threads increase beyond 16, the performance gain moderately ceases for all algorithms due to the increased synchronization costs without additional parallel computing power. For the used low-cardinality domain $[2, 2, 2, 2, 100]$ we observed that our algorithms ARL-S A and HPL-S A outperform the competitors until 128 threads. Beyond that, the parallel computing power decreases and ARL-S A and HPL-S

A become worser. This is due to the fact of the high number of locks on the BTG nodes, in particular when using 512 threads. The Skyline has 37971 objects.

In Figure 5f we measure the speed-up of each algorithm on anti-correlated data using $n = 5 \cdot 10^6$ and 5 dimensions ($[2, 3, 5, 10, 100]$). We observed that all algorithms have nearly linear speed-up up to 8 threads. From the ninth thread on, the performance only marginally increases and beyond 16 thread it gradually decreases. This can be explained with decreasing cache locality and increasing communication costs as our test systems uses two quad-core processors with Hyper-Threading (8 cores per CPU). Starting with the ninth core, the second processor must constantly communicate with the first.

Figure 5g presents the behavior of the algorithms for increased data size. APSkyline is better than pBNL and pSkyline as mentioned in [7]. However, the domain derived from $[2, 3, 5, 10, 100]$, which is typical for Skyline computation (a few small attributes together with a large attribute) [8], is best suited for our algorithms, which significantly outperform all others.

Figure 5h shows the obtained results when increasing the number of dimensions: $d = 3$ ($[2,2,100]$) to $d = 7$ ($[2,2,2,2,2,2,100]$). The number of input tuples (anti) was fixed to $n = 10^6$ and $c = 16$. pSkyline and APSkyline are quiet similar for all dimensions, whereas pBNL is better for $d = 3$. It should be mentioned that the size of the Skyline set normally increases on anti-correlated data with the dimensionality of the dataset [25] (138 Skyline objects for $d = 3$, 4125 objects for $d = 7$). This makes Skyline processing for algorithms relying on tuple-to-tuple comparison more demanding. This experiments verifies the advantage of our algorithms based on the lattice structure and not on a tuple comparison, in particular for higher values of dimensionality.

Real Dataset. In Figure 5i we show the obtained results for the real-world Zillow dataset. The parallel BNL algorithm is outperformed in an order of magnitude by all other algorithms. APSkyline is outperformed by pSkyline because of an unfair data partitioning as mentioned in [7]. Our lattice based algorithms do not rely on any partitioning scheme and are independent from data distribution. Therefore, the best performing algorithms are ARL-S A and HPL-S A. Thereby the latter one slightly performs better. For both algorithms the adding phase is the most time consuming part. Note that we skipped the HashMap and SkipList implementations for an better overview. They are also outperformed by ARL-S A and HPL-S A. There are 95 objects in the Skyline.

6   Conclusion

In this paper we presented two algorithms for high-performance parallel Skyline computation on shared-memory multi-processor systems. Both algorithms are based on the lattice structure constructed by a Skyline query over low-cardinality domains, and do not rely on any data partitioning. Our algorithms have linear runtime complexity and a memory requirement which is linear w.r.t. the size of the lattice. In our extensive experiments on synthetic and real data, we showed the superior characteristics of these algorithms in different settings. Exploiting the parallelization on the lattice structure we are able to outperform state-of-the-art approaches for Skyline computation on modern hardware architectures. As future work we want to extend our algorithms to handle high-cardinality domains, which could be a challenging task.

Acknowledgement. We want to thank Selke et al. [5] for providing us with the source code of the parallel BNL and pSkyline. The implementation of APSkyline is based on the source code made available by Liknes et al. [7].

References

1. Börzsönyi, S., Kossmann, D., Stocker, K.: The Skyline Operator. In: Proc. of ICDE 2001, pp. 421–430. IEEE, Washington, DC (2001)
2. Chomicki, J., Ciaccia, P., Meneghetti, N.: Skyline Queries, Front and Back. SIGMOD Rec. 42(3), 6–18 (2013)
3. Mattson, T., Wrinn, M.: Parallel Programming: Can we PLEASE get it right this time? In: Fix, L. (ed.) DAC, pp. 7–11. ACM (2008)
4. Gustafson, J.L.: Reevaluating Amdahl's law. Commun. ACM 31(5), 532–533 (1988)
5. Selke, J., Lofi, C., Balke, W.-T.: Highly Scalable Multiprocessing Algorithms for Preference-Based Database Retrieval. In: Kitagawa, H., Ishikawa, Y., Li, Q., Watanabe, C. (eds.) DASFAA 2010. LNCS, vol. 5982, pp. 246–260. Springer, Heidelberg (2010)
6. Im, H., Park, J., Park, S.: Parallel Skyline Computation on Multicore Architectures. Inf. Syst. 36(4), 808–823 (2011)
7. Liknes, S., Vlachou, A., Doulkeridis, C., Nørvåg, K.: APSkyline: Improved Skyline Computation for Multicore Architectures. In: Bhowmick, S.S., Dyreson, C.E., Jensen, C.S., Lee, M.L., Muliantara, A., Thalheim, B. (eds.) DASFAA 2014, Part I. LNCS, vol. 8421, pp. 312–326. Springer, Heidelberg (2014)
8. Morse, M., Patel, J.M., Jagadish, H.V.: Efficient Skyline Computation over Low-Cardinality Domains. In: Proc. of VLDB 2007, pp. 267–278. (2007)
9. Preisinger, T., Kießling, W.: The Hexagon Algorithm for Evaluating Pareto Preference Queries. In: Proc. of MPref 2007 (2007)
10. Godfrey, P., Shipley, R., Gryz, J.: Algorithms and Analyses for Maximal Vector Computation. The VLDB Journal 16(1), 5–28 (2007)
11. Park, S., Kim, T., Park, J., Kim, J., Im, H.: Parallel Skyline Computation on Multicore Architectures. In: Proc. of ICDE 2009, pp. 760–771 (2009)
12. Lo, E., Yip, K.Y., Lin, K.-I., Cheung, D.W.: Progressive Skylining over Web-accessible Databases. IEEE TKDE 57(2), 122–147 (2006)
13. Wu, P., Zhang, C., Feng, Y., Zhao, B.Y., Agrawal, D.P., El Abbadi, A.: Parallelizing Skyline Queries for Scalable Distribution. In: Ioannidis, Y., et al. (eds.) EDBT 2006. LNCS, vol. 3896, pp. 112–130. Springer, Heidelberg (2006)
14. Hose, K., Vlachou, A.: A Survey of Skyline Processing in Highly Distributed Environments. The VLDB Journal 21(3), 359–384 (2012)
15. Afrati, F.N., Koutris, P., Suciu, D., Ullman, J.D.: Parallel Skyline Queries. In: Proc. of ICDT 2012, pp. 274–284. ACM, New York (2012)
16. Cosgaya-Lozano, A., Rau-Chaplin, A., Zeh, N.: Parallel Computation of Skyline Queries. In: Proc. of HPCS 2007, p. 12 (2007)
17. Park, Y., Min, J.-K., Shim, K.: Parallel Computation of Skyline and Reverse Skyline Queries Using MapReduce. PVLDB 6(14), 2002–(2013)

18. Vlachou, A., Doulkeridis, C., Kotidis, Y.: Angle-based Space Partitioning for Efficient Parallel Skyline Computation. In: Proc. of SIGMOD 2008, pp. 227–238 (2008)
19. Köhler, H., Yang, J., Zhou, X.: Efficient Parallel Skyline Processing using Hyperplane Projections. In: Proc. of SIGMOD 2011, pp. 85–96. ACM (2011)
20. Heller, S., Herlihy, M., Luchangco, V., Moir, M., Scherer III, W.N., Shavit, N.: A Lazy Concurrent List-Based Set Algorithm. In: Anderson, J.H., Prencipe, G., Wattenhofer, R. (eds.) OPODIS 2005. LNCS, vol. 3974, pp. 3–16. Springer, Heidelberg (2006)
21. Bøgh, K.S., Assent, I., Magnani, M.: Efficient GPU-based Skyline computation. In: Proc. of DaMoN, pp. 5:1–5:6. ACM, New York (2013)
22. Woods, L., Alonso, G., Teubner, J.: Parallel Computation of Skyline Queries. In: Proc. of the FCCM, pp. 1–8. IEEE, Washington, DC (2013)
23. Pugh, W.: Skip Lists: A Probabilistic Alternative to Balanced Trees. Commun. ACM 33(6), 668–676 (1990)
24. Glück, R., Köppl, D., Wirsching, G.: Computational Aspects of Ordered Integer Partition with Upper Bounds. In: Bonifaci, V., Demetrescu, C., Marchetti-Spaccamela, A. (eds.) SEA 2013. LNCS, vol. 7933, pp. 79–90. Springer, Heidelberg (2013)
25. Shang, H., Kitsuregawa, M.: Skyline Operator on Anti-correlated Distributions. In: Proc. of VLDB 2013 (2013)

Top-k Differential Queries in Graph Databases

Elena Vasilyeva[1], Maik Thiele[2], Christof Bornhövd[3], and Wolfgang Lehner[2]

[1] SAP AG, Chemnitzer Str. 48, 01187 Dresden, Germany
elena.vasilyeva@sap.com
[2] Technische Universität Dresden, Database Technology Group
Nöthnitzer Str. 46, 01187 Dresden, Germany
{maik.thiele,wolfgang.lehner}@tu-dresden.de
[3] SAP Labs, LLC, Palo Alto, USA
christof.bornhoevd@sap.com

Abstract. The sheer volume as well as the schema complexity of today's graph databases impede the users in formulating queries against these databases and often cause queries to "fail" by delivering empty answers. To support users in such situations, the concept of differential queries can be used to bridge the gap between an unexpected result (e.g. an empty result set) and the query intention of users. These queries deliver missing parts of a query graph and, therefore, work with such scenarios that require users to specify a query graph. Based on the discovered information about a missing query subgraph, users may understand which vertices and edges are the reasons for queries that unexpectedly return empty answers, and thus can reformulate the queries if needed. A study showed that the result sets of differential queries are often too large to be manually introspected by users and thus a reduction of the number of results and their ranking is required. To address these issues, we extend the concept of differential queries and introduce top-k differential queries that calculate the ranking based on users' preferences and therefore significantly support the users' understanding of query database management systems. The idea consists of assigning relevance weights to vertices or edges of a query graph by users that steer the graph search and are used in the scoring function for top-k differential results. Along with the novel concept of the top-k differential queries, we further propose a strategy for propagating relevance weights and we model the search along the most relevant paths.

Keywords: Graph databases, Top-k Differential Queries, Flooding.

1 Introduction

Following the principle "data comes first, schema comes second", graph databases allow to store data without having a predefined, rigid schema and enable a gradual evolution of data together with its schema. Unfortunately, schema flexibility impedes the formulation of queries. Due to the agile flavor of integration and interpretation processes, users very often do not possess deep knowledge of the data and its evolving schema. As a consequence, issued queries might return

Y. Manolopoulos et al. (Eds.): ADBIS 2014, LNCS 8716, pp. 112–125, 2014.
© Springer International Publishing Switzerland 2014

unexpected result sets, especially empty results. To support users to understand the reasons of an empty answer, we already proposed the notion of a differential query [15]; a graph query (for example see Figure 1(a)) that has a result consisting of two parts: (1) a discovered subgraph that is a part of a data graph isomorphic to a query subgraph like in Figure 1(b), and (2) a difference graph reflecting the remaining part of a query like in Figure 1(c). Differential queries work in scenarios, where users need to specify a query in the form of a graph, such as subgraph matching queries. Although the approach in [15] already supports users in the query answering process, it still has some limitations: the number of intermediate results can be very large, e.g. it can reach up to 150K subgraphs for a data graph consisting of 100K edges and a query graph with 10 edges. Optimization strategies reducing the number of traversals for a query based on cardinality and degree of a query's vertices could prune intermediate results, but, as a side effect, they also could remove important subgraphs, since these strategies do not consider the users' intention.

Contributions

To cope with this issue, we extend the concept of differential queries with a top-k semantic, resulting in so-called top-k differential queries that are the main contribution of this paper. These queries allow the user to mark vertices, edges, or entire subgraphs of a query graph with relevance weights showing how important specified graph elements are within a query. To make the search of a top-k differential query with multiple relevance weights possible, we present an algorithm for the propagation of relevance weights: relevance flooding. Based on the propagated weights, the system decides automatically how to conduct the search in order to deliver only the most relevant subgraphs to a user as an alternative result set of the original query. The initial weights are used to rank the results. The concept of top-k differential queries allows us to reduce processing efforts on the one side and allows to rank individual answers according to the user's interest on the other side.

The rest of the paper is structured as follows. In Section 2 we present the state of the art related work. The property graph model and differential queries are introduced in Section 3. Section 4 describes the relevance-based search and its application to top-k differential queries. We evaluate our approach in Section 5.

2   Related Work

In this section we present solutions for "Why Not?" queries and for the empty-answer problem, ranking of query results, and flexible query answering.

"Why Not?" Queries and Empty-Answer Problem

The problem of unexpected answers is generally addressed by "Why Not?" queries [3] determining why items of interest are not in the result set. It is assumed that the size and complexity of data prevent a user from manually studying the reasons in a feasible way. A user specifies the items of interest with attributes or key values and conducts a "Why Not?" query. The answers to such a query can be (1) an operator of a query tree, removing the item from processing [3], (2) a data source in provenance-based "Why Not?" queries like for example in [6], or (3) a refined query that contains the items of interest in the result set like in [14]. In contrast to approaches tailored for relational databases, we do not operate on a query tree constructed for a query execution plan, but we deal with a query graph, for which we search corresponding data subgraphs by a breadth-first or depth-first traversal considering user-defined restrictions with respect to vertices and edges based on their attribute values. It is important to understand, which query edges and vertices are responsible for the delivery of an empty result set.

Query rewriting for the empty-answer problem can also be enhanced by user interaction [10]. This interactive query relaxation framework for conjunctive queries [10] constructs a query relaxation tree from all possible combinations of attributes' relaxations. Following the tree top-down, a user receives proposals for query relaxations and selects preferred ones. This approach [10] has only a single objective function. In our settings, it would be only a single vertex of interest. To model multiple relevant elements and to detect the optimal path between them cannot be achieved by this approach proposed in [10].

Ranking of Query Results

The concept of top-k queries derives from relational database management systems, where the results are calculated and sorted according to a scoring function. In graph databases top-k queries are used for ranking (sub)graph matchers [16,17]. These ranking strategies differ in regard to how a data graph is stored in a graph database. If a database maintains multiple data graphs, for example chemical structures, then a similarity measure based on a maximum common subgraph between a query and an individual data graph can be used as a scoring function [16]. If a database maintains a single large data graph, then the approach of top-k subgraph matchers [17] can be applied. In this context, it is assumed that a data graph has naturally a hierarchical structure that can be used for index construction and clustering of data subgraphs enabling effective pruning. These solutions do not consider any relevance function for a query graph which is paramount in our setup.

To rank the results, an "interesting" function [5], relevance and distance functions [4], or estimation of confidence, informativeness, and compactness [7] can be used. In the first case [5], such an "interesting" function is defined in advance by a use case, for example, it can be a data transfer rate between computers in a network. Up front, we do not have any "interesting" function in a data graph. In the second case [4], the matching problem is revised by the concept of "output node", which presents the main part of a query answer to be delivered to a user. In our settings, this approach could be compared to a single vertex with a user-specified relevance weight. In the third case [7], additional semantic information is used to estimate scoring functions. In contrast, we assume that the data graph has the maximal confidence, our user is interested in subgraph matching queries without accounting for additional semantic information. The compactness of answers is not considered in our work, because we deal with exact matching, and the answers containing more relevant parts matching to the initial query are ranked higher. Our approach can be further improved by estimating the informativeness, which should be based on a user's preferences. This question is left for future work.

In [1] the top-k processing for XML repositories is presented. The authors relax the original query, calculate the score of a new query based on its content- and structure-based modifications, and search for the matches. While Amer-Yahia et al. relax the query and search for a matching document, we process a data graph without any changes to the original query. Instead we do search for exact subgraph matches. Subgraphs can also be matched and ranked by approximate matching and simulation-based algorithms, which can result in inaccurate answers with a wrong graph shape or non matching vertices. Since we provide exact matches, the class of inexact algorithms is not considered in our work.

Flexible Query Answering

A different approach tackling the problem of overspecified queries can be modeled by the SPARQL language [11]. SPARQL provides the OPTIONAL clause, which allows to process a query graph if a statement or an entire subgraph is missing in a data graph. The UNION clause allows to specify alternative patterns. Defining a flexible query is not straight-forward: a user has to produce all possible combinations of missing edges and vertices in a query graph to derive results, this requires good knowledge of SPARQL. Moreover, this language does not support relevance weights on a query graph directly, and a user cannot have a direct impact on the search within the database. Furthermore, it does not support the calculation of difference graphs.

3   Preliminaries

In this section we present a general overview on the used graph model and differential queries in graph databases.

Property Graph

A graph database stores data as vertices and edges. Any query to a graph database and corresponding results may be understood as graphs themselves. As an underlying graph model we use the property graph model [12], a very general model, describing a graph as a directed multigraph. It models entities via vertices and relationships between them via edges. Each graph element can be characterized by attributes and their values, allowing the combination of data with different structures and semantics. The mathematical definition and the comparison of this model with other graph models are provided in [12,15].

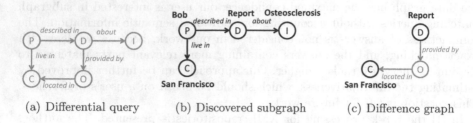

(a) Differential query   (b) Discovered subgraph   (c) Difference graph

Fig. 1. Differential query and its results

Differential Queries

If a user receives an empty result set from a graph database, a differential query can be launched that investigates the reasons of an empty result [15]. The differential query is the initial graph query delivering an empty answer, marked by a specified keyword. In order to provide some insights into the "failure" of a query, a user receives intermediate results of the query processing consisting of two parts: a data subgraph and a missing part of the original query graph. The first part consists of a maximum common subgraph between a data graph and the query graph that was discovered by any maximum common subgraph algorithm suitable for property graphs. This can be for example the McGregor maximum common subgraph algorithm [8]. The second part reflects a difference graph - a "difference" between a query graph and a discovered maximum common subgraph. It shows the part of a differential query that is missing from a data graph and therefore displays the reason why the original query "failed". The difference graph is also annotated with additional constraints at the vertices, which are adjacent to the discovered subgraph as connecting points.

As an example, imagine a data graph derived from text documents that contains information about patients, their diagnoses, and medical institutions. We store the data graph together with a source description in a graph database to allow its collaborative use by several doctors. Assume a doctor is interested in names of all patients (P), their diseases (I), their cities of residence (C), medical institutions (O), and information documents (D) like in Figure 1(a). If the query does not deliver any answer, the doctor launches the query as a differential query and receives the following results:

- The discovered subgraph in Figure 1(b): A person, called Bob, living in San Francisco, whose information was described in "Report", which is about osteosclerosis.
- The difference graph in Figure 1(c): There is no information about any medical institution located in San Francisco, which provided the "Report".

Differential Query Processing

The processing of a differential query is based on the discovery of maximum common subgraphs between a query and a data graph as well as on the computation of difference graphs. Firstly, the system selects a starting vertex and edge from a query graph. Secondly, it searches a corresponding data subgraph in a breadth-first or depth-first manner. If a maximum possible data subgraph for a chosen starting vertex is found, then the system stores this intermediate result, chooses a next starting vertex, and searches again. This process is repeated with every vertex as a starting point. If the search is done only from a single starting vertex, then the largest maximum common subgraph might be missing, because not all edges exist in a data graph. In a final step, the system selects the maximum common subgraphs from all intermediate results, computes the corresponding difference graphs, and returns them to a user.

Due to the nature of the differential queries, redundant intermediate subgraphs and their multiple processing create a potentially significant processing overhead. The number of intermediate results can reach up to 150K subgraphs (Figure 5(d)) for a data graph of 100K edges. In order to cope with this issue, we already proposed different strategies for the selection of a starting vertex [15]: based on cardinality or degree of vertices. Although the number of answers is reduced, it can still remain large to be processed manually. As a side effect, some subgraphs, which are potentially relevant for a user, might be excluded from a search, because the strategies do not take a user's intention into account. To avoid this, we propose an extended concept of differential queries – top-k differential queries, which process a query graph and rank results according to user-defined relevance weights.

4   Top-k Differential Query Processing

In this section we describe the core of our approach – the relevance-based search with relevance flooding and the detection of an optimal traversal path through a differential query, and ranking of results.

4.1   Top-k Differential Queries

We define a **top-k differential query** as a directed graph $G_q^k = (V, E, u, f, g, k)$ over attribute space $A = A_V \dot\cup A_E$, where: (1) V, E are finite sets of N vertices and M edges, respectively; (2) $u : E \to V^2$ is a mapping between edges and vertices; (3) $f(V)$ and $g(E)$ are attribute functions for vertices and edges; (4) A_V and A_E are their attribute space, and (5) k is a number of required results.

The goal of a top-k differential query is to search subgraphs based on relevance weights and to rank the discovered subgraphs according to a relevance-based scoring function. For this, we introduce so-called *relevance weights* for vertices $\omega(v_i)$ and edges $\omega(e_j)$ in a query graph, which annotate graph elements, vertices and/or edges, in a query graph with float numbers $\in [0; 1]$. A weight $\omega = 0$ denotes low relevance and thus reflects the default of a vertex and an edge. In our work we do not concern negative evidence, because if a graph element is not interesting to a user, then it would not be included in the query. Graph elements with higher relevance weights in a query are more important to a user than those with lower values. The introduction of relevance weights does

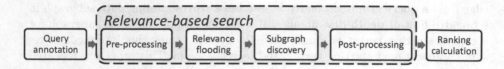

Fig. 2. Top-k differential query processing

not affect the definition of top-k differential queries, this is just an additional property for edges and vertices: $\omega(V) \subseteq f(V)$ and $\omega(E) \subseteq g(E)$.

The relevance weights are used for several purposes, e.g. (1) for steering our search in a more relevant direction, (2) for earlier processing of elements with higher relevance, and most importantly (3) in a scoring function for the ranking itself. The values facilitate the discovery of such subgraphs that are more interesting to a user, and the elimination of less relevant subgraphs.

The processing of top-k differential queries is performed as depicted in Figure 2. After a user has annotated a query with the relevance weights, a relevance-based search is started. When no new data subgraphs can be found, the system stops the search, calculates the rank of discovered subgraphs, and returns results to a user. In the following, we describe all these processing steps in more detail.

4.2   User and Application Origin of Relevance Weights

Relevance weights described in the previous paragraph can be determined based on a user's preferences or based on a particular use case. If relevance weights are assigned by a user, then the more important graph elements get higher weights. With reference to our running example (Figure 1(a)), if a doctor is more interested in the names of patients and their diseases, then he provides the highest relevance to corresponding vertices: $\omega(v_P) = 1$ and $\omega(v_I) = 1$.

If relevance weights are determined by a particular use case, then they are defined considering specific features of the use case – an *objective function*, which the use case tries to minimize or maximize. Some examples of objective functions would be the data transfer rate in networks of hubs or traffic in the road networks. If a user aims to maximize the objective function, then graph elements with higher values of the objective function are annotated by higher relevance weights. In our approach, we do not assume any specific use case and expect that relevance weights are defined by a user.

4.3   Relevance-Based Search

After a user has annotated the query graph, the relevance-based search is conducted, which is outlined in the dashed box in Figure 2. At the stage of pre-processing, the relevance weights are transformed into the format required by the relevance flooding: edge relevance weights are converted into the relevance weights of incident vertices. Afterwards, the relevance flooding propagates the weights along the query graph, if at least one vertex does not have a user-defined relevance weight. Then, relevant subgraphs are searched in a data graph. After the search the post-processing is executed over relevance weights to prepare them for the further ranking.

Pre-processing and Post-processing of Relevance Weights. Relevance flooding considers relevance weights only on vertices. To account for the relevance weights on edges, we transform them into the weights of incident vertices before the flooding. The pre-processing consists of two steps: assignment of missing relevance weights and transformation of relevance weights. If a graph element is not annotated by a relevance weight, then the default value is assigned to it. Afterwards, the system distributes the relevance weights of edges to their incident vertices as follows: (1) The user-defined relevance weight of an edge is distributed equally across its ends: the source and target vertices. (2) Given a set of K incident edges to a vertex v_i, the relevance weight of a vertex $\omega(v_i)$ is the sum of the square root of edges' relevance weights $\omega(e_j)$, which are incident to the vertex v_i, and its initial relevance weight $\omega^{init}(v_i)$ (if any) like in Equation 1.

The post-processing is conducted after the subgraph search; it prepares the weights for the ranking. By default, the user-defined weights are used in the ranking, therefore, the weights changed during the relevance flooding have to be reset to values derived at the pre-processing step. Non-annotated graph elements are specified by the minimal weights (see Equations 2 – 4). If we want to use the relevance flooding weights for the ranking, we have to derive the weights for edges by multiplying the weights of their sources and targets (Equation 3).

$$\omega(v_i) = \sum_{j=1}^{K} \sqrt{\omega(e_j)} + \omega^{init}(v_i) \quad (1) \qquad\qquad \omega^{min}(e_j) = 1/M \qquad (2)$$

$$\omega(e_i) = \omega(e_i^{source}) * \omega(e_i^{target}) \quad (3) \qquad\qquad \omega^{min}(v_i) = 1/N \qquad (4)$$

Relevance Flooding. The goal of relevance flooding is to annotate all vertices in a query graph by relevance weights. It takes place if not all vertices of a query graph have user-defined relevance weights. This is necessary to allow the subgraph search based on relevance weights and to facilitate the early detection of the most relevant parts of a query graph, which are specified by relevance weights. The algorithm for relevance flooding is based on similarity flooding [9], where two schemes are matched by comparing the similarity of their vertices. We extend this algorithm to propagate the relevance weights to all vertices in a query graph and to keep the initial user-defined relevance weights.

The relevance flooding takes several observations into account: locality and stability of relevance. The locality assigns higher relevance weights to the direct neighbors and lower relevance weights to remote vertices. The stability keeps the relevance weights provided by a user and prevents the system from reducing them during the flooding.

Algorithm 1. Relevance Flooding

1: **for all** vertex v_i in query graph G_q **do**
2:     **if** $v_i.getWeight() > 0$ **then**   ▷ if a vertex has a weight
3:         $\omega = v_i.getWeight()$ ▷ store a weight in ω
4:         $neighbors = getNeighbors(v_i)$ ▷ take all direct neighbors
5:         $\Delta\omega = \omega/neighbors.size()$ ▷ calculate a propagation weight
6:         **for all** $neighbors_j$ in $neighbors$ **do**
7:             $neighbors_j.addPropWeight(\Delta\omega)$ ▷ store a propagation weight
8: **for all** vertex v_i in query graph G_q **do**
9:     $v_i.increaseWeight()$ ▷ increase all weights with propagation weights
10: $max(\omega) = 0$
11: **for all** vertex v_i in query graph G_q **do**
12:     **if** $max(\omega) < v_i.getWeight()$ **then**
13:         $max(\omega) = v_i.getWeight()$ ▷ find a vertex with the maximal weight
14: **for all** vertex v_i in query graph G_q **do**
15:     **if** $v_i.getInitWeight() > 0$ **then**   ▷ if a vertex has a user-defined weight
16:         $v_i.setWeight(v_i.getInitWeight())$ ▷ reset to an initial weight
17:     **else**
18:         $v_i.setWeight(v_i.getWeight()/max(\omega))$ ▷ normalize a weight
19: $sum = 0$
20: **for all** vertices v_i in G_q **do**   ▷ calculate a difference between iterations
21:     $sum = sum + (v_i.getPrevWeight() - v_i.getWeight())^2$
22: **if** $sum <= \epsilon$ OR $\kappa >= longestPath$ **then**
23:     $terminateFlooding()$ ▷ check termination conditions

Relevance flooding works as described in Algorithm 1. In the main part at lines 1- 9, each vertex broadcasts its value to direct neighbors according to the locality property. Afterwards, the values are normalized to the highest value at line 18 and user-defined relevance weights are set back to ensure the stability of given relevance weights at line 16. If a termination condition is satisfied, the propagation is interrupted at line 23. As the termination condition we can use a threshold ϵ for the difference of relevance weights of two subsequent iterations or the number of iterations κ, which corresponds to the size of the longest path between two vertices in a query graph.

Following our example in Figure 1(a) and assigned relevance weights $\omega(v_P) = \omega(v_I) = 1$, at each iteration we propagate the equal relevance weights to all direct neighbors (an exemplary weight propagation during the second iteration is shown in Figure 3(b)). During the flooding we do not consider the direction of edges, because the processing of a graph can easily be done in both directions without any additional efforts. After the first iteration, vertices D, C get the propagated relevance weights from P, I according to the locality property (Figure 3(a)). Vertex O still remains without relevance weight. After each iteration, we normalize the relevance weights to the highest value and set those of them

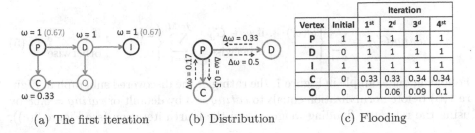

Vertex	Initial	Iteration			
		1st	2d	3d	4st
P	1	1	1	1	1
D	0	1	1	1	1
I	1	1	1	1	1
C	0	0.33	0.33	0.34	0.34
O	0	0	0.06	0.09	0.1

(a) The first iteration   (b) Distribution   (c) Flooding

Fig. 3. Relevance flooding: gray relevance weights show the case, where the initial relevance weights are not set back

back to initial values that have weights defined by the user. The gray relevance weights in brackets for vertices P, I show the weights without reset. We repeat the process, until it converges according to the specified threshold ϵ or when the number of iterations κ has exceeded the longest path between two vertices ($\kappa = 4$). The results of relevance flooding are presented in Figure 3(c).

Maximum Common Subgraph Discovery with Relevance Weights. User-defined relevance weights represent an interest of a user in dedicated graph elements: such elements have to be processed first. We treat a traversal path between all relevant elements in a query graph as a cost-based optimization, where we maximize the relevance of a path.

The search of subgraphs is modeled by the GraphMCS algorithm [15], a depth-first search for property graphs, discovering maximum common subgraphs between a query and a data graph. First, we choose the first vertex to process. The vertices with highest relevance weights are prioritized and processed first. Second, we process such an incident edge of the selected starting vertex that has a target vertex defined by the highest relevance weight. Finally, this process continues till all vertices and edges in a query graph are processed. If a query edge is missing in a data graph, then the system adjusts the search dynamically: it selects the incident edge with the next highest relevance weight or revises the search from all possible target vertices.

The relevance-based search chooses a next edge to process dynamically based on relevance weights of edges' ends. If several vertices have the same weight, then the edge that has a vertex with minimal cardinality or minimal degree is chosen to be processed. The proposed strategy steers the search in the most relevant direction first, guaranteeing the early discovery of the most relevant parts.

4.4 Rank Calculation

The ranking is based only on the discovered subgraphs, the difference graph does not influence the rating score. The answers with higher relevance weights are ranked higher. A rating score is calculated based on the values of edges and vertices a result comprises. After ratings of all results are computed, they are normalized to the highest discovered rating score. Given N vertices and M edges in a query graph G_q, the rating of discovered subgraph G'_d is calculated as follows

$$rating(G'_d) = \sum_{i=1}^{i=N} \begin{cases} \omega(v_i) & \text{, if } v_i \in G'_d \\ 0 & \text{, otherwise} \end{cases} + \sum_{j=1}^{j=M} \begin{cases} \omega(e_j) & \text{, if } e_i \in G'_d \\ 0 & \text{, otherwise} \end{cases} \quad (5)$$

Following our example in Figure 1, the rating of the discovered subgraph in Figure 1(b) before normalization equals to $rating = 3$ by default or $rating = 5.68$ by using the relevance flooding weights from the fourth iteration (see Figure 3(c)).

5   Evaluation

In this section, we compare top-k differential queries and unranked differential queries. We describe the evaluation setup in Section 5.1 and compare both approaches in Section 5.2. Then, we present and interpret the scalability of the top-k differential queries in Section 5.3.

5.1   Evaluation Setup

We implemented a property graph model on the top of an in-memory column database system with separate tables for vertices and edges, where vertices are represented by a set of columns for their attributes, and edges are simplified adjacency lists with attributes in a table. Both edges and vertices have unique identifiers. To enable efficient graph processing, the database provides optimized flexible tables (new attributes can efficiently be added and removed) and compression for sparsely populated columns like in [2,13]. This enables schema-flexible storage of data without a predefined rigid schema. Our prototypical graph database supports insert, delete, update, filter based on attribute values, aggregation, and graph traversal in a breadth-first manner in backward and forward directions with the same performance.

Data and queries are specified as property graphs. In a query, each graph element can be described with predicates for attribute values. To specify a dedicated vertex, we use its unique identifier.

As a data set, we use a property graph constructed from DBpedia RDF triples, where labels represent attribute values of entities. This graph consists of about $20K$ vertices and $100K$ edges. We have tested each case for each query ten times and have taken the average runtime as a measure.

5.2   General Comparison

We constructed an exemplary query shown in Figure 4(a) and marked three edges of the type "deathPlace" with relevance weight $\omega = 1$. The unranked differential query delivers results with a lower maximal rating and exhibits longer response times than the top-k differential query (Figure 4(b)). The top-k differential query discovers more subgraphs of higher ratings than the unranked differential query (Figure 4(c)). The unranked query also discovers the graphs with low ratings.

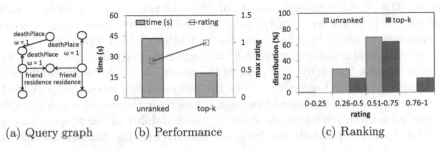

(a) Query graph   (b) Performance   (c) Ranking

Fig. 4. Evaluation of unranked and top-k differential queries

5.3 Performance Evaluation

We evaluate two kinds of query graphs, one for the path topology and one for the zigzag topology. The query for the path topology consists of edges of the same type "successor", and the first edge is marked by a relevance weight. The

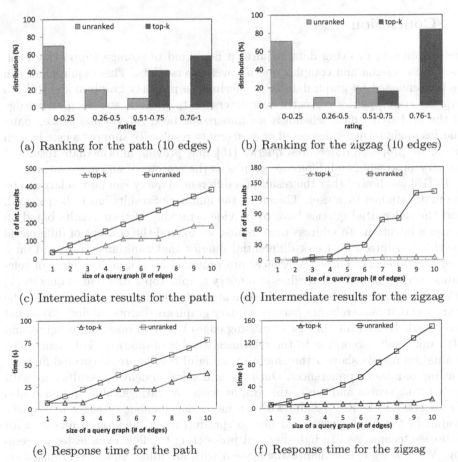

(a) Ranking for the path (10 edges)   (b) Ranking for the zigzag (10 edges)

(c) Intermediate results for the path   (d) Intermediate results for the zigzag

(e) Response time for the path   (f) Response time for the zigzag

Fig. 5. Performance evaluation for the differential query and top-k differential query

query for the zigzag topology consists of edges of two types "birthPlace" and "deathPlace". It starts with "birthPlace" marked by a relevance weight and is extended incrementally by a new edge for "deathPlace", then "birthPlace" etc.

We compare rating distributions for the largest query (ten edges in a query graph) in Figures 5(a)-5(b). The most of the results delivered by the unranked differential query have low ratings, while the proposed solution provides at least 60% of its results with the highest ratings. We increase the size of a query graph from one edge up to ten edges and evaluate the scalability of the proposed solution. The size of intermediate results grows linearly with the number of edges in a query graph (Figures 5(c)-5(d)), and it is lower than at least one order of magnitude for the top-k differential query. This can be explained by the elimination of low-rated subgraphs from the search. The response time evaluation exhibits the steep decrease for the top-k differential query (Figures 5(e)-5(f)). From this we can conclude, the top-k differential query is more efficient than the unranked differential query: it delivers results with a higher rating score, omits low-rated subgraphs, and consumes less processing time.

6   Conclusion

Heterogeneous, evolving data requires a new kind of storage supporting evolving data schema and complex queries over diverse data. This requirement can be implemented by graph databases offering the property graph model [12]. To express graph queries correctly over diverse data without any deep knowledge of the underlying data schema is a cumbersome task. As a consequence, many queries might return unexpected or even empty results. To support a user in such cases, we proposed differential queries [15] that provide intermediate results of a query processing and difference graphs as the reasons of an empty answer.

In [15] we showed that the result of a differential query can be too large to be manually studied by a user. Therefore, the number of results has to be reduced, and the differential queries have to provide a ranking of their results based on a user's intention. To address these issues, we extend the concept of differential queries and introduce top-k differential queries that rank answers based on a user's preferences. These preferences are provided by a user in a form of relevance weights to vertices or edges of a query graph. Top-k differential queries (1) allow marking more relevant graph elements with relevance weights, (2) steer the search so that more relevant parts of a query graph are discovered first, (3) adjust the search dynamically in case of missing edges based on relevance weights, and (4) rank results according to the relevance weights of discovered elements. The evaluation results showed that more meaningful results are discovered first according to a user's preferences. Our proposed solution delivers results only with high rating scores and omits the graphs with low ratings. Our approach also shows good scalability results with an increasing number of edges in a query graph. In the future, we would like to speedup top-k differential queries with database techniques like indexing and pre-sorting to allow even faster processing. We also want to enhance the system with an online adaptive propagation of relevance weights based on a user's feedback.

Acknowledgment. This work has been supported by the FP7 EU project LinkedDesign (grant agreement no. 284613).

References

1. Amer-Yahia, S., Koudas, N., Marian, A., Srivastava, D., Toman, D.: Structure and Content Scoring for XML. Proc. of VLDB Endow., 361–372 (2005)
2. Bornhövd, C., Kubis, R., Lehner, W., Voigt, H., Werner, H.: Flexible Information Management, Exploration and Analysis in SAP HANA. In: DATA, pp. 15–28 (2012)
3. Chapman, A., Jagadish, H.V.: Why not? In: Proc. of ACM SIGMOD, pp. 523–534. ACM, New York (2009)
4. Fan, W., Wang, X., Wu, Y.: Diversified top-k graph pattern matching. Proc. of VLDB Endow. 6, 1510–1521 (2013)
5. Gupta, M., Gao, J., Yan, X., Cam, H., Han, J.: Top-k interesting subgraph discovery in information networks. In: Proc. of ICDE, pp. 820–831. IEEE (2014)
6. Huang, J., Chen, T., Doan, A., Naughton, J.F.: On the provenance of non-answers to queries over extracted data. Proc. of VLDB Endow. 1, 736–747 (2008)
7. Kasneci, G., Suchanek, F., Ifrim, G., Ramanath, M., Weikum, G.: Naga: Searching and ranking knowledge. In: Proc. of ICDE, pp. 953–962. IEEE (2008)
8. McGregor, J.J.: Backtrack search algorithms and the maximal common subgraph problem. Software: Practice and Experience 12(1), 23–34 (1982)
9. Melnik, S., Garcia-Molina, H., Rahm, E.: Similarity flooding: A versatile graph matching algorithm and its application to schema matching. In: Proc. of ICDE, pp. 117–128. IEEE (2002)
10. Mottin, D., Marascu, A., Roy, S.B., Das, G., Palpanas, T., Velegrakis, Y.: A probabilistic optimization framework for the empty-answer problem. Proc. of VLDB Endow. 6, 1762–1773 (2013)
11. Prud'hommeaux, E., Seaborne, A.: SPARQL Query Language for RDF. W3C Recommendation (2008)
12. Rodriguez, M.A., Neubauer, P.: Constructions from dots and lines. Bulletin of the American Society for Inf. Science and Technology 36(6), 35–41 (2010)
13. Rudolf, M., Paradies, M., Bornhövd, C., Lehner, W.: The Graph Story of the SAP HANA Database. In: BTW, pp. 403–420 (2013)
14. Tran, Q.T., Chan, C.Y.: How to conquer why-not questions. In: Proc of ACM SIGMOD, pp. 15–26. ACM, New York (2010)
15. Vasilyeva, E., Thiele, M., Bornhövd, C., Lehner, W.: GraphMCS: Discover the Unknown in Large Data Graphs. In: EDBT/ICDT Workshops, pp. 200–207 (2014)
16. Zhu, Y., Qin, L., Yu, J.X., Cheng, H.: Finding top-k similar graphs in graph databases. In: Proc. of EDBT, pp. 456–467. ACM (2012)
17. Zou, L., Chen, L., Lu, Y.: Top-k subgraph matching query in a large graph. In: Proc. of the ACM First Ph.D. Workshop in CIKM, pp. 139–146. ACM (2007)

Static Integration of SQL Queries in C++ Programs

Maciej Sysak, Bartosz Zieliński, Piotr Kruszyński, Ścibór Sobieski,
and Paweł Maślanka

Department of Computer Science,
Faculty of Physics and Applied Informatics,
University of Łódź,
ul. Pomorska nr 149/153, 90-236 Łódź, Poland
{maciej.sysak,bzielinski,piotr,scibor,pmaslan}@uni.lodz.pl

Abstract. Contemporary frameworks offer essentially two methods of accessing data in relational databases. The one using plain SQL requires writing a lot of boilerplate code and storing the SQL in a string, which is error prone and denies the benefits of static query analysis. The other one enforces the use of an additional (usually object oriented) abstraction layer which incurs additional runtime costs and hinders the use of advanced SQL capabilities. In this paper we present a working implementation of a radically different approach. Our tool uses the database engine to analyze the native SQL queries prepared by the user, and generates all the necessary classes representing query responses, single result rows and database connections. The use of native queries allows to utilize advanced and highly optimized SQL features. On the other hand, the use of the generated classes ensures that data access happens in a statically checked, type-safe way.

1 Introduction

There are well known difficulties in accessing data stored in relational databases from application programs, especially if good object orientation is required:

1. A query is formulated in SQL, which is stored in the program code as string. Such strings are opaque to the compiler, which defers the error discovery until the query is actually sent to database during program execution. In case of C/C++ this situation is further aggravated by the low quality of runtime exception mechanisms. Also, writing the query as a string requires the programmer to properly escape all the special characters which might appear in the query (such as quotes). This garbles the query string stored in the program code and increases the error rate even more.
2. Iteration through the query result and binding prepared statements parameters requires a lot of unpleasant boilerplate code which is not type-safe — the number of columns in the result set, their types and names cannot be

Y. Manolopoulos et al. (Eds.): ADBIS 2014, LNCS 8716, pp. 126–138, 2014.

checked by the compiler. It is also the task of a programmer to perform all the necessary type conversions between SQL and host language types.

A number of solutions is known and used in practice. Those fond of SQL can use the embedded SQL precompilers supplied for all major databases and languages. Unfortunately this solution has some disadvantages. For one thing, it mixes the SQL code with host language code and some auxiliary glue syntax. This confuses the syntax coloring tools — a seemingly minor inconvenience, which nevertheless might negatively influence the programmer's productivity. Also, it requires the coder to learn the aforementioned glue syntax and the actual implementations do not handle the type conversions as smoothly as one might expect. Finally, the precompilation stage introduces issues with some compilers.

A more common approach is to use ORM frameworks such as *Hibernate* or *Java Persistence Api*. The obvious advantages are that they are fully object-oriented. In particular, queries return objects, which can be made persistent in order to simplify the data modification management in program code. The frameworks often decrease the amount of boilerplate code development by advanced code generation features which makes the programmer task less error prone. As an additional boon the use of their own object oriented query language (such as HQL or JPQL) permits developing database vendor agnostic code. Additionally, many of the ORM frameworks (like JPA or Hibernate) allow the transparent data caching where the updates and inserts can be collected locally and actually sent to database as the single batch at the end of transaction. Similarly, the queries can first examine the cache for the presence of requested data.

Unfortunately the ORM frameworks have also well known drawbacks. A common and not entirely unjustified charge against ORM's is that the quality of generated SQL statements sent to database is very low and it is hard to get the performance right (see e.g. [17]). Less obviously, the same features mentioned above as beneficial can also be considered as a downsides from another point of view. Special query language isolates the developer from the particulars of the given SQL dialect so much that it precludes the use of advantageous features of the dialect. Moreover, some more advanced elements of SQL standard might not be implemented in the custom query language of the framework (consider e.g. window clauses of analytic functions). Finally, the developer is required to learn one more query language only superficially resembling SQL. Let us add that the OQL (HQL, JPQL, etc.) queries are passed to the library functions as strings, which, just like in case of SQL queries, delays the query syntax and semantics checking to the time of program execution. This is why many ORM frameworks include query building DSL's (Domain Specific Languages) utilizing the host language syntax — well known examples include LINQ and Java EE Criteria API. This takes care of compile time syntax checking but still leaves semantics checking to the run time.

More generally, ORM's often impose an active, object oriented view of the data in which objects representing entities manage their own updates leading to an impedance mismatch with the relational model ([13], see eg. [18] for a mathematical treatment of this mismatch). This object oriented view, despite the claims to the contrary [9] the authors (like some others [8]) do not perceive as the most natural way of conceptualizing the database application.

Another fundamental flaw of most of the existing ORM frameworks is that they perform during runtime many of the activities, such as translating OQL to SQL, which can be done during compilation as well. The impact on the performance is hard to assess, but might be significant in some applications. More important, however, is the loss of the benefits of static code analysis.

The dbQcc framework presented in this paper takes a completely different approach. We start with the plain, native SQL query as the source of all needed information. The external tool we created uses the analyzed query to generate classes representing the result row, and the query response. The application code uses the generated classes together with the featured database connection library to access data in a statically checked, type-safe way. The only query used in program runtime is the native one explicitly specified by the programmer in the beginning — we do not modify the query. Note that this is the opposite approach to the one taken by the ORM frameworks, which produces SQL queries from the annotated class definitions. Also it is worthy to emphasize that the class generating tool checks both the syntactic and semantic correctness of the query as well. Hence the successful class creation guarantees the lack of query syntax and data model mismatch errors in the runtime.

1.1   Comparison with Previous Work

Most of the existing approaches to supporting compile time verification of correctness of SQL statements involves introducing special syntax (often an internal DSL [10]) instead of using the native SQL like in our approach — see e.g. [11] for a native C++ example making use of template metaprogramming. An interesting exception is [12] in which a verifier of dynamically constructed SQL is presented. Another example of internal DSL is [14] in which a C# library is introduced which allows to construct SQL strings correct by construction (similarly as XML libraries construct correct XML files). Both ([11] and [14]) utilize an external tool, which examines the database schema and generates the appropriate access classes from table metadata. Note that in our approach we generate the access classes for each statement, rather than for each table. Also note that unlike our framework, which is meant to support the execution of verified queries, the tool presented in [12] only verifies the SQL statements, and does not include any provisions for type safe interface between SQL and the host program. Similarly, neither of the frameworks [11] or [14] supports actual execution of the queries generated.

The ORM frameworks utilize their own query languages for more complicated tasks. This introduces the same problems as dynamic SQL statements. Safe *Query Objects* framework [7] allows to construct compile time checked JDOQL queries (for *Java Data Objects* [1]). The framework is interesting, because unlike e.g., JDO's own JDO Typesafe, it allows to specify filter conditions not with special objects which represent them, but with native Java boolean expressions. The actual queries are generated by the external tool from the bytecode.

A somewhat different approach to generation of SQL queries correct by construction was taken by the authors of [16], in which the SQL queries are created from the specification in term rewriting system [6]. The language of specification includes a base sublanguage, which is made to resemble SQL — modulo some quirks imposed by the Maude's own syntax. The system allows to extend the SQL-like base with the ability to factorize common SQL fragments or to introduce some higher level features like named joins. In effect it can be seen as an SQL metaprogramming system. The framework outputs actual vendor specific queries, which can then be fed to a system like the one developed in this paper.

A system which partially inspired our framework and the approach of which bears the most similarity to ours is *Web4j* Java Web Application Framework [5]. Web4j uses a custom file format containing specially tagged SQL statements. At runtime the statements can be fetched by identifier and then processed, after supplying, if applicable, the parameter values. The iteration through the query result is enabled through the method which accepts the list of arguments consisting of the `SomeType.class` object of the desired class of rows, the identifier of the query, and the query parameters if any. The method returns a list of `SomeType` objects. The object construction process utilizes reflection to determine which constructor to use, matching the constructor with the same number of arguments that there are columns in the rows returned by the query. Note that, while the framework allows, as an option, to check the validity of the query at application startup by preparing the query (for the databases that support it), but, similarly as the matching of row columns to constructor arguments, it happens during runtime, and does not, unlike our solution, offer the compile time type safety for the queries.

When speaking about SQL metaprogramming it is worthwhile to mention *Metagen* [15] — a tool for generating database schemas from vendor independent descriptions in a special language.

Let us note that while object relational mappers isolate the user from some vendor specific database capabilities, presenting instead the generic interface, the support for advanced features in this generic interface is already considerable and is rising fast, see eg. [19], which enriches Hibernates HQL with recursive queries.

Finally, note that our framework strongly supports implementing the domain logic using the transaction script pattern [9], whereas most of the higher level, object oriented frameworks are geared towards either the domain model [9] — e.g., the Hibernate framework, or the table model [9] — e.g. ADO Net.

2   System Architecture and Internals

This section provides a detailed account of our system. The data flow diagram in Figure 1 presents the overview of the system architecture. In particular:

- The sql statements used in the application utilizing the DbQcc framework ought to be placed in the separate uqf file(s) (for each database used) which are essentially SQL files with some metadata placed in specially formatted comments. Subsection 2.1 contains a detailed description of the file format.

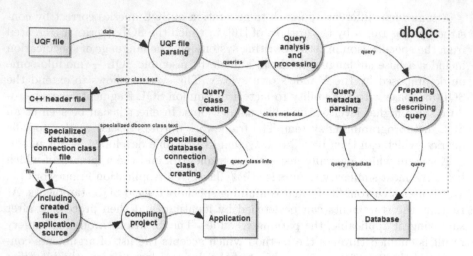

Fig. 1. Data flow in the system

- We use a C++ database access abstraction layer responsible for database connection management and the actual query execution at runtime. This abstraction level is not meant to be visible to the user but to serve as an implementation layer for the classes actually used by the programmer. A transparent support for standard database operations in different database engines is provided through the driver mechanism. Currently, only the driver for the PostgreSQL database (based on *libpq* [2]) is available. In order to furnish the required functionalities in an efficient way we use some of the new features introduced in C++11 like shared pointers and move semantics.
- Moving data from the database into the application necessitates converting data from the representation used by the DBMS to the native C++ types. In Subsection 2.2 we describe the mechanisms supporting the conversions.
- The uqf file is processed by the separate tool which can be integrated into compilation toolchain, and which generates for each statement classes representing: query record and query response. Additionally, for each uqf file the database access class is generated. The Subsection 2.3 contains the account of various stages of processing of user queries extracted from uqf file(s).

2.1   User Query File

Database queries used by the application ought to be placed in the separate `.sql` file (see Figure 2), which we will refer to as an **uqf** (user query) file, collecting all queries referring to the particular database. The file employs the special key-value format of comments, which permits adding some auxiliary information necessary for the correct execution of the class generating tool. The project may contain many **uqf** files, each one associated with a different database.

```
/* DBEngine = PostgreSQL
   DBName = dbqcc
   DBHost = ***.uni.lodz.pl
   DBUser = dbqcc_tester
   DBPassword = SQLorNothing
   DBFileName = Blog */
-- PREPARABLE_SELECT = ClientRanking
SELECT c.email, s.name, r.client_rank
FROM (
    SELECT pcnt.*, rank() OVER(
        PARTITION BY pcnt.section_id ORDER BY pcnt.cnt DESC
    ) AS client_rank
    FROM (SELECT count(*) AS cnt, p1.owner_id, p1.section_id
          FROM page p1 GROUP BY p1.owner_id, p1.section_id) pcnt
) r JOIN client c ON (r.owner_id = c.id)
        JOIN section s ON (r.section_id = s.id)
WHERE r.client_rank <= $1 ORDER BY s.name, r.client_rank
```

Fig. 2. User query SQL file sample containing parametrized query

The **uqf** file starts with the information about the database engine vendor
which is used by the class generator to choose the appropriate driver (if available). Next, the file contains the connection data — it is utilized by the generator
to connect to the database server during query analysis phase. For security reasons one may omit some of the authorization information — the missing data
can be supplied at the generator's execution time. In addition, the authorization
data provide default values in the generated database connection class.

The last section of **uqf** file contains a sequence of SQL statements present in
the application (intended to be executed in the database the **uqf** file is associated with). Queries are to be separated with empty lines. Each statement should
start with a specially formatted SQL comment specifying exact statement type
(**PREPARABLE_SELECT** in the query in Figure 2) and assigning a unique (within
the file) statement identifier (**ClientRanking** in the query in Figure 2), which
must be a valid C++ type name. This identifier will be utilized as the statement
class name. The statement type declares whether it is a SELECT query, DML
statement or a stored subprogram as well as whether the statement should be
prepared prior to the first execution, and thus stored in prepared form in the
database server for the duration of the session. Preparing statements may bring
significant performance benefits, specifically for frequently executed and particularly complex statements [4] — their parsing, rewriting and creation of their
execution plans happen only once, when they are prepared.

After the statement header follows the statement in the native form. There is
no need to escape special characters, the statement will be converted to a valid
C++ string placed in the query class body during class generation. The query
may be parametrized, where the parameters are denoted by consecutive numbers
prefixed with the dollar character (see Figure 2). Note that the **uqf** file is a valid

```
template <typename T> class ValueConverter{};

template <> class ValueConverter<uint32_t> {
public:
   static bool fromDB(char * rawValue, uint32_t & value) {
       uint32_t tmp;
       memcpy(&tmp, rawValue, sizeof(tmp));
       value = ntohl(tmp);
       return true;
   }
   static bool toDB(const uint32_t & value, std::vector<char> & rawValue,
                    int & rawValueFormat) {
       rawValue.resize(sizeof(uint32_t));
       uint32_t tmp = htonl(value);
       memcpy(&rawValue[0], &tmp, rawValue.size());
       rawValueFormat = DBManagerBase::BINARY_FORMAT;
       return true;
   }
};
```

Fig. 3. Value converters for uint32_t (PostgreSQL oid)

SQL file and any editor with SQL syntax coloring capabilities can be used for simple editing.

2.2   Data Conversion Mechanism

Data access layers for C++ of some of the databases (PostgreSQL in particular) allow to exchange data both in binary and text formats. Using the text format is not applicable for all types and clearly less efficient (both computationally and in terms of compactness of representation) than binary format. Data conversion to C++ types is necessary for performing operations on data regardless of format used for communication with database. To facilitate data conversion we provide a mechanism based on C++ templates. The ValueConverter class template (see Figure 3) has its specializations for all supported data types. For an unsupported data type (one not having a template specialization) compilation time error occurs. Note that the libpq library representations of database integer types are not completely determined, e.g, oid type employs C/C++ unsigned int which has no strict definition of size and endianness in language standards. Therefore, our conversion mechanism provides converters (template specializations) for all C++11 standard fixed size integer types (e.g., uint32_t) and the compiler matches the specialization for an appropriate type (e.g, uint32_t for unsigned int on the majority of 32-bit platforms). The endianness problem is solved in a platform-specific way, taking into account that PostgreSQL DBMS provides binary data (for integer representations) in network byte order.

2.3   Query Analysis and Processing

The operation of the query analyzer module is based on using the existing database to verify both the query syntax and the validity of references to schema objects, and, when the query is valid, to obtain the metadata associated with the query. The metadata is used later to generate appropriate classes representing the SQL statement, and, if applicable, the result rows. Hence, the statement analysis stage starts with the query analyzer module connecting with the database server on which the analyzed statements are supposed to be eventually executed.

The present system implementation provides a prototype plugin for the PostgreSQL database engine based on the native *libpq* [2] database access library for C which provides advanced functionalities to obtain rich metadata describing the SQL statement and its parameters. Query analyzing module operation, described in detail below, is largely depending on those *libpq* metadata capabilities.

```
class ClientRankingRecord: public QueryRecord {
public:
    DBValue<std::string> client_email, section_name;
    DBValue<long long int> client_rank;
    ClientRankingRecord(QueryResult * queryResult, int rowNumber) :
        QueryRecord(queryResult, rowNumber),
        client_email(queryResult, rowNumber, 0),
        section_name(queryResult, rowNumber, 1),
        client_rank(queryResult, rowNumber, 2) {}
};
```

Fig. 4. Generated query row class example

The SQL statement extracted from uqf file is not executed on the database server. Instead, it is sent to the database as a prepared statement. In the majority of relational database engines (and in particular in PostgreSQL) the prepared statement is parsed, and the query plan is generated (with some slots reserved for filling by parameters, if any) and stored in the database, ready for execution (perhaps multiple times) until explicitly closed. Most relevant for our purpose is that one does not need to execute the prepared statement nor supply the values of the parameters in order to be able to receive all the available statement metadata (in particular we can work with the database containing empty tables). After preparing the statement, such as the one depicted in Figure 2, it suffices to send the request *describe prepared*. In reply we receive the information about:

- names and types of columns of the result set (in case of queries). For columns defined by simple table column reference we also get the source table name,
- types of parameters, in case the statement was parametrized.

The query analyzer module uses this metadata together with the statement itself and the statement identifier extracted from the query header in the uqf file to

```cpp
class ClientRanking: public QueryResult{
  static constexpr const char * queryName = "ClientRanking";
  static constexpr const char * query = "SELECT [...]";
  static bool prepared;
public:
  long long int param1;
  ClientRanking(DBManagerBase * dbManager,
                const long long int & _param1): param1(_param1){
    const char * paramValues[1];
    int paramLengths[1]; int paramFormats[1];
    std::vector<char> param1RawValue;
    ValueConverter<long long int>::toDB(
        param1, param1RawValue, paramFormats[0]);
    paramValues[0] = &param1RawValue[0];
    paramLengths[0] = param1RawValue.size();
    if(!prepared){
      auto result = dbManager->prepareStatement(
          queryName,query,1,nullptr);
      if(result->bad()) throw std::runtime_error(dbManager->getError());
      prepared = true;
    }
    dbResult = dbManager->executePreparedStatement(
        queryName,1,paramValues,paramLengths,paramFormats,
        DBManagerBase::BINARY_FORMAT);
    if(dbResult->bad()) throw std::runtime_error(dbManager->getError());
  }
  std::shared_ptr<ClientRankingRecord> operator[](int rowNumber){
    if(rowNumber >= dbResult->getRowsNumber()) return nullptr;
    return std::make_shared<ClientRankingRecord>(this,rowNumber);
  }//[...]
};
```

Fig. 5. Generated query response class example

assemble, using the class builder internal tool, header files holding definitions of the classes associated with the statement, such as:

- (In case of SELECT statement) the class representing a single result set row, with a field (and possibly accessor methods) of appropriate type for each result set column (e.g., see Figure 4). The field and accessor names are based on column names, qualified with table name, if applicable (with some necessary conversions, like substituting underscores for spaces). Fields are constructed using the DBValue class template which uses the mechanism described in Subsection 2.2 to convert data received from the database. Because DBValue<T> overloads the conversion operator to type T it follows that it preserves the const T& semantics. The class provides also conversion operators supporting structural type equivalence of database records with the same types of columns (c.f. [11]). In our approach the type of a query result set row class identifies the query, but the conversion operators permit transparent combination or substitution of rows from different queries, provided that they have the same number and types of columns.

- The class representing the query result, which also encapsulates the statement itself (e.g., see Figure 5). This class also provides the functionality to execute the statement (and supply the arguments if applicable), and in the case of the SELECT statement also to iterate through the result rows. The statement, if declared as preparable, is prepared during the construction of the first object of a given class. In this case, executions refer to the parsed statement stored in the database server.

```
class DBManagerBlog : public DBManagerBase {
    DBManagerBlog(){};
public:
    typedef const std::string & csr;
    static std::shared_ptr<DBManagerBlog> create(csr host, csr dbName,
        csr user, csr password, csr errorMessage) {/* [...] */}
    // [...]
    std::shared_ptr<ClientRanking>
    SELECT_ClientRanking(const long long & param1) {
        return std::make_shared<ClientRanking>(this, param1);
    }
};
```

Fig. 6. Generated database connection class

The last stage of query analysis module operation is the generation of specialized database connection classes (for each database server application connects to). The classes inherit from the generic database access driver for the particular DBMS. It extends the driver with methods for executing SQL statements and stored procedures specified in the processed uqf file (see, e.g., Figure 6).

After all the queries are analyzed and all necessary classes are generated, the query analyzer module cleans up the database, removing all prepared statements (in case of PostgreSQL it suffices to close the connection, as the prepared statements are associated with the session and die with it).

3   Simple Example of DbQcc Usage

The following code presents basic example usage of our framework:

```
std::string error;
auto dbManager = DBManagerBlog::create("***.uni.lodz.pl",
    "dbqcc", "dbqcc_tester", "SQLorNothing", error);
if(!dbManager) {throw std::runtime_error(error);}
auto ranking = dbManager->SELECT_ClientRanking(10);
for(auto i = 0; i < ranking->getRowCount(); i++) {
    auto row = ranking->getRow(i);
    // [...]
}
```

Classes generated by dbQcc provide simple API for programmer to establish database connection, execute queries described in user queries file and access response data. Our example schema may be treated as an extremely simplified blog's database. It consists of three tables: PAGE, CLIENT and SECTION connected with referential constraints. PAGE stores the title and content of blog pages. Each page is authored by the unique client and belongs to the unique section. In our sample UQF we placed the `ClientRanking` query (Figure 2) defined as preparable, containing one parameter. This select is supposed to return, for each section, n most prolific authors ranked with respect to the number of the authored pages, where n is passed as a parameter. Note that `ClientRanking`, in addition to being a complex query with joins and multiply nested subqueries, makes use of a window (or analytic) functions (the `rank()`), which, despite being in the standard, are not widely supported by ORM's.

We have performed tests to compare our framework with the directly used PostgreSQL API (libpq). The blog schema described above was filled with random data: 100000 rows in the CLIENT table, 200000 rows in the PAGE table, and, finally, 10 rows in the SECTION table. We prepared two equivalent implementations of the same application which executes the query from Figure 2 one hundred times with the sole parameter set to 7500 and collects the result rows, performing all the necessary data conversions. The two implementations, compiled with the same (default) optimization level, utilize, respectively:

- The dbQcc framework (**dbQcc**),
- Directly used PostgreSQL API (**libpq**).

Both implementations utilize statement preparation. Note that statement is prepared only once. The durations of each query execution and data conversion were measured separately, and for each implementation averages and standard deviations were computed. The results are presented in Table 1. Note that the differences in times between query executions in both implementations are less than the standard deviation, and hence it follows that our framework does not add any significant overhead.

Table 1. Execution times for the two alternative implementations of the same application executing query from Figure 2

	dbQcc	libpq
Query execution	6.8 ± 0.2	6.7 ± 0.2
Row parsing	0.058 ± 0.007	0.005 ± 0.002

4   Conclusion

In the paper we presented an alternative approach to programming database applications, which is more natural and effective for SQL oriented people (and, potentially, also leads to a better performance). We developed a working tool generating C++ classes from plain native SQL statements. Those generated classes allow us to execute the statements in the application program and to iterate in a semantically correct way through the result sets (in case the statement was a query). Thus, we effectively couple native SQL with a C++ code in a way which is statically checked for correctness, excluding, in particular, the possibility of runtime type mismatch errors. Unlike in the case of embedded SQL, the queries and application program is kept separate, which simplifies application development by programmers with specialized skills — the C++ programmer has no need to see the SQL queries developed by database wizards, and conversely, database specialists are happily separated from C++.

Using plain SQL in its native form makes it possible to write specialized and highly optimized queries even with DBMS vendor specific features. For a simple example see Section 3, which presents a non trivial SQL query, hardly supported by typical approaches, but potentially useful and non-artificial.

Because we utilize mechanisms such as a lazy evaluation of values, the current implementation of dbQcc is not thread safe. This means that the use of generated code in threaded application requires protecting all database operations with a mutex associated with a given database session. Presently we work on effective synchronization of generated code using atomic types and non-blocking algorithms.

The present implementation supports only PostgreSQL, and it will be worthwhile to develop drivers for some other database engines. Despite a well-layered architecture of our system, it might not be entirely trivial, as we assume the availability of certain PostgreSQL features which are not required by the standard and might not be available from other vendors.

Moreover, using the database server to parse SQL statements, while not without merit (it is simple to implement and, if the database is the same as the target one, we can be sure that the query will be accepted also during the application runtime), it has some downsides as well. For one thing, sometimes the database might not exist (even in the form of a schema with empty tables) at program design time. Arguably, it might be better (and more elegant) to parse the statement directly in the tool with the option of supplying the schema description in the specialized format. Moreover, once the tool does the parsing, and hence understands the SQL statement syntax, new possibilities appear, e.g., of extending the SQL syntax, introducing, in particular, special annotations for automated BLOB conversion into specified class objects during query execution.

Finally, as we can create connection objects to many databases and database servers in the same application, it is only natural for the need for the distributed commit mechanisms to appear. Therefore the support for two-phase commit is currently under active development.

Acknowledgements. We would like to thank the reviewers for their helpful remarks and to Łukasz Krawczyk for his help in implementing the tests.

References

1. Java Data Objects, http://db.apache.org/jdo/
2. Libpq — C Library, http://www.postgresql.org/docs/9.3/static/libpq.html, chapter in [3]

3. PostgreSQL 9.3.2 Documentation, http://www.postgresql.org/docs/9.3
4. PREPARE command,
   http://www.postgresql.org/docs/9.3/static/sql-prepare.html
5. Web4j java web application framework,
   http://www.web4j.com/Java_Web_Application_Framework.jsp
6. Clavel, M., Durán, F., Eker, S., Lincoln, P., Martí-Oliet, N., Meseguer, J., Talcott, C.: The maude 2.0 system. In: Nieuwenhuis, R. (ed.) RTA 2003. LNCS, vol. 2706, pp. 76–87. Springer, Heidelberg (2003)
7. Cook, W., Rai, S.: Safe query objects: statically typed objects as remotely executable queries. In: Proceedings of 27th International Conference on Software Engineering, ICSE 2005, pp. 97–106 (May 2005)
8. Date, C.: An Introduction to Database Systems. Addison-Wesley (2003)
9. Fowler, M.: Patterns of Enterprise Application Architecture. A Martin Fowler signature book. Addison-Wesley (2003)
10. Fowler, M.: Domain-Specific Languages. Addison-Wesley Signature Series (Fowler). Pearson Education (2010)
11. Gil, J.Y., Lenz, K.: Simple and safe SQL queries with C++ templates. In: Proceedings of the 6th International Conference on Generative Programming and Component Engineering, GPCE 2007, pp. 13–24. ACM, New York (2007)
12. Gould, C., Su, Z., Devanbu, P.: Static checking of dynamically generated queries in database applications. In: Proceedings of 26th International Conference on Software Engineering, ICSE 2004, pp. 645–654. IEEE (2004)
13. Maier, D.: Representing Database Programs As Objects. In: Advances in Database Programming Languages, pp. 377–386. ACM, New York (1990)
14. McClure, R., Kruger, I.: Sql dom: compile time checking of dynamic sql statements. In: Proceedings of 27th International Conference on Software Engineering, ICSE 2005, pp. 88–96 (May 2005)
15. Pustelnik, J., Sobieski, Ś.: Metagen — the text tool for generating sql database descriptions from ER diagrams (in polish). In: Bazy Danych - Modele, Technologie, Narzędzia, pp. 309–314. WKL Gliwice (2005)
16. Sobieski, S., Zieliński, B.: Using maude rewriting system to modularize and extend sql. In: Proceedings of the 28th Annual ACM Symposium on Applied Computing, SAC 2013, pp. 853–858. ACM, New York (2013)
17. Wegrzynowicz, P.: Performance antipatterns of one to many association in hibernate. In: 2013 Federated Conference on Computer Science and Information Systems (FedCSIS), pp. 1475–1481 (September 2013)
18. Wiśniewski, P., Burzańska, M., Stencel, K.: The impedance mismatch in light of the unified state model. Fundamenta Informaticae 120(3), 359–374 (2012)
19. Wiśniewski, P., Szumowska, A., Burzańska, M., Boniewicz, A.: Hibernate the recursive queries - defining the recursive queries using hibernate orm. In: ADBIS (2), pp. 190–199 (2011)

A Demand-Driven Bulk Loading Scheme
for Large-Scale Social Graphs

Weiping Qu and Stefan Dessloch

University of Kaiserslautern
Heterogeneous Information Systems Group
Kaiserslautern, Germany
{qu,dessloch}@informatik.uni-kl.de

Abstract. Migrating large-scale data sets (e.g. social graphs) from cluster to cluster and meanwhile providing high system uptime is a challenge task. It requires fast bulk import speed. We address this problem by introducing our "Demand-driven Bulk Loading" scheme based on the data/query distributions tracked from Facebook's social graphs. A client-side coordinator and a hybrid store which consists of both MySQL and HBase engines work together to deliver fast availability to small, "hot" data in MySQL and incremental availability to massive, "cold" data in HBase on demand. The experimental results show that our approach enables the fastest system's starting time while guaranteeing high query throughputs.

Keywords: Bulk loading, HBase, MySQL.

1   Introduction

As the biggest social network company, Facebook's social graph system nowadays serves tens of billions of nodes and trillions of links at scale [1]. Billions of daily queries demand low-latency response times. Recently, a social graph benchmark called LinkBench [2] was presented by Facebook which traces distributions on both data and queries on Facebook's social graph stores.

Two main tables node(id, type, data) and link(id1, link_type, id2, data) are used to build the social graph at Facebook (primary keys are underlined). Nodes represent objects like user, photo, video, etc. while links are connections between the objects and have types like "post", "like" and "friend_of". We learned several interesting facts from LinkBench, for example, one observation on access patterns and distributions states that there is always some "hot" data that is frequently accessed while massive amounts of "cold" data is seldom used. With a 6-day trace, 91.3% of the data is cold. In addition, hot data often exists around social graph nodes with high outdegrees, which means the access likelihood grows along with the node outdegrees. As an example, a video with high *like* rates will be recommended more widely than others. Based on another observation on social graph operations, an operation called *get_link_list* occurs frequently and constitutes 50.7% of the overall workload. The signature of this

Y. Manolopoulos et al. (Eds.): ADBIS 2014, LNCS 8716, pp. 139–152, 2014.
© Springer International Publishing Switzerland 2014

operation is `get_link_list(link_type, id1, max_time, limit, ...)` where id1 is the starting node id of this link. Given the type of a link (e.g. *like*) and the id of the starting node (e.g. a user *id1*), *get_link_list* returns a list of links to answer certain types of queries like "what objects have been recently *liked* by user *id1*". This *get_link_list* performs a short range scan in massive amounts of rows in graph stores based on (`link_type` and `id1`), a subset of the composite key.

As more and more applications are added to Facebook, like Facebook Messaging and Facebook Insights [3], and workloads evolve on graph stores, the change of the underlying data infrastructure or software requires migrating existing data from cluster to cluster while high data availability must be guaranteed. To provide 24-h system uptime, normally hot-standby clusters are used to store replicas of source data and serve query workloads during data migration. This incurs high data redundancy and the new system still cannot start until a long-running bulk import is finished.

Our work investigates the problem of migrating large-scale social graphs based on their data distributions and access patterns introduced above. To guarantee high system uptime, a trade-off between data availability and query latency is utilized in this work. The "hotness and coldness" of migrated data is balanced by a hybrid graph store which is composed of a traditional index-based relational MySQL (http://www.mysql.com) database and an Apache HBase (http://hbase.apache.org) cluster. Both systems have received high attention as a backend storage system for realtime operations on big data. The debate on MySQL and HBase began in 2010 in terms of multiple metrics like read/write throughput, I/O, etc. In this work, we will first compare these two systems regarding their bulk load speed and short range scan latency and then introduce our "demand-driven bulk loading" scheme.

The comparison of MySQL and HBases' load and scan performance is given in Section 2 and the motivation of this work is explained there. In Section 3, we introduce the architecture of our "demand-driven bulk loading" scheme. The experimental results are analyzed in Section 4. We discuss related work in Section 5, and Section 6 concludes our work.

2   Bulk Loading in MySQL and HBase

In [4,5], the performance of sequential/random read/write access has been compared among Cassandra, HBase and (sharded) MySQL. The results show that, due to their different architectures, HBase has the highest write throughput when the insertions fit in memory, while MySQL is best for read access. Both engines provide extra bulk load utilities in addition to the interfaces for individual, row-level writes/updates. In this section, we compare their bulk load mechanisms by analyzing their architectural differences. Based on the comparison result, we describe our motivation of providing incremental availability to external queries during bulk loading large-scale social graphs.

2.1 MySQL

Like other traditional databases, MySQL uses B-trees as an index structure for fast read and write on large tables. One crucial step of generic bulk loading in traditional databases is an index construction process. Using a classical sort-based bulk loading approach, the entire data set is pre-sorted ($O(n\log(n))$) and grouped in file blocks as index leaf nodes. A B-tree index can be easily built from this set of sorted leaf nodes in a bottom-up fashion from scratch. In contrast, inserting the tuples from the same data sets once at a time in a top-down fashion without pre-sorting incurs overhead i.e. a lot of splits on index's internal nodes and a large number of disk seeks with random I/O. There are other approaches for building indices during bulk loading, like buffer-based/sample-based bulk loading [6] which will not be detailed here.

To import large amounts of data in MySQL, there are two primitive approaches: batch insert and bulk loading. By including multiple tuples in one INSERT statement and inserting these tuples in batch, batch insert results in fewer operations and less locking/commit overhead. Yet the bulk load command LOAD DATA INFILE is usually 20 times faster than using the INSERT statement because of its less overhead for parsing [7]. However, the user has to ensure that the tuples to be inserted won't violate integrity constraints. Before bulk loading, the use of indices is normally disabled to avoid disk seeks for updating on-disk index blocks at load time. After bulk loading, indices are enabled again and created in memory before writing them to disk. However, when bulk-loading non-empty tables where indices are already in use, a performance impact of bulk-loading on concurrent reads occurs.

As mentioned in Section 1, the frequently used *get_link_list* operation performs a short range scan on a subset of the composite key. Therefore, using MySQL as the graph storage backend, very low scan latency can be achieved by traversing the leaf nodes of primary key index sequentially after the starting block has been found.

2.2 HBase

Apache HBase^{TM} is the Hadoop database built directly on the Hadoop Distributed File System (HDFS) [8]. HDFS is an open-source version of Google File Systems (GFS), which inherently provides batch processing on large distributed files using MapReduce jobs. HBase was modeled after Google's Bigtable [9] and provides random, realtime read/write access to HDFS. This is done by directing client requests to specific region, with each server handling only a fraction of large files within a certain key range.

Both Bigtable and HBase use an "append-only" log-structured merge tree (LSM-tree) [10] structure. In HBase, inserted tuples are first buffered in an in-memory structure called MemStore and sorted very fast in memory. Once the size of the MemStore exceeds a certain threshold, it is transformed to an immutable structure called HFile and flushed onto disk. A new MemStore is then created to further buffer new incoming rows. Updates on existing rows are treated as new insertions appended to existing files instead of in-place modification, which needs random disk seeks for reading updated blocks. In addition, no auxiliary index structure needs to be maintained. In this way, high write throughput can be achieved in HBase. However, a single row can appear many times in multiple HFiles or MemStore. To retrieve a single row, HBase has to scan those HFiles or MemStore that contain the copies of this row and merge them to return the final results. Sequential scans are carried out on sorted HFiles. Thus the read speed is dominated by the number of files in a region server. In order to improve read performance, a compaction process runs periodically to merge HFiles and reduce the number of row copies. Furthermore, Bloom filters can be used to skip a large number of HFiles during reading.

To insert large amounts of files into HBase, an efficient MapReduce-based bulk loading approach can be used to directly transform HDFS files into HFiles without going through the write path introduced above. Each row appears only once in all HFiles. The map tasks will transform each text line to a Put object (a HBase-specific insert object) and send it to a specific region server. The reduce tasks sort these Put objects and generate final HFiles. This process is much faster than HBase writes as it exploits batch processing on HDFS. The *get_link_list* operation can benefit from this bulk loading approach as well since the number of HFiles to read is small. By setting up Bloom filters, target HFiles can be found very fast.

2.3   Motivation

As introduced in Section 1, 91.3% of the social graph is rarely used while 8.7% of the data sets are frequently accessed. When migrating such a social graph to new clusters, the availability of hot data is delayed since the system downtime will only end when all data has been loaded. For frequently emerging queries, system uptime could start earlier if there was a mechanism that can tell whether all the relevant data is already available before the remaining data is loaded. As data is loaded chunk by chunk, it would be enough to have a global view of the key ranges of all the chunks before starting loading. This can be seen as an index construction process. In this way, a query can identify its relevant chunks by comparing its queried key with the key ranges of all the chunks. Once its relevant chunks have been loaded, this query can start to run over the current storage state. In addition, as the small amount of hot data can be arbitrarily distributed among all the chunks, faster availability to frequent queries can be achieved by prioritizing the loading of the chunks which contain hot data.

MySQL's bulk loading builds indices either during data loading or after loading. Building indices upfront is impossible. Using HBase's bulk loading, source files must be first copied to HDFS and then transformed to HFiles. Comparing the bulk loading techniques in both systems, similar performance can be expected, since both techniques have to sort files either to build B-tree index in MySQL or to generate HFiles in HBase. But HDFS's batch processing feature can be exploited in HBase's two-phase bulk loading approach to build indices on copied chunks in batch upfront before going to the second transformation phase, which is more desirable.

However, according to the experimental results in [2], MySQL slightly outperforms HBase in latency and MySQL executes *get_link_list* operations 2x faster than HBase. We see a trade-off between fast availability and low query latency here. Loading all data in HBase can have fast availability by creating indices upfront. Loading all data in MySQL leads to low query latency after long-time bulk loading ends. It makes sense to load only a small amount of hot data into MySQL in a smaller time window for fast processing while copying massive amounts of cold data into HBase where its query latency is still acceptable for cold data. But the cost of identifying the hotness and coldness of tuples could be a large overhead for bulk loading. Hence, we introduce our "demand-driven bulk loading" scheme to address these considerations.

3   Demand-Driven Bulk Loading

In this section, we introduce the architecture of our demand-driven bulk loading scheme. According to the modification timestamps of files, the whole input data set is separated into two parts: small number of recent files and large, historical files. Recently changed files are imported into a MySQL table called "link table" using MySQL's own fast bulk load utility. These files are used for answering queries on hot data and providing partial results for all other queries. Meanwhile, massive amounts of historical files are first split to chunks with pre-defined size and then copied into HDFS in batch. With parallel loading of recent and historical files into MySQL and HDFS, respectively, the latency of loading HDFS is normally higher than that of MySQL's bulk load due to the input size, thus dominating the overall load speed. After loading in MySQL completes, hot data is available for querying. The HDFS files will be gradually loaded into HBase to complement the MySQL query results for answering queries that involve cold/historical data. Figure 1 illustrates the architecture of our hybrid-storage approach.

Two main processes are involved in this hybrid-storage architecture: offline index building and online bulk load coordination. In contrast to traditional bulk load approaches, client requests are allowed to query link data before the historical data sets are completely available in HBase. To determine the completeness of query results on the client side, a so-called *bucket index table* is used. At the HBase layer (left) side of this architecture, an offline MapReduce job called *distribute chunk* (*dist_ch*) job is batch processed on each file chunk in HDFS once copied from the remote server by a *HDFS loader*. The implementation of this job is based on a hash function that maps and writes each text line in a chunk to a specific "bucket" (HDFS file) with a unique id and a key range disjoint from others. A new bucket will be created if needed and its *bucket index* and *key range* will be captured by the bucket index table at the (right-side) MySQL layer. These steps form the *offline index building* process. More details will be provided in Subsection 3.1. With the completion of the last *dist_ch* job, all cold

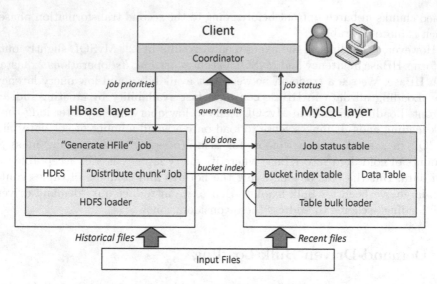

Fig. 1. Architecture of Demand-driven Bulk Loading

files have been copied into HDFS and clustered into multiple buckets with disjoint key ranges. The key ranges and index information of all the buckets are contained in the bucket index table in MySQL.

At this time, system uptime begins and query requests are allowed to run on our hybrid link data storages through a client-side *coordinator* component. At the same time, another online MapReduce job called *generate HFiles* (*gen_HF*) job is triggered to continuously transform buckets in HDFS to HFiles so that they can be randomly read from HBase. The transformed HFiles incrementally build a "link table" in HBase which can be seen as an external "table partition" of the "link table" in MySQL. Tuples stored in both engines share the same logical table schema. With a given key (the id1 of the starting node of a link) specified in a query, the coordinator checks whether the HBase layer should take part in this query execution by asking the MySQL-side bucket index table. If so, another *job status table* tracks the availability of the required tuples in HBase. For tuples available in HBase, the query request is offloaded to HBase by the coordinator. In case there are tuples that are not available yet because they reside in buckets that wait in the *gen_HF* queue, the query is marked as "incomplete" and buffered by the coordinator. As more and more incomplete queries occur at the coordinator side, the coordinator makes the online MapReduce job prioritize the job execution sequence for specific buckets, delivering fast availability on demand. Once the buckets are transformed to the portions of HBase's "link table", corresponding buffered queries are released. This process is called *online bulk load coordination*. The implementation of this process will be detailed in Subsection 3.2.

3.1 Offline HDFS Load and Index Construction

We use a dedicated Hadoop cluster to take over the job of loading massive amounts of cold/historical link data from remote servers to MySQL. HDFS's free copy/load speed and batch processing (using MapReduce) natures are exploited here to provide only indices on clustered file groups (buckets) using the *dist_ch* jobs, as introduced above. A *dist_ch* job writes text lines in each chunk in HDFS to different buckets (HDFS directories) and outputs indices of new buckets to MySQL's bucket index table (Hadoop's MultipleOutputs is used here to include TextOutputFormat and DBOutputFormat for writing lines to buckets and writing indices to MySQL, respectively).

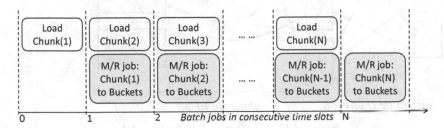

Fig. 2. Execution Pipelines in the HDFS loader

Instead of running one big MapReduce-based *dist_ch* job after all files have been completely copied from a remote server, large historical files are split to several chunks and multiple small *dist_ch* jobs are executed in parallel with copying small file chunks to HDFS. As shown in Figure 2, chunk copying and *dist_ch* job run simultaneously in each time slot except the first and the last one which copies the first chunk and builds the indices for the last chunk, respectively. The resource contention is low since chunk copying does not use any MapReduce job and each *dist_ch* job runs individually. As the chunk copying pipeline overlaps the *dist_ch* job pipeline, the overall latency is derived from loading all chunks plus running the last *dist_ch* job. The chunk size is selected in a way that the latency of loading a chunk of this size is higher than running one *dist_ch* job on that chunk. If this requirement can be guaranteed, the chunk size should be defined as small as possible so that the time running the last *dist_ch* job is the shortest. Therefore, the overall system downtime is similar to the latency of copying massive amounts of cold data from a remote server to HDFS.

As tuples belonging to a specific key might be arbitrarily distributed in all chunks, we introduce a simple hash function to cluster tuples into buckets according to disjoint key ranges. If we have numeric keys (e.g. id1 for links) and a key range of 1K (0..1K; 1K..2K; ...), the bucket index for each tuple is derived from b_index=round(id1/1K). With bucket index and key range, the coordinator can tell exactly which bucket should be available in HBase to complete the results for an incomplete query. As shown on the left side of Figure 3, *dist_ch* jobs take fix-sized chunks as inputs and generate a set of buckets of dynamic

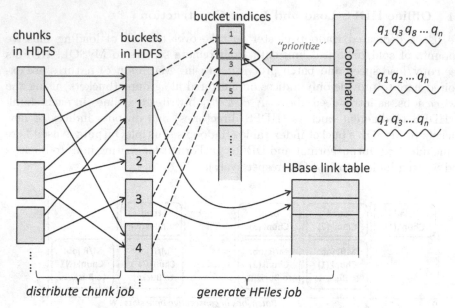

Fig. 3. Offline & Online MapReduce Jobs

sizes based on id1. This can be explained by the LinkBench's observation on the distribution of nodes' outdegrees described in Section 1. Here, a bucket might contain a large number of links which belong to a node with very high outdegree.

3.2   Online HBase Bulk Load and Query Coordination

Once the remote files are copied to local HDFS, our system starts to accept queries from the client side and a *gen_HF* job runs continuously to finish the remaining bulk load work. Three components are involved here: the client-side coordinator, two tables in MySQL (bucket index table and job status table) and the online *gen_HF* job in HBase layer (see Figure 1).

A *gen_HF* job is a MapReduce-based bulk loading job that is executed in several runs in the HBase layer. In each run, it takes HDFS files in "bucket" directories (directed by bucket indices) as input and generates HFiles as results (see Subsection 2.2). Tuples in HFiles can be randomly read without batch processing. The cost of the *gen_HF* job is dominated by sorting. When the local memory on each region server cannot hold the entire set of Put objects for in-memory sorting, expensive disk-based external sorting occurs. Hence, a *gen_HF* job each time will take only the top two buckets (included in the red rectangle in Figure 3) as input to avoid external sorting on local region servers.

The coordinator plays an important role in our demand-driven bulk loading scheme. It maintains a set of four-element tuples (b_index, j_stat, k_range, q_list) at runtime. The b_index is the bucket index which directs the *gen_HF* job to the input files in this bucket. The k_range represents the key range of these input files which will be further checked by an incoming query whether this bucket can contain required tuples. The b_index and k_range of all the buckets are initially read from the MySQL-side bucket index table at once. Note that, after the hot link data has been bulk loaded into MySQL, a special bucket will be created to contain the k_range of MySQL-side "link table".

Furthermore, before the *gen_HF* job starts a run, it registers its two input bucket indices in the job status table in MySQL. When the job is done, it updates its status in the job status table, whose content will be periodically pulled by the coordinator to maintain the j_stat elements for all buckets. At the beginning of system uptime, files in most of the buckets have not been transformed to HFiles and thus are not available in HBase. It's much likely that the incoming queries at that moment cannot be completely executed and are further pushed into the query list q_list of certain buckets. The coordinator will release the queries in a q_list once the j_stat states that this bucket is readable. Moreover, the coordinator will also sort the four-element tuples according to the size of q_list due to emergency so that the *gen_HF* job will always work on transforming the top two buckets with the largest number of waiting queries.

As an example of demand-driven bulk loading shown in Figure 3, three client-side threads keep sending queries to the coordinator. Most of them contain queries that would access tuples in the key ranges of bucket 1 and 3. The coordinator checks the size of the query waiting list and prioritizes the *gen_HF* job execution sequence for bucket 1 and 3. Hence, after the next run, query q_1 and q_3 will be released by the coordinator since the required tuples now can be found in the HBase "link table".

4   Experiments

In Section 2, we mentioned our motivation of partitioning and loading large-scale link sets of a social graph to a hybrid storage system (consisting of a MySQL database and a HBase cluster) based on the observation that only a fraction of links is frequently accessed while most of the data is seldom used. To enable fast availability (i.e. fast load speed) of the entire social graph system, we introduced our "Demand-driven Bulk Loading (DeBL)" scheme in Section 3. In this section, we validate our approach by analyzing the experimental results. The performance difference in terms of load speed and query latency is shown by comparing the results using a single MySQL database, using a single HBase cluster or using our DeBL approach. Our approach serves as a compromise between these two systems and outperforms both of them when loading large-scale social graphs.

We used a logical *link* table with its schema (id1, link_type, id2, ..., data) to represent links stored in MySQL, HBase or both systems (as links occupy the largest portion in a graph, we ignored loading graph nodes in our test). The test query is the get_link_list operation which performs a short range scan to fetch links with given id1s and link_types and constitutes 50% of the whole workload. We think that this test setup is general and representative.

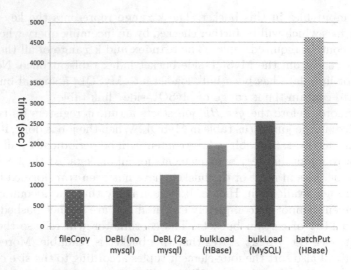

Fig. 4. Loading Time of Different Approaches

(For write operations, both MySQL and HBase can provide fast random write facility. However, we excluded these operations to simplify our test.)

We extended the LinkBench [2] program for our test purpose which is based on a client/server architecture. In the original LinkBench implementation, multiple threads run on the client side to either load links (**load phase**) or send requests (**query phase**) to a server-side "link table" (using MySQL INNODB engine) in a MySQL database. To compare the load performance of different approaches, we first recorded the latency of bulk loading a CSV input file (100M links, 10GB) from a remote client (through 100Mb/s Ethernet) into a link table in a MySQL instance running on a single-node machine (2 Quad-Core Intel Xeon Processor E5335, 4×2.00 GHz, 8GB RAM, 1TB SATA-II disk) using LOAD DATA INFILE command (the primary key index was first disabled and re-enabled after the file was loaded) [bulkLoad (MySQL)]. Since we were only comparing short range scan performance in the query phase later, a faster "link table" using MySQL MYISAM engine instead was used which is optimized for heavy read operations (MySQL's batch insert was excluded in this test as MYISAM engine uses table-level locking on tables which is very slow for massive, concurrent inserts).

In the second case, we tested the bulk load performance on a HBase cluster - a 6-node HBase cluster (version 0.94.4) with a master running on the same node as the MySQL instance and 5 region servers (2 Quad-Core Intel Xeon Processor X3440, 4×2.53GHz, 4GB RAM, 1TB SATA-II disk) connected by Gigabit Ethernet. A big MapReduce job (the *gen_HF* job) was implemented to generate HFiles and populate a HBase link table after the same input file was copied from remote to local HDFS [bulkLoad (HBase)]. Since HBase also provides high write throughput, we also tested the performance of writing links to HBase in batch using HBase's Put method [batchPut (HBase)]. To improve performance, the link table was first pre-split evenly in the cluster according to its key distribution to avoid "hotspots" during loading and querying.

Two variants of the load phases were tested using our DeBL approach. The first variant was composed of bulk loading 2GB, recently changed, remote link subsets into the MySQL table, copying the rest 8GB link files in batch from remote client to HDFS (in our HBase cluster) and meanwhile running multiple small MapReduce jobs (the *dist_ch* jobs) in parallel [DeBL (2g mysql)]. Another extreme case was shown by the second variant where no files were loaded into MySQL and the entire input file was copied to HDFS [DeBL (no mysql)]. In this case, the "'hotness and coldness" in input data was not pre-defined manually but was captured automatically by our coordinator during *gen_HF* phase according to incoming query distribution. To indicate fast availability of DeBL approach, we attached the time taken to simply copy the test input file to server's local file system [fileCopy] to the final results as the bottom line as well.

Table 1. Detailed Latencies (sec) in Load Phase

DeBL (no mysql)		DeBL (2g mysql)		bulkLoad (MySQL)		bulkLoad (HBase)	
chunk load:	24.33	mysql load:	463.26	bulk load:	1674.76	HDFS copy:	895.31
dist_ch job:	24.95	hbase load:	793.98	gen. index:	890.57	gen. HFiles:	1082.81
total:	952.17	total:	1257.23	total:	2565.34	total:	1978.12

The results of load latencies are shown in Figure 4 and detailed in Table 1. The latency of fileCopy is the bottom line which is 893 seconds and cannot be improved anymore. The result of DeBL (no mysql) is 952.17s and very closed to fileCopy's latency. The input file was transferred chunk by chunk and each chunk has 256MB size. With this chunk size, both chunk load job and *dist_ch* job took similar time (24~25s). As both jobs ran in parallel, the result of DeBL (no mysql) could be derived from the sum of fileCopy's time and the latency of the last *dist_ch* job. It provides the fastest starting time of system uptime with near wire-speed. However, incoming queries still have to wait until their files are available in HBase. Another variant DeBL (2g mysql) took a little bit longer for bulk loading 2GB hot data into MySQL (including index construction) which is 463.26s and its total latency is 1257.23s.

Latency gets higher when using traditional bulk loading approaches. Using bulkLoad (MySQL), the LOAD DATA INFILE command took 1674.76s while re-enabling primary key index spent 890.57s. Using bulkLoad (HBase), copying remote files to HDFS had the same latency as fileCopy and generating HFiles reached similar cost (1082.81s) as MySQL's index construction since both processes required sorting on large input files. However, bulkLoad (HBase) is faster than bulkLoad (MySQL) since HBase is a distributed system where MapReduce's batch processing feature can be exploited. Apart from this difference, both bulk loading approaches still outperforms HBase's fast writes where some overheads like compaction occurred due to HBase's implementation as mentioned in Subsection 2.2.

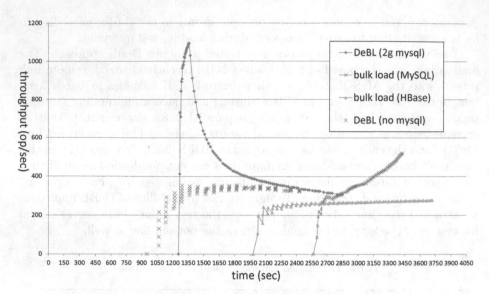

Fig. 5. System Uptime & Query Throughput in four Approaches

The end of load phase indicates the start of system uptime and lower load latency means faster data availability. After the load phase, we ran 50 threads on the client side to continuously send 500,000 *get_link_list* requests to the server. Both MySQL and HBases' bulk loading approaches led to complete data availability and we tracked the query throughputs starting from their system uptime. In this case, the difference of the query throughputs represents the difference of query latencies on MySQL and HBase as well. We tried our best to tune our HBase cluster for example, by enabling block caching, setting up Bloom filters and pre-splitting regions. For 10GB test data, our in-house HBase cluster yet could not cope with a single-node MySQL engine in terms of query throughput as shown in Figure 5. However, as HBase cluster took less time to ingest the input data, its throughput curve started earlier (from 1978.12s, the end of its load phase) to rise and converged at 278.7 op/sec throughput and 3.6 ms query latency in average for each *get_link_list* request. The bulkLoad (MySQL) approach took the longest time until all links are available in link table. Its throughput was rising rapidly (till 521.8 op/sec, 1.9 ms) and all the queries were finished in a small time window.

In contrast to traditional bulk loading approaches, our DeBL approach trades complete data availability for fast system uptime. It provides incremental availability to files stored in HDFS on demand. It can be seen in the DeBL (no mysql) variant that the system started the earliest at 952.17s but the query throughputs occurred intermittently as relevant file partitions continuously got available in HBase. The throughputs were higher than bulkLoad (HBase) at the beginning since less available files needed to be scanned in HBase. Along with growing data size, the throughputs kept close to the highest value in bulkLoad

(HBase) (332 op/sec). Using DeBL (2g mysql), the system uptime had 300s delay whereas its throughput curve first climbed up drastically and reached its peak 1092.7 op/sec. After that, it began to fall and finally converged with bulkLoad (HBase)'s curve. The reason is that a big portion of frequently emerging queries were immediately answered by the 2GB hot data in MySQL at first. As the size of data in MySQL was much smaller, their query latencies were also faster than those on 10GB data in bulkLoad (MySQL). The rest of the queries that could not be answered by MySQL were buffered by the coordinator and released as soon as the data was available in HBase. Important to mention, both DeBL variants were able to digest the entire 500,000 requests before the system uptime began in bulkLoad (MySQL).

5 Related Work

Bulk loading techniques normally serve the loading phase in Extract-Transform-Load (ETL) processes which handle massive data volumes at regular time intervals. A middleware system called "Right-Time ETL (RiTE)" [11] provides ETL processes with INSERT-like data availability, but with bulk-load speeds. Instead of loading entire input data directly into the target table on a server, a server-side, in-memory buffer is used to hold partial rows of this table before they are materialized on disk. Since loading data in memory is much faster, the time window of data loading is shrunk. A logical view is defined to enable query execution on rows stored in both locations. However, problems will occur when large-scale social graphs cannot fit into memory. In our case, we let a distributed file system take over partial load/query jobs on large files from databases.

In [12], Goetz Graefe proposed his idea of fast loads and online B-tree index optimization as a side-effect of query processing. Instead of building complete indexes during data loading, a small, auxiliary structure called *partition filters* (similar to small materialized aggregates [14]) is created for each new loaded partition. With this information, the indexes are optimized incrementally and efficiently on demand according to queries' predicates. This inspired us to use a bucket index table as auxiliary information to identify required file buckets to be available in HBase for incoming queries.

With the advent of "Big Data" and its emerging Hadoop/MapReduce techniques, database vendors now have lots of solutions that integrate open-source Hadoop with their products. IBM's InfoSphere BigInsights and Big SQL [13] is one of them. Big SQL provides a SQL interface to files stored in Hadoop, BigInsights distributed file systems or external relational databases. It allows companies with existing large relational data warehouses to offload "cold" data to cheap Hadoop clusters in a manner that still allows for query access. In this context, our approach exploits the features of underlying MySQL and HBase engines to balance the availability between "hot" and "cold" data.

6 Conclusion

In this work, we first introduced the bulk loading techniques used for MySQL and HBase and then proposed our demand-driven bulk loading scheme. This scheme utilizes a hybrid storage platform consisting of a fast-load/slow-query HBase and a slow-load/fast-query MySQL to accommodate large-scale social graphs, which is a compromise as fast available "hot" graph data and slowly accessible "cold" data. Our experimental results show that our approach provides fast system uptime and incremental availability to "cold" data on demand.

We do not assume that the data partition stored in MySQL is always hot since the "hotness" of files that resides in HBase can still be discovered in our approach. The limitation is that the query latency of HBase is not satisfactory if the data partition stored in HBase gets frequently accessed in the future. Our future work is to remove this limitation by online data re-balancing in the hybrid storage cluster.

References

1. Curtiss, M., Becker, I., Bosman, T., Doroshenko, S., Grijincu, L., Jackson, T., Zhang, N.: Unicorn: a system for searching the social graph. VLDB, 1150–1161 (2013)
2. Armstrong, T.G., Ponnekanti, V., Borthakur, D., Callaghan, M.: Linkbench: a database benchmark based on the facebook social graph, pp. 1185–1196. ACM (2013)
3. Borthakur, D., Gray, J., Sarma, J.S., Muthukkaruppan, K., Spiegelberg, N., Kuang, H., Aiyer, A.: Apache Hadoop goes realtime at Facebook. In: SIGMOD, pp. 1071–1080 (2011)
4. Cooper, B.F., Silberstein, A., Tam, E., Ramakrishnan, R., Sears, R.: Benchmarking cloud serving systems with YCSB, pp. 143–154. ACM (2010)
5. Rabl, T., Gómez-Villamor, S., Sadoghi, M., Muntés-Mulero, V., Jacobsen, H.A., Mankovskii, S.: Solving big data challenges for enterprise application performance management. VLDB, 1724–1735 (2012)
6. Bercken, J., Seeger, B.: An evaluation of generic bulk loading techniques. VLDB, 461–470 (2001)
7. https://dev.mysql.com/doc/refman/5.0/en/insert-speed.html
8. White, T.: Hadoop: The definitive guide. O'Reilly Media, Inc. (2012)
9. Chang, F., Dean, J., Ghemawat, S., Hsieh, W.C., Wallach, D.A., Burrows, M., Gruber, R.E.: Bigtable: A distributed storage system for structured data. In: TOCS (2008)
10. O'Neil, P., Cheng, E., Gawlick, D., O'Neil, E.: The log-structured merge-tree (LSM-tree). Acta Informatica, 351–385 (1996)
11. Thomsen, C., Pedersen, T.B., Lehner, W.: RiTE: Providing on-demand data for right-time data warehousing. In: ICDE, pp. 456–465 (2008)
12. Graefe, G., Kuno, H.: Fast loads and queries. Transactions on Large-Scale Data-and Knowledge-Centered Systems II, 31–72 (2010)
13. http://www.ibm.com/developerworks/library/bd-bigsql/
14. Moerkotte, G.: Small materialized aggregates: A light weight index structure for data warehousing. VLDB, 476–487 (1998)

Open Source Is a Continual Bugfixing by a Few

Mikołaj Fejzer, Michał Wojtyna, Marta Burzańska, Piotr Wiśniewski,
and Krzysztof Stencel

Faculty of Mathematics and Computer Science,
Nicolaus Copernicus University,
Toruń, Poland
{mfejzer,goobar,quintria,pikonrad,stencel}@mat.umk.pl

Abstract. Github is one of the most popular repository sites. It is a
place where contributors come together to share code, ideas, thoughts
and report issues. By using topic modelling applied to comments we are
able to mine plentiful interesting information. Three aspects of an open
source project mostly attracted our attention: the existence of a "Core
Team"' - small number of developers that have the most contributions,
the prevailing popularity of topics related to bug fixing and the continuous development of project without significant iteration phases.

Keywords: Bug fixing, Developers behavioural patterns, Development
phases, Github, LDA, Topic analysis, Team work.

1 Introduction

Today, the the most popular code repository sites for open source projects are
Github and Sourceforge. They gather massive data on users, their activities and
the code they produce. From 2014 MSR Mining Challenge [1] we have obtained
a portion of repositories stored in Github. After a careful analysis of the provided data structure, we have decided to study the influence of committers on
their projects. We focused our attention on mining the information from commit
messages and issue comments.

In order to generalize information about each of the studied commits and issues and to gather statistics, we used topic modelling [2]. Each comment has been
treated as a single document. To obtain topics we applied the Latent Dirichlet
allocation [3] using the Mallet topic modelling toolkit [4]. We trained our topic
model per project, to capture each project's unique history and the context of
programmers interaction. As a next step we have aggregated the data by a number of attributes - among which the most helpful aggregations were by date and
by author.

Based on the series of empirical tests of training Mallet with different parameters, we have finally decided that 50 topics and 1000 iterations gives us the
best generalization without losing too many specific details about the studied
data. We have also added custom stop words that matches the Github context to
clean up committers' messages. Mostly we had to deal with numerous comments

Y. Manolopoulos et al. (Eds.): ADBIS 2014, LNCS 8716, pp. 153–162, 2014.

containing "thank you", "good job" and other such praising that fell out of our scope of interest.

Unfortunately for our approach, out of 92 projects provided, only 52 had a sufficient number of commit messages allowing to extract reliable topic, and of those 52 only 43 projects had enough issue comments to be studied. We have examined the resulting data to verify a number of hypotheses, sometimes discovering new ones worth investigating. The first thing we have noticed was that repeatedly the bug fixing topic was always either the most popular topic, or among the top 5. We shall address this issue in Section 2. Also our initial observations indicated that most of the projects have a small hard working "core developers" group comprising of specialists in one or two topics and people who contribute to almost every topic. More detailed information about the corresponding hypothesis, gathered data and results can be found in Section 3.

Of course among many interesting questions some remained unanswered, while some turned out with a negative answer although our initial intuition suggested that they should validate positively. Two most noteworthy of such hypotheses are the close correlation between issue topics and following commit topics and that the open source projects are created iteratively. Those hypotheses are addressed in the Section 4.

This report deals mostly with three hypotheses:

- In most of the projects the contributors can be divided into two main groups: a great number of contributors is responsible for only a small portion of code, since their input is minor. On the other hand - a small group of contributors, let us call them the "Core team", is responsible for substantial developments through majority of commits to the code.
- Despite advancements of software engineering open source projects are not developed according to a methodology based on any form of cycles. In fact, in general the development process is iterative.

In following section we take a closer look at the aforementioned hypotheses.

2   "The Core Team"

Our goal was to check how do people involved with a project contribute to its development. Do they mostly assume a role of a specialists - local domain specific experts - or perhaps they are generalists who contribute to different parts of their project? Or maybe there are other who fall out of those two categories. Furthermore, how big (in percentages) are those groups and can we find some universal trends on this matter?

In our terminology a specialist is a committer, whose number of commits matching specific topic is larger than half of the maximum number of commits to the most popular topic of a project. A generalist is a committer whose number of commits matching multiple topics is larger than average number of topics per committer.

After analysing the data of the chosen 43 projects we found out that in each project the majority of committers behave like partisans, using hit and run tactics. They create only one commit matching specific topic and disappear, never to be seen again. The average number of committers per project is 35.9 contributors which is equal to 70.93% of average project's committers. Specialists usually concentrate on the most significant topics. Our calculations show that they constitute average 2.07 (12.06%) committers per project. They are also competent or willing enough to create numerous commits matching other topics, so many of them are also counted among generalists, whose average is 9.59 (28.12%) of committers per project.

In order to illustrate the issue we have selected two charts generated from data aggregated by comments' author. Figures 1 and 2 visualize the number of contributions to a topic by an author. For a project 3583 presented in Figure 1 the global number of committers is 84 and there are 5 specialists and 27 multi-topic generalists. The project 107534 from Figure 2 has 34 contributors in total, among which there are 3 specialists and 10 generalists.

Similar research has been conducted by the author of [5]. They have studied the Apache and Mozilla projects and their contributors. Despite slight differences in percentages (their results were closer to 20%-80% ratio of the number of core team developers to others), they came to a similar conclusion. A minority of contributors is responsible for the majority of work.

Table 1. Statistics of committers groups

Specialists	Generalists	Other	Definition
12,06%	28,12%	70,93%	Average percentage per project
2,07	9,59	35,9	Average number per project
5,77%	26,70%	72,96%	Percentage of all
85	393	1074	Number of all

3 Bugfixes

The analysis of the charts containing popularity of topics has lead us to our second hypothesis: bug fix commits are notably popular in open source projects. Our first intuition was that bug fixing related topics should have significant popularity. We analysed data fetched from Github database using our generic solution based on Mallet as follows. We took all comments of each commit and created a list of topics describing it. Every topic consists of top 10 words with frequencies. We classified a topic as bug fix related if it contained at least 2 words from the list of bug fix keywords, such as: fix, bug, solve. We took all commits of a project and generated a reduced list of 10 topics. In a 50 topic list we have often seen a number of bug fix related topics, thus the need for reduction.

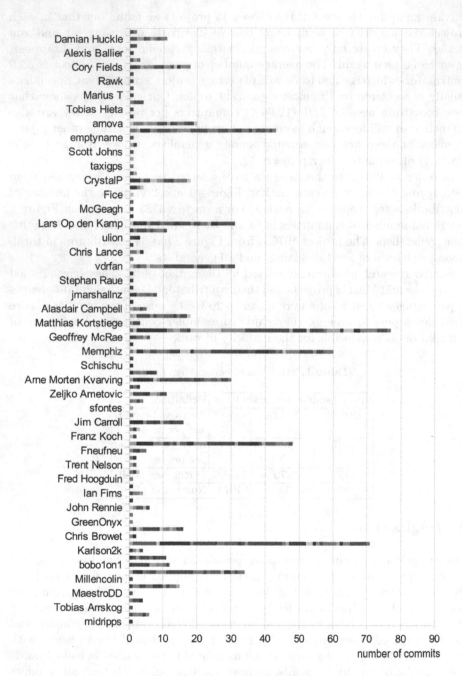

Fig. 1. Topic aggregation by committer's name on project 3583

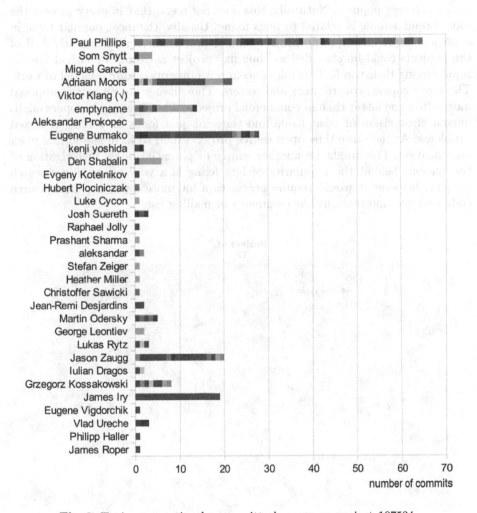

Fig. 2. Topic aggregation by committer's name on project 107534

Now, we have assumed that developers' work is focused mainly around fixing bugs if a project contains at least one bug fix related topic within top 5 topics list. We have found out that most of the chosen projects share topics which we can classify as related to bugs fixing. Only one of 44 projects did not have a bug fixing related topic. Also, bug fix topics are often equally distributed over time. Bug fix commits usually span over many months or even years. Moreover, these topics are very popular. Naturally, this does not mean that in every project the most popular topic is related to bugs fixing. Usually, the most popular topic in a project represents specific project-related problems. Actually, nearly half of the projects could be classified as "bug-fix" project as 20 out of 44 had bug-fix topic among their top 5. This might seem reflecting open source model of work. There are copious contributors and testers. Thus, issues are reported/proposed more often and faster than in commercial projects. That is why developers might have a clear vision of what to do and they can just focus on solving reported problems. At the same time open source projects tend to have less formal work organization. This might be another source of potential bugs. Investigation of the reasons behind the popularity of bug fixing is a very interesting research subject, however it would require access to a lot more data, e.g. both source code and user interactions like comments or mailing lists.

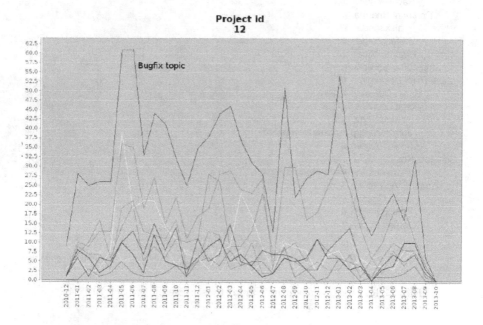

Fig. 3. Topic aggregation by commit date on project 12

Figures 3 and 4 present two example charts showing distribution of commit topics over time in two projects. The first chart (project id 12) represents topics of TrinityCore Open Source MMO Framework. The second chart (project

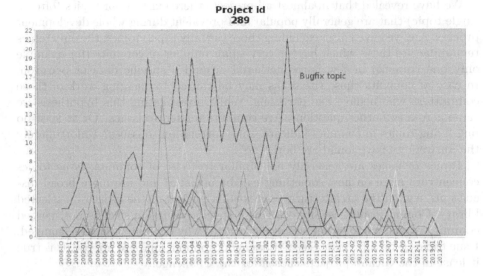

Fig. 4. Topic aggregation by commit date on project 289

id 289) concerns MaNGOS, a MMO server suite. Both of them have plentiful contributors and commits over time. Therefore they are suitable to show some tendencies. Both projects have also rich histories (unlike many other projects), ranging from 2008 to 2013 (project 289) and from 2010 to 2013 (project 12). We can see that in both projects bug fix commits are not only notably popular, but also spanned across the entire chart. This leads us to conclusion that in both projects developers' main efforts are focused on solving issues/bugs reported by the community or other developers.

4 Project's Development Phases

Before we began to study the data provided by the 2014 MSR Mining Challenge [1] we strongly believed that the majority of open source projects is being developed in cycles loosely corresponding to either classical prototype methodologies, or sprints used in Scrum. To verify this hypothesis we have studied topics of issue comments and commit messages separately, in both cases with and without the most significant topics among. We have also tried reducing the initial number of 50 topics and then eliminating the most significant one. However, much to our surprise, in most projects we were unable to detect cycles of topics (of either issues or commits), such as a regular increase and decrease of interest in particular topic in specific months. Only one project (51669) had a form of topics' cycle between 2012-07 and 2013-03. Numerous projects probably spanned over a too small period of time, or had too few contributions to detect cycles even if committers worked according to them.

We have revealed that in almost all projects there exist major topics (often a single topic) that are generally popular and prevalent during whole development process. And yet, even after excluding those (usually top 3) topics, the only topics remaining are those which have a very small number of commits (for example only one commit) or few issues gathered around a specific date or occasion. In case of commits those situations may be caused by merging work of those committers who behave like partisans. While investigating this hypothesis we came across two other questions. Are commits related to issues? Or at least do bug fixing topics in commits indicate some correlation to issues? Unfortunately the answers we have found are both "no".

Topics of issues are generally not similar to those of commits. Issue topics concentrate more on how something can be achieved and are more broad discussions, e.g. on the architectural context of a project or the usage of a selected library. Commit topics usually describe a specific situation such as not merged commit, a failed compilation or specific reasons why commit should be changed. Issue topics generally do not seem influencing commit topics to appear. It is true it least in the Github projects chosen for the Challenge.

5   Research Limitations

Our results are subject to a number of limitations threatening the accuracy or even the correctness of our assumptions. Our biggest concern is that we did not have access to the source code. Therefore, we were unable to verify if the specialists actually produce a majority of code, or they simply make lots of small corrections - possibly even insignificant. The same concerns generalists. Their work may additionally be triggered not by the desire to improve their projects but more by the reputation and contribution score. On the other hand "partisans" may contribute to a project not by Github itself but through other means like forums - where they may post helpful code, hints or suggestions.

Another problem that should be clearly stated here is that we analysed a very small number of projects. As mentioned earlier only 92 projects were provided, out of which only 43 were subjected to our method of investigation. And even those projects usually had a very small number of committers and comments making it difficult to generalize based on them. In general people tend to leave commit messages empty or they include a short, nearly meaningless sentence.Thus, without looking at the source code it is impossible to say anything about such commit. One of our concerns here is that we were unable to distinguish between commits that are meaningful to a project and commits that were cancelled or overwritten.

Our approach itself left a space for additional work. The fact, that we have eliminated topics related to praising and thanking others for their work might have negatively influenced our results. The same goes for parameters we used when training the LDA. For topics generated in some other way for example there may exist software development cycles.

6   Conclusions

In this paper we have shown that the majority of open source projects have a strong "Core Team", i.e. a group of developers that are either strongly involved in the development of a chosen topic, or they browse through the project bringing together committers and their contributions. The work of open source developers, no matter what is their role in a project, usually involves a notable amount bug fixing. Moreover, while investigating this problem we have found no correlation between bug fixing and Test Driven Development, or even simple testing phases. This leads to our third main conclusion that in most open source projects the work is continuous and cannot be clearly divided into stages or phases.

There are still numerous interesting hypotheses to be researched. As we have mentioned earlier, as we dug down the data, more and more questions arose. For example, is a project's popularity related somehow to the main technology? Or what are the trends behind the open source development? One of the hypotheses that we were unable to verify due to the small sample of projects is connected with the existence of the Core Team. Specifically, how big should be the leading team for a project to succeed? Or is a small core team a guarantee of a failure? This questions have been also formed within the research paper on Apache and Mozilla projects [5]. But we may also try to analyse means of interaction between the "Core Team" and other members, bearing in mind the research in [6]. How open are the core team members to other participants, and does it corelate to the amount of bug-reports or commits done by them. Going further on this subject we may want to ask what other project's features are directly linked to the project's success or failure. The authors of [7] have attempted to assess the main bug-types and quality of bug reports for selected Android Apps. This research could be also expanded with our findings and lead to more in-depth analysis of general trends in bug-reporting for open-source projects. In particular, the work on identifying key bug-fixing patches for Linux kernel [8] could be enhanced with our approach for more general topic classification.

Last but not least, we have to remember that more work should also address the limitations described in Section 5.

References

1. Gousios, G.: The GHTorrent dataset and tool suite. In: Proceedings of the 10th Working Conference on Mining Software Repositories, MSR 2013, pp. 233–236 (2013)
2. Hofmann, T.: Probabilistic latent semantic indexing. In: Proceedings of the 22nd Annual International ACM SIGIR Conference on Research and Development in Information Retrieval, pp. 50–57. ACM (1999)
3. Blei, D.M., Ng, A.Y., Jordan, M.I.: Latent Dirichlet allocation. The Journal of Machine Learning Research 3, 993–1022 (2003)
4. McCallum, A.K.: MALLET: A Machine Learning for Language Toolkit (2002), http://mallet.cs.umass.edu

5. Mockus, A., Fielding, R.T., Herbsleb, J.D.: Two case studies of open source software development: Apache and Mozilla. ACM Transactions on Software Engineering and Methodology (TOSEM) 11, 309–346 (2002)
6. Scialdone, M.J., Li, N., Heckman, R., Crowston, K.: Group maintenance behaviors of core and peripherial members of free/Libre open source software teams. In: Boldyreff, C., Crowston, K., Lundell, B., Wasserman, A.I. (eds.) OSS 2009. IFIP AICT, vol. 299, pp. 298–309. Springer, Heidelberg (2009)
7. Bhattacharya, P., Ulanova, L., Neamtiu, I., Koduru, S.C.: An empirical analysis of bug reports and bug fixing in open source android apps. In: Proceedings of the 2013 17th European Conference on Software Maintenance and Reengineering, CSMR 2013, pp. 133–143. IEEE Computer Society, Washington, DC (2013)
8. Tian, Y., Lawall, J., Lo, D.: Identifying linux bug fixing patches. In: Proceedings of the 34th International Conference on Software Engineering, ICSE 2012, pp. 386–396. IEEE Press, Piscataway (2012)

Materialized View Selection Considering the Diversity of Semantic Web Databases

Bery Mbaiossoum[1,2], Ladjel Bellatreche[1], and Stéphane Jean[1]

[1] LIAS/ENSMA, University of Poitiers 86960, Futuroscope Cedex, France
{mbaiossb,bellatreche,jean}@ensma.fr
[2] University of N'Djamena, Chad Republic

Abstract. With the extensive use of ontologies in various domains, Semantic Web Databases ($\mathcal{SWDBs}$) have appeared in the database landscape. Materialized views are one of the most popular optimization structures in advanced databases. Queries represent the most important input of the problem of selecting materialized views. In the context of $\mathcal{SWDB}$, queries are expressed using the SPARQL language. A SPARQL query consists of a set of triple patterns executed on a set of triples representing the logical level of the $\mathcal{SWDB}$. But a $\mathcal{SWDB}$ may have several deployments according to the used storage layout (vertical, horizontal, binary). As a consequence the process of selecting materialized views has to consider this diversity. In this paper, we first present the difficulty of the process of materializing views in the context of $\mathcal{SWDB}$ considering the diversity of storage layouts. Secondly, we define two approaches to select materialized views. The first approach hides the implementation aspects and views are selected at the ontological level using a rule-based approach. In the second approach, views are selected at the logical level and the view selection is guided by a cost model which considers the diverse storage layouts that can be used. Finally, intensive experiments are conducted by the means of the Lehigh University Benchmark and we empirically compare our finding with state-of-the-art algorithms.

1 Introduction

Materialized views ($\mathcal{MV}$) are one of the most popular optimization structures used in many fields such as data warehousing [1], data mining [2], XML databases [3], caching in mediator databases [4], cloud computing [5], etc. $\mathcal{MV}$s are used to pre-compute and store aggregated data, which is particularly the case of big data analytical workload. Once materialized views are selected, queries are rewritten using materialized views (*query rewriting*). Two major problems related to materialized views are: (a) the view selection problem and (b) the view maintenance problem.

View Selection Problem. The database administrator (DBA) can not materialize all possible views, as he/she is constrained by limited resources such as disk space, computation time, maintenance overhead and cost required for the query rewriting process [1]. Hence, the DBA needs to select an appropriate set of

Y. Manolopoulos et al. (Eds.): ADBIS 2014, LNCS 8716, pp. 163–176, 2014.

views to materialize under some resource constraints. Formally, the view selection problem $(\mathcal{VSP})$ is defined as follows. Given a set of most frequently used queries $Q = \{Q_1, Q_2, ..., Q_n\}$, where each query Q_i has an access frequency f_i $(1 \leq i \leq n)$ and a set of resource constraints M. The $\mathcal{VSP}$ consists in selecting a set of materialized views that minimizes one or more objectives, possibly subject to one or more constraints. Many variants of this problem have been studied considering several objective functions and resource constraints: (i) minimizing the query processing cost subject to storage size constraint [1], (ii) minimizing query cost and maintenance cost subject to storage space constraint [6], (iii) minimizing query cost under a maintenance constraint [1], etc. This problem is known to be NP-hard [1]. For more details, the interested reader can refer to the survey papers [6, 7].

View Maintenance Problem. Materialized views store data from base tables. In order to keep the views in the database up to date, it is necessary to maintain the materialized views in response to the modifications of the base tables. This process of updating views is called view maintenance which has been of great interest in the past years. Views can either be recomputed from scratch, or incrementally maintained by propagating the base data changes onto the views. As re-computing the views can be prohibitively expensive, the incremental maintenance of views is of significant value [1].

The above description shows the great interest that the database community gives to materialized views. In the last two decades, a new type of databases has emerged: the Semantic Web Databases (or ontology-based databases) $(\mathcal{SWDB})$. Contrary to traditional databases, a $\mathcal{SWDB}$ brings new dimensions [8]:

(a) *the diversity of ontology formalism*: each $\mathcal{SWDB}$ uses a particular formalism to define its ontologies (e.g., OWL [9], RDFS [10] or PLIB [11]),

(b) *the diversity of storage layouts*: in a $\mathcal{SWDB}$, several storage layouts (horizontal, vertical, binary) are used to store ontologies and their data,

(c) *the diversity of architectures*: three main architectures of database management system (DBMS) managing $\mathcal{SWDB}$ are distinguished. In the first type (that we called $Type_1$), the traditional database architecture was reused to store both the data and the ontologies referencing them. The ontology and its associated data are stored in a unique part. To separate the ontology from the data, a second type $(Type_2)$ was proposed, where the ontology and its ontological data are stored independently into two different schemes. Therefore, the management of ontology and data parts is different. The $Type_1$ and $Type_2$ architectures hardcoded the ontology model (OWL or RDFS, etc.). To enable the evolution of ontology models, a third architecture $(Type_3)$ extends the second one by adding a new part called the meta-schema.

(d) *the most popular query language* for $\mathcal{SWDB}$ is SPARQL. Its particularity is that it is defined at the logical level of the $\mathcal{SWDB}$ where ontologies and their instances are considered as a set of triples and thus, it hides the deployment structure. Therefore, it can be applied to any type of $\mathcal{SWDB}$.

The problem of selecting materialized views in the context of $\mathcal{SWDB}$ is formalized as follows [12]: given a SPARQL query workload Q and a storage constraint S, it consists in selecting a set of views that minimizes the SPARQL queries processing cost and the total size of materialized views must not exceed the imposed storage constraint S. Several recent studies have proposed algorithms for selecting views to optimize SPARQL queries [12–14]. These studies focused mainly on the vertical storage layout which consists in storing the triples in a three-column table *(subject, predicate, object)*.

To abstract the storage layout used by a $\mathcal{SWDB}$, the data can be seen either as a set of triples (logical level) or as instances of ontologies (conceptual level). As a consequence, we present in this paper two approaches to select materialized views in the context of the diversity of $\mathcal{SWDB}$ (Fig. 1): (i) the first approach, defined at the conceptual level, hides the implementation aspects of the triples storage. Consequently, materialized views are selected according to the usage of ontology classes by SPARQL queries. The second approach, defined at the logical level, is similar to the one used in relational data warehouses [15], where plans of individual queries are merged into one unified global query plan that captures the interaction between queries. The intermediate nodes of this plan are candidates to be materialized. The quality of the selected views is guided by a cost model estimating the number of inputs/outputs and considering the deployment of the $\mathcal{SWDB}$. These two approaches allow DBAs to select materialized views to optimize SPARQL queries without wondering about the storage layout of their databases. The interest of our approach is evaluated on the Lehigh University Benchmark with $\mathcal{SWDB}$s that use different storage layouts. Moreover, we show that our second approach is competitive with the the state-of-the-art approach for the vertical storage layout [12].

This paper is organized as follows. In Section 2, our first approach for materializing views at the conceptual level is described. Section 3 details our second

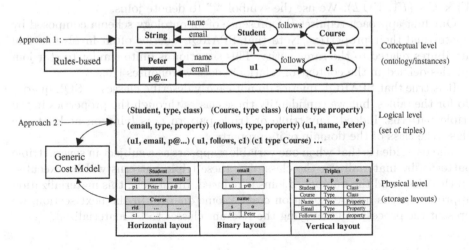

Fig. 1. Our Two Approaches

approach in which a genetic algorithm is presented to perform the selection at the logical level. Section 4 exposes our validations tests. Section 5 discusses the related work. Finally we conclude this work with a summary and outlook in section 6.

2   Materialized View Selection at Ontological Level

Before describing our first algorithm to materialize views at the conceptual level, some concepts and definitions are needed.

Definition 1. *An RDF triple is a 3-tuple (subject, predicate, object) $\in (U \cup B) \times U \times (U \cup B \cup L)$ where U is a set of URIs, B is a set of blank nodes and L is a set of literals.*

Three main storage layouts are used:

- *vertical*: a triple table (*subject, predicate, object*);
- *binary*: a two-columns table (*subject, object*) for each property;
- *horizontal*: a table per class $\mathcal{C}(p_1, \ldots, p_n)$ where $p_1,\ldots, p_n$ are the single-valued properties used at least by one instance of the class. Multi-valued properties are represented by a two-columns table (*subject, object*).

Definition 2. *A $\mathcal{SWDB}$ is said to be saturated if it contains initial instances and inferred instances, otherwise it is called unsaturated.*

For this work, we use saturated $\mathcal{SWDB}$, which means that we do not need to perform inference mechanism at querying processing time. An RDF triple pattern t is a triple (*subject, predicate, object*) $\in (U \cup V) \times (U \cup V) \times (U \cup V \cup L)$, where V is a set of variables disjoint from the sets U, B and L.

We consider RDF queries defined as a conjunction of triple patterns $t \in (U \cup V) \times U \times (U \cup V \cup L)$. We use the symbol "^" to denote joins.

Our first approach requires the presence of an ontology schema composed by classes and their properties. Note that SPARQL queries often involve several attributes related to the same entity. This situation is quite similar to star join queries defined in the context of relational data warehouses [16].

It is true that SPARQL queries do not clearly describe classes as SQL queries do for the tables, but we can identify these classes through the properties in the triple patterns. Indeed, any triple or triple pattern t_i is lying on at least one class: the class of the domain of its property.

The basic idea is that when one variable x appears as a subject in several triple patterns, by materializing the class of x, these triple patterns will be executed on this class, and the number of joins is reduced. Intuitively, this materialization approach speeds up the execution cost of some queries. In the next section, we present the process of identifying the relevant classes to be materialized.

Identification of View Candidates. We begin with the algebraic tree of each query. For each triple pattern, we consider the domain of its property (it is the class of its *subject*). It is obvious that if this class is materialized, the branch of the query tree built on this triple pattern will have this class as leaf. So, by replacing each triple pattern by the class of its property, and merging the same classes, a class-based query tree is obtained. Merging the classes is performed by computing the union of their properties. In other words, we group all triple patterns with the same domain. Let t_i and t_j be two triple patterns and $C(p)$ a class derived from a triple pattern t with p as property. If t_i and t_j have the same domain, the combination of t_i and t_j is translated by creating a class with properties pi and pj ($C(pi,pj)$). Figures 2 and 3 present an illustration of class-based query trees for the query $q = t1\hat{\ }t2$.

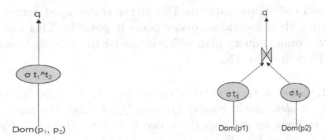

Fig. 2. t_1 and t_2 on a same domain   **Fig. 3.** t_1 and t_2 on different domains

Note that a property can have many classes as domain. We denote $Dom(p)$, the set of these classes for a property p. During the identification process, for each query, all triple patterns must be considered. For each variable, the selected classes are those that are common to the sets of classes obtained for each triple pattern. But to avoid having a large number of these classes, a predominance is granted to the "*rdf:type*" property. If in a given query, a variable x is involved in several triple patterns as subject, and if among these triple patterns properties, the "*rdf:type*" predicate is present, then the class of x is the one specified by "rdf:type". During the classes identification, we keep only the properties which are present in the query.

Usage Matrices. By using the ontology schema, query workload and identification of classes, two usage matrices are defined: *class usage matrix (MUC)* and *property usage matrix (MUP)*. MUC rows represent queries, whereas MUC columns represent classes. $MUC[i][j] = 1$, with $1 \leq i \leq n$ and $1 \leq j \leq m$, if the query q_i uses the class C_j, otherwise $MUC[i][j] = 0$. MUP is filled using the same principle, where properties replace classes.

These two matrices are exploited to select materialized views. If a class is used by a query, then it is considered as a view and the properties used by at least one query are associated to the view.

3   Materialized View Selection at the Logical Level

This approach is based on triple patterns of queries and exploits the interaction between queries that we describe in the next sections.

3.1   Generic Global Query Plan

A SPARQL query can be represented by an algebraic expression tree [17, 18]. The root corresponds to the query (i.e., the query result). Intermediate nodes correspond to algebraic operations such as selection, join or cross product. The leaves correspond to the entities used by a given query. For instance, in a $\mathcal{SWDB}$ which uses a vertical storage layout, leaves are the triples table. In binary and horizontal storage layout, leaves represent property tables and class tables, respectively. In our study and to hide the implementation aspects of a given $\mathcal{SWDB}$, the leaf nodes represent only triple patterns. Due to the star-shaped nature of SPARQL queries merging their individual query plans is possible. This merging process gives raise to a unified query plan which is similar to Multiple Views Processing Plan (MVPP) defined in [15].

Example 1. *Let $q_0 = t_1$, $q_1 = t_1 \hat{} t_2 \hat{} t_3$ and $q_2 = t_1 \hat{} t_2 \hat{} t_4$ be three queries. The branch $t_1 \hat{} t_2$ is identical in queries Q_1 and Q_2. Pooling the plans of these two queries can be done by merging these two branches. The unified plan of these queries is depicted in Fig. 4.*

As in [15], we consider each intermediate node as a potential view. We consider the unified plan as a search space for the views selection problem. The leaves of our graph which represent triple patterns result from the selection operation made on the underlying storage layout of the $\mathcal{SWDB}$. The selection nodes $(\sigma(.))$ which clearly express the meaning of triple patterns execution results are included in the unified plan (see Fig.5). For a triple pattern t, a selection node is defined as follows according to the used storage layout:

$$\sigma(t) = \begin{cases} select \ * \ from \ TT \ where \ condV \ \text{if vertical} \ \mathcal{SWDB} \\ select \ * \ from \ \ TabProp(t) \ where \ condB \ \ \text{if binary} \ \mathcal{SWDB} \\ select \ * \ from \ TabClass(t) \ where \ condH \ \text{if horizontal} \ \mathcal{SWDB} \end{cases} \qquad (1)$$

where TT, $TabProp(t)$ and $TabClass(t)$ represent respectively the triples table, the property table and the class table corresponding to the triple pattern t. $condV$, $condB$ and $condH$ represent respectively the selection predicates corresponding to the triple pattern t in vertical, binary and horizontal layout. In our work, we discard the projection operation, as a consequence it is not represented in our graph. The unified graph of the queries in example 1 is shown in Fig.5.

Due to the properties of algebraic operations [19], there exists a multitude of unified query plans. As a consequence, we propose the following procedure to build our plan. First we begin by optimizing each query individually. Several techniques are available for this task, we used the selectivity triple patterns

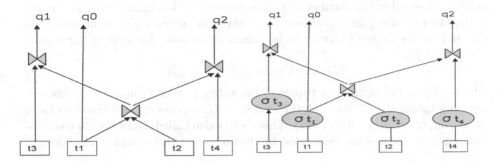

Fig. 4. MVPP of SPARQL queries   **Fig. 5.** Global Query Plan with selection nodes

[20, 21]. The triple patterns are ordered according to their selectivity increasingly. For the workload, queries are ordered according to their *cost*frequency* decreasingly.

The construction of the global plan involves the following steps:

- *Extraction of the Triple Patterns*: for each query of the workload, all triple patterns are extracted and kept in a set. Each triple pattern is stored only once. Variables can have different names but they can refer to the same triples, so we define a mapping that aligns variables on the whole workload.
- *Extraction of the Selection Nodes*: this is the first operation on the data. It represents the execution triple pattern on a $\mathcal{SWDB}$ (equation 1). The corresponding nodes are a set of triples corresponding to the triple patterns.
- *Extraction of the Join Nodes*: for each query, we identify the join nodes and extract them. We recall that a join is a combination of two triple patterns sharing at least one variable. These nodes are created on selection nodes of the triple patterns involved.
- *Definition of the Projection Nodes*: we consider these nodes as the results of queries.

Our approach for building a global plan follows the one proposed by [15]. Since our global plan depends on query order, then we propose to order the queries according to the *cost*frequency* criterion. The unified plan is created by a circular permutation until the first query of the list returns to the head of the list, and finally one global graph that has a minimum cost is chosen.

Once the optimal global plan is obtained, we proceed to the selection of views. This step takes into account the cost of construction and the cost of access of each node, the frequency of each query, the frequency of updating tables. These different costs are defined in the cost model below.

Cost Model. Each intermediate node is characterized by two costs: the cost of construction and the cost of access.

Construction Cost. It is the computational cost of the node. For a selection node, it is equal to the number of tuples of the selection table. For a join node, it depends on the join algorithm. Only the cost of hash join is presented in this paper. We denote $C\_cout(v)$, the construction cost of a node v. For a node $v = TP_1 \hat{\ } TP_2$.

$$C\_cost(v) = 3 * (||T_1|| + ||T_2||) \qquad (2)$$

where $||T||$ is the number of tuples of the table T; and T_1 and T_2 are the underlying tables of TP_1 and TP_2 (i.e., triples tables, property table or class table according to the type of $SWDB$). This cost is also called maintenance cost (assuming that updates are performed by dropping and (re)creating the nodes).

Access Cost. The access cost of a node is defined as the node size. For simplicity, we consider it as the number of tuples of the node. We denote $A\_cost(v)$ the access cost of a node v. Computing this cost depends on the type of node (selection node or join node).

Selection Node

$$A\_cost(v) = \begin{cases} sel(TP) * nbrTuple(TT) \text{ if vertical } SWDB \\ nbrTuple(TabProp(TP)) \text{ if binary } SWDB \\ sel(TP) * nbrTuple(TabClass(TP) \text{ if horizontal } SWDB \end{cases} \qquad (3)$$

sel(TP) is the selectivity of the triple pattern TP when the vertical storage layout is defined as in [20]. In the case of horizontal storage layout, the selectivity of the predicate is computed by translating triple pattern TP into the SQL language.

  Join Node

$$A\_cout(v) = jsel * nbrtuple(TP_1) * nbrTuple(TP_2) \qquad (4)$$

where *nbrTuple* (TP_1) and *nbrTuple* (TP_2) are the number of tuples in each relation corresponding to the results of processing triple patterns TP_1 and TP_2 and *jsel* is their join selectivity. Stocker et al. [20] have defined the selectivity of two conjunctive triple patterns TP_1 and TP_2 as the number of triples resulting from the join of TP_1 and TP_2 normalized by the total number of triples square. To take into account the diversity of storage layouts, we have adapted the classical method of relational databases [19]. Let's suppose that R_1 and R_2 are relations resulting from processing triple patterns TP_1 and TP_2 separately.

  Let $card(R_1 \bowtie R_2)$ be the size of join node of R_1 and R_2.

$$card(R_1 \bowtie R_2) = \frac{nbrTuple(TP_1) * nbrTuple(TP_2)}{max(V(R_1, ?x), V(R_2, ?x))} \qquad (5)$$

where nbrTuple (TP_1) and nbrTuple (TP_2) are numbers of tuples of R_1 and R_2 respectively, $?x$ is a variable shared by TP_1 and TP_2, and $V(R_i, ?x)$ is the size of the domain of the attribute $?x$ in the relation R_i. In other words, *nbrTuple(TP_i)* is the number of triples which are solutions of the triple pattern TP_i and $V(R_i, ?x)$ is the number of distinct values that the variable $?x$ can take in the solutions of TP_i. For more details on these sizes computing, the interested reader can refer to [21].

Query Cost. The query cost is based on materialized views. If the views are used, the query cost is the construction cost of the query root node. Else it is equal to the sum of access costs of the used views. We denote it $cost(q, M)$ for a query q, where M is a materialized views set.

Total Cost of Workload. The total cost of a query workload is the sum of the cost of each query multiplied by the frequency of the query.

$$cost\_total = \sum_{q \in Q} freq(q) * cost(q, M) \tag{6}$$

where $freq(q)$ is the frequency of the query q.

Materialized Views Selection. Recall that each node is a potential view. Therefore, we can define our selection algorithm that minimizes the total cost of the query workload and meets the storage constraint.

If the storage constraint does not exist, the algorithm for views selection used in [15] can be used to have the set of optimal views. Since we are constrained by the storage capacity, we propose the use of genetic algorithm that has shown its efficiency in the context of relational data warehouses [22].

Genetic Algorithm. Our genetic algorithm is based on the API JeneGA [23]. We consider for each view, the space it occupies and the profit obtained if it is the only view that is materialized. The occupied space is equal to the access cost as mentioned in section 3.1. The profit of a view v is defined as the difference between the total cost of the workload without views and the total cost of the workload if the view v is only materialized.

$$profit(v) = \sum_{q \in Q} freq(q) * cost(q, \emptyset) - \sum_{q \in Q} freq(q) * cost(q, M) \tag{7}$$

where *freq(q)* is the frequency of the query q and $M = \{v\}$ is a materialized view set.

The problem is to find a set of views whose the sum of the occupied space is less than or equal to the storage constraint S and which maximizes profits.

We define our chromosome as an array of bits (0 or 1). All intermediate nodes of the global plan are mapped into the bit array. If a bit at a position is 1, it means that the mapped node is selected to be materialized. A random solution is initially generated and improved to be a good one. Our fitness function is based on the maximum profit of chromosome. Indeed, for each chromosome (solution), the sum of its nodes benefit and the sum of the occupied space are calculated. If the sum of the occupied space is less than the storage constraint, the chromosome is declared valid and can be improved to be a solution else it is not interesting. We used the following parameters because they are often used in view selection: crossover probability: 0.8, mutation probability: 0.02, population size 1000 and maximum generation number 200. This algorithm provides us some interesting results seeing the experimental results presented in the next section.

4   Experimentations

4.1   Tests Datasets

We have performed several experiments to test our approaches on a 3.10 GHZ Intel Xeon DELL personal computer with 4GB of RAM and 500GB of hard disk. We used the benchmark of Lehigh University named LUBM [24] that creates instances of an university domain ontology. We generated ontology-based data for 100 universities denoted *Lubm100* with 12.674.100 triples. We used the 14 queries provided by the benchmark. For each query, we made four measures of processing time and we took the average. We did this for the initial queries without views and for the queries with views. We used PostgreSQL 8.2 as the underlying DBMS for all used *SWDB*s.

4.2   Different Tests

We call *TBA* (Triple-Based Approach), our approach defined at the logical level which is based on triple patterns, and *CBA* the Class-Based Approach defined at the conceptual (ontological) level. Our experiments are performed on existing *SWDB*s (Jena and Sesame) and on a *SWDB* created especially for comparing our approach with the one of [12] that we call native *SWDB*.

Experimentations on Existing *SWDB*s. In order to evaluate the generic aspect of our approaches, we have used the Jena *SWDB* which uses a vertical storage layout and Sesame *SWDB* (binary layout). We have also made experiments on OntoDB which uses an horizontal storage layout. But for conciseness, we only report experiments made on Jena and Sesame. The considered workload of queries is the set of the 14 LUBM benchmark queries. Our approaches were used to choose a set of views to optimize this workload. Then we measured the query processing time of the queries with the created views and without them (denoted *Sparql*).

Results are presented in Fig.6 for Jena *SWDB* and Fig.7 for Sesame binary *SWDB*.

Interpretation of the Results. *TBA* provides good results for all queries using views. Queries that can not be rewritten with appropriate views are executed on the native storage layout. Otherwise, their results would be worse than if they were performed with a SPARQL engine. Indeed, the SPARQL engine uses some optimization techniques and is faster than a flat SQL engine. Castillo and Leser [13] used SPARQL-SQL rewriting and made the same remark. *CBA* comes after *TBA*. Some of its results are worse than the one of the SPARQL engine mainly when the views are very large. This is the case of queries 1 and 3 which target the *GraduateStudent* and *Publication* classes. However, the *CBA* approach gives an acceptable cumulative time despite the fact that for individual queries, the results are mitigated.

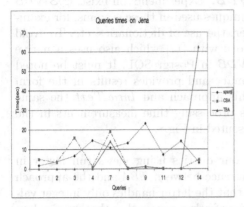

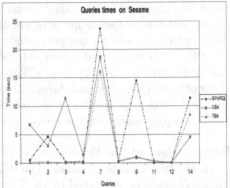

Fig. 6. Queries times on vertical $\mathcal{SWDB}$

Fig. 7. Queries times on binary $\mathcal{SWDB}$

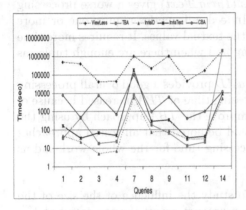

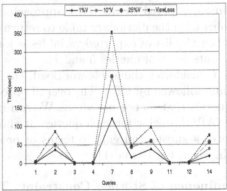

Fig. 8. Queries times on Native $\mathcal{SWDB}$

Fig. 9. Influence of storage constraint

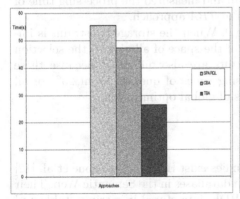

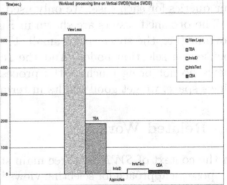

Fig. 10. Query workload times on binary $\mathcal{SWDB}$

Fig. 11. Query workload times on vertical $\mathcal{SWDB}$

Experimentations on a Native $\mathcal{SWDB}$. Experiments on existing $\mathcal{SWDB}$ could be influenced by optimization techniques used in these systems, for example the existence of some index or cluster, the use of dictionaries, etc. To avoid these influences and to compare our work with [12] which also uses a native $\mathcal{SWDB}$, we have created a vertical $\mathcal{SWDB}$ in PostgreSQL. It must be noted that the approach in [12] uses a dictionary and provides results in the form of identifiers (ID). We call *Inria_ID* this approach and *Inria_Text* the same approach with text results. The queries processing time measurements in the different approaches give the results presented in Fig.8.

Interpretation of the Results. Most of our queries using views run faster in *Inria_ID* and *TBA* approaches. The difference between *TBA* and the approach *Inria_ID* could be explained by the fact that the latter handles only integer values, so its operations are faster than those of other approaches that manipulate strings.

INRIA[1] approach with textual results (*Inria_Text*) gives a worse processing time compared to the *TBA* approach. Indeed, all queries require one or more joins with the dictionary table to return the textual values. It requires more time when the list of exported variables is long and when there are enough tuples as results (case of queries 5 and 7).

Considering the whole workload, *Inria_ID* provides a good overall processing time followed by *CBA*. The overall processing time of *TBA* increased because of queries that do not use views. We can improve the *TBA* approach by using the SPARQL engine for such queries. It should be noted that all tested approaches have a substantial gain in terms of processing time for the whole workload as shown in Fig.10 and 11.

Influence of Storage Constraint. To study the influence of the size of the storage space for MV on the view selection process, we set the constraint at 1%, 10% and 25% of the size of all views. Then we make the selection of views and created them for each case. Finally, we ran and measured the processing time of our queries for each case. We only used the *TBA* approach.

The obtained results are shown in Fig.9. When the storage constraint is important (i.e., the space size is small, 1% of the space of all views), the selection focuses on selection nodes. And the results are also mitigated because these views do not bring much to the processing times of queries. From 25% of all views space, we get good results in terms of global optimization.

5   Related Work

In the context of $\mathcal{SWDB}$s, three main studies exist [12–14]. Goasdoué et al. [12] propose an approach for selecting views in databases in the Semantic Web. Their approach uses a vertical $\mathcal{SWDB}$, where RDF data stored in a triples table and based on the concept of state inspired by [25]. A state is a pair (V, R) where V

---

[1] Institut National de Recherche en Informatique et en Automatique.

is a set of views and a set R of rewriting queries on these views. The objective is to find a state that minimizes the cost of executing queries, the storage space and the maintenance cost views.

In [13], the authors propose an approach for selecting views on RDF data for a given workload of SPARQL queries. The RDF data used are stored in a triples table. The idea is to discover shared patterns to be used as indexes to improve queries processing times. They consider the given set of queries as the initial space of views, extend this space by analyzing each query of the workload and identifying all combinations of connected triples patterns of some length and add them to the space of view candidates.

Dritsou et al. [14] proposed a materialization of paths frequently accessed (called shortcut) as a solution to reduce the RDF queries processing cost. The basic idea is that given the prevalence of the path expression in RDF data, they can be materialized to reduce the time of queries processing. A shortcut is a query fragment with at least two triples patterns. Having defined the set of candidate nodes, they develop the set of candidate shortcuts by considering all valid combinations between candidate shortcut nodes.

Unlike these approaches which were defined for vertical $SWDB$, our approaches have been defined and evaluated for different storage layouts used by $SWDBs$

6   Conclusion

In this paper, we identify a fundamental aspect related to the process of selecting materialized views in the context of $SWDBs$ which is the diversity of storage layouts. By exploring the main important state-of-art approaches, we figured out the existence of algorithms and approaches for selecting materialized views considering a unique storage layout: the triple table. To cover different types of $SWDB$, we have proposed in this paper two approaches: one hides the implementation aspects of $SWDBs$ in which materialized views are selected at the conceptual level by exploiting the usage of classes by the SPARQL queries. The second approach proposes a selection of materialized views at the logical level and takes into account the storage layout of the target database. This flexibility is offered thanks to a cost model defined to quantify the quality of the final solution. Intensive experiments were conducted to show the interest of our approach using $SWDBs$ that use different storage layouts. Moreover, our experiments on the vertical storage layout show that our second approach is competitive with the main approach defined for this type of $SWDB$. Currently, we are conducting other experiments by considering other benchmarks involving interactive queries.

References

1. Gupta, H.: Selection and maintenance of views in a data warehouse. PhD thesis (1999)
2. Morzy, T., Wojciechowski, M., Zakrzewicz, M.: Materialized data mining views. In: Zighed, D.A., Komorowski, J., Żytkow, J.M. (eds.) PKDD 2000. LNCS (LNAI), vol. 1910, pp. 65–74. Springer, Heidelberg (2000)
3. Arion, A., Benzaken, V., Manolescu, I., Papakonstantinou, Y.: Structured materialized views for xml queries. In: VLDB, pp. 87–98 (2007)

4. Adali, S., Candan, K.S., Papakonstantinou, Y., Subrahmanian, V.S.: Query caching and optimization in distributed mediator systems. ACM SIGMOD, 137–148 (1996)
5. Upadhyaya, P., Balazinska, M., Suciu, D.: How to price shared optimizations in the cloud. VLDB 5(6), 562–573 (2012)
6. Mami, I., Bellahsene, Z.: A survey of view selection methods. SIGMOD Record 41(1), 20–29 (2012)
7. Dhote, C., Ali, M.: Materialized view selection in data warehousing: A survey. Journal of Applied Sciences 9(1), 401–414 (2009)
8. Mbaiossoum, B., Bellatreche, L., Jean, S.: Towards performance evaluation of semantic databases management systems. In: Gottlob, G., Grasso, G., Olteanu, D., Schallhart, C. (eds.) BNCOD 2013. LNCS, vol. 7968, pp. 107–120. Springer, Heidelberg (2013)
9. Bechhofer, S., van Harmelen, F., Hendler, J., Horrocks, I., McGuinness, D., Patel-Schneider, P., Stein, L.: Owl web ontology language reference. W3C (2004), http://www.w3.org/TR/owl-ref/
10. Brickley, D., Guha, R.: Rdf vocabulary description language 1.0: Rdf schema. W3C (2002), http://www.w3.org/TR/rdf-schema/
11. Pierra, G.: Context representation in domain ontologies and its use for semantic integration of data. Journal of Data Semantics (JoDS) 10, 174–211 (2008)
12. Goasdoué, F., Karanasos, K., Leblay, J., Manolescu, I.: View selection in semantic web databases. VLDB 5(2), 97–108 (2011)
13. Castillo, R., Leser, U.: Selecting materialized views for RDF data. In: Daniel, F., Facca, F.M. (eds.) ICWE 2010. LNCS, vol. 6385, pp. 126–137. Springer, Heidelberg (2010)
14. Dritsou, V., Constantopoulos, P., Deligiannakis, A., Kotidis, Y.: Optimizing query shortcuts in RDF databases. In: Antoniou, G., Grobelnik, M., Simperl, E., Parsia, B., Plexousakis, D., De Leenheer, P., Pan, J. (eds.) ESWC 2011, Part II. LNCS, vol. 6644, pp. 77–92. Springer, Heidelberg (2011)
15. Yang, J., Karlapalem, K., Li, Q.: Algorithms for materialized view design in data warehousing environment. In: VLDB, pp. 136–145 (1997)
16. Arias, M., Fernández, J.D., Martínez-Prieto, M.A., de la Fuente, P.: An empirical study of real-world sparql queries. CoRR abs/1103.5043 (2011)
17. Frasincar, F., Houben, G.J., Vdovjak, R., Barna, P.: Ral: An algebra for querying RDF. World Wide Web 7(1), 83–109 (2004)
18. Cyganiak, R.: A relational algebra for SPARQL. Technical report, Digital Media Systems Laboratory, HP Laboratories Bristol (2005)
19. Garcia-Molina, H., Ullman, J.D., Widom, J.: Database Systems: The Complete Book, 2nd edn. Prentice Hall Press, Upper Saddle River (2008)
20. Stocker, M., Seaborne, A., Bernstein, A., Kiefer, C., Reynolds, D.: Sparql basic graph pattern optimization using selectivity estimation. In: WWW, pp. 595–604 (2008)
21. Kaoudi, Z., Kyzirakos, K., Koubarakis, M.: SPARQL query optimization on top of DHTs. In: Patel-Schneider, P.F., Pan, Y., Hitzler, P., Mika, P., Zhang, L., Pan, J.Z., Horrocks, I., Glimm, B. (eds.) ISWC 2010, Part I. LNCS, vol. 6496, pp. 418–435. Springer, Heidelberg (2010)
22. Hylock, R., Currim, F.: A maintenance centric approach to the view selection problem. Information Systems 38(7), 971–987 (2013)
23. Troiano, L., Pasquale, D.D.: A java library for genetic algorithms addressing memory and time issues. In: NaBIC, pp. 642–647 (2009)
24. Guo, Y., Pan, Z., Heflin, J.: Lubm: A benchmark for owl knowledge base systems. Web Semantics: Science, Services and Agents on the World Wide Web 3(2-3) (2011)
25. Theodoratos, D., Sellis, T.: Designing data warehouses (1999)

A Robust Skip-Till-Next-Match Selection Strategy for Event Pattern Matching

Bruno Cadonna[1], Johann Gamper[2], and Michael H. Böhlen[3]

[1] Humboldt-Universität zu Berlin, Germany
cadonna@informatik.hu-berlin.de
[2] Free University of Bozen-Bolzano, Italy
gamper@inf.unibz.it
[3] University of Zurich, Switzerland
boehlen@ifi.uzh.ch

Abstract. In event pattern matching, various selection strategies have been proposed to impose additional constraints on the events that participate in a match. The skip-till-next-match selection strategy is used in scenarios where some incoming events are noise and therefore should be ignored. Skip-till-next-match is prone to blocking noise, i.e., noise that prevents the detection of matches. In this paper, we propose the robust skip-till-next-match selection strategy, which is robust against noise and finds matches that are missed by skip-till-next-match when blocking noise occurs in the input stream. To implement the new strategy in automaton-based pattern matching algorithms, we propose a backtracking mechanism. Extensive experiments using real-world data and different event pattern matching algorithms show that with skip-till-next-match the number of matches not detected due to blocking noise can be substantial, and that our backtracking mechanism outperforms alternative solutions that first produce a superset of the result followed by a post processing step to filter out non-compliant matches.

1 Introduction

Event pattern matching finds matches of a pattern in an incoming stream of events, where a pattern specifies constraints on extent, order, values, and quantification of matching events. The result is a set of matches. Besides the pattern, a *selection strategy* [2] (or selection policy [4]) imposes additional constraints on the events that participate in a match. Various selection strategies have been proposed to adapt to application-specific needs. The strategies range from finding matches whose events need to be strictly contiguous in the stream [2, 6, 10, 12] to finding all possible matches by not imposing any additional constraints [2, 7–9].

In many applications for event pattern matching some incoming events are noise to a match, i.e., they occur between events that participate in a match but are not themselves part of the match. In such scenarios, the *skip-till-next-match (STNM)* [2] selection strategy has been used [2, 3, 5, 11]. STNM ignores all events that do not match with the pattern until the next matching event is read. As the following example reveals, STNM distinguishes two types of noise: noise that is ignored and does not affect the detection of matches, and noise that is ignored but blocks the detection of some matches.

Y. Manolopoulos et al. (Eds.): ADBIS 2014, LNCS 8716, pp. 177–191, 2014.
© Springer International Publishing Switzerland 2014

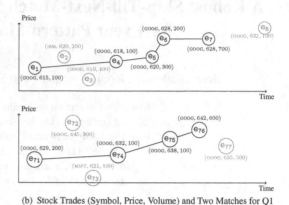

Stocks			
	S	P	V	T

	S	P	V	T
e_1	GOOG	615	100	9:32:344
e_2	IBM	620	200	9:32:357
e_3	GOOG	610	400	9:32:368
e_4	GOOG	618	100	9:32:380
e_5	GOOG	620	300	9:32:396
e_6	GOOG	628	200	9:32:401
e_7	GOOG	628	700	9:32:421
e_8	GOOG	632	100	9:32:450
⋮	⋮	⋮	⋮	⋮
e_{71}	GOOG	629	200	14:15:555
e_{72}	GOOG	645	300	14:15:572
e_{73}	MSFT	621	100	14:15:581
e_{74}	GOOG	632	100	14:15:592
e_{75}	GOOG	638	100	14:15:613
e_{76}	GOOG	642	600	14:15:628
e_{77}	GOOG	635	500	14:15:640

(a) Event Stream Stocks

(b) Stock Trades (Symbol, Price, Volume) and Two Matches for Q1

Fig. 1. Stock Trade Events

Consider the event stream Stocks in Fig. 1(a) that records stock trades (stock symbol (S), price per stock (P), volume of the trade (V), and occurrence time (T)). For instance, event e_1 represents the trade of 100 Google stocks at a price of \$615 at time 9:32:344. To analyze Google stock trades, consider the following pattern query Q1: *Within a period of 100 ms, find three or more Google (GOOG) stock trades with strictly monotonic increasing prices, followed by one Google stock trade with volume larger than each of the price-increasing trades.* Figure 1(b) shows two matches for query Q1 plotted over time. Events are represented as circles and labeled with stock symbol, stock price and trade volume; noise events are marked in gray. The first match consists of the events e_1, e_4, e_5, e_6, and e_7. Event e_1 starts the match, e_4, e_5, e_6 are subsequent events with strictly increasing prices (\$615 < \$618 < \$620 < \$628), and e_7 is the final event that has a larger trade volume than e_1, e_4, e_5, e_6 (100 < 700, 100 < 700, 300 < 700, 200 < 700). The events e_2, e_3, and e_8 are not part of the match since e_2 is an IBM stock trade, e_3 has a lower price than e_1, and e_8 exceeds the 100 ms period. The second match consists of e_{71}, e_{74}, e_{75}, and e_{76}. Events e_{72} and e_{73} are noise and therefore ignored: though e_{72} has a higher price than e_{71}, there are no other Google events that continue this trend in order to complete the match; and e_{73} is a Microsoft stock trade.

STNM finds the first match, but misses the second since only the current partial match is considered to decide whether an event is matched or ignored as noise. After matching event e_{71}, STNM reads and matches e_{72}, since e_{72} is the next Google stock trade with a price greater than e_{71}. The Microsoft stock trade e_{73} is ignored as noise as well as the subsequent Google stock trades, e_{74}, e_{75}, and e_{76}, since none of them continues the positive price trend of e_{71} and e_{72}. If e_{72} were ignored, e_{74}, e_{75}, and e_{76} would continue the positive trend, and the second match would be detected. STNM implicitly distinguishes between two types of noise. Noise events e_2, e_3 and e_{73} are ignored, and we would get exactly the same match if these events were not present in the stream. In contrast, noise event e_{72} blocks the detection of the second match. If e_{72} were not present in the stream, the second match would be found. We call e_{72} *blocking noise*. The result set detected with STNM depends on the presence of blocking noise.

In this paper, we propose a selection strategy, called *robust skip-till-next-match (RSTNM)*, that ignores noise events and does not allow them to block the detection of matches, i.e., RSTNM is robust against any noise. With RSTNM, the result set does not depend on noise. RSTNM considers all constraints in the pattern together with a complete match to determine whether events participate in a match or are noise. It finds all matches that are found with STNM plus those matches that are not found by STNM due to blocking noise. The matches found with RSTNM are a subset of all possible matches of the pattern in the stream, namely the matches that contain the events occurring earliest after the start of the match. We argue that RSTNM is safer to use than STNM because the presence of blocking noise in the stream cannot be foreseen. To compute pattern queries using RSTNM, we propose an efficient backtracking mechanism that can be integrated into automaton-based pattern matching algorithms. The key idea of our solution is to identify blocking noise events and to avoid the blockage. To summarize, the technical contributions are as follows:

– We formally define STNM and RSTNM and compare them.
– We present a backtracking mechanism that can be integrated into automaton-based event pattern matching algorithms to support RSTNM.
– We experimentally show that the difference in the amount of matches between STNM and RSTNM in real-world data can be substantial and that our backtracking solution clearly outperforms two alternative solutions.

The rest of the paper is organized as follows. Section 2 discusses related work. In Section 3, we define event pattern matching. In Section 4, we formally define RSTNM and STNM. Section 5 presents the backtracking mechanism for automaton-based event pattern matching algorithms. In Section 6 we report experimental results. Section 7 concludes the paper and points to future work.

2   Related Work

SASE+ [2] is an automaton-based event pattern matching algorithm. It introduces various selection strategies including STNM. However, the selection strategies are defined through the automaton-based evaluation model rather than in a formal declarative way. Other event pattern matching approaches that find matches using STNM are sequenced event set (SES) pattern matching [3], Cayuga [5], and the NEXT CEP system [11]. All three approaches are based on automata.

T-REX [4] is a middleware for event pattern matching. It provides the selection strategy `first-within` that finds the earliest matching event in a stream and can be individually applied to each event specified in the pattern. If `first-within` is applied to all events specified in a pattern, T-REX, similar to RSTNM, treats all noise events equally and ignores them. The presence of noise in the stream does not affect the detection of a match. Though the set of matches found with T-REX overlaps with the set of matches that comply with RSTNM, they are generally not equal. The reason for the inequality is that T-REX, unlike other event pattern matching algorithms, specifies matches starting from the latest event in the match, whereas RSTNM is defined for patterns that specify matches starting from the earliest event in the match. The evaluation of T-REX, which is based on automata, first computes a superset of the matches that conform to `first-within` and then discards the non-compliant matches.

The Amit [1] middleware detects matches of patterns in a stream by evaluating a combination of specialized operators. Similar to T-REX, it defines a selection strategy `first` that can be individually applied to each event specified in the pattern. If `first` is applied to all events in the sequence operator, Amit finds the matches that comply with RSTNM. However, Amit does not support trends over unbounded numbers of events, such as the price trend in our example. Such trends represent a particular challenge for event pattern matching with RSTNM as shown in Sect. 5.2.

ZStream [9], NEEL [8], and Event Analyzer [7] find all possible matches of a pattern in a stream. ZStream is a cost-based query processor using join trees. NEEL is a pattern matching engine for nested patterns with an automaton-based evaluation. Event Analyzer is a data warehouse component to analyze event sequences.

DejaVu [6] and SQL-TS [10] find only contiguous matches, i.e., no noise events are allowed between events in a match. DejaVu is based on automata. SQL-TS adopts a solution based on the Knuth-Morris-Pratt string matching algorithm.

3   Background

In this section, we give a definition of event pattern matching that incorporates common properties of existing event pattern matching algorithms.

An *event* is represented as a tuple with schema $\mathbf{E} = (A_1, \ldots, A_k, T)$, where T is a temporal attribute that stores the occurrence time of an event. For T we assume a totally ordered discrete time domain. An *event stream*, E, is a set of events with a total order given by attribute T. A *chronologically ordered sequence of events* is represented as $\vec{e} = \langle e_1, \ldots, e_n \rangle$, where e_1 and e_n specify the first and the last event in $\vec{e}$, respectively.

Definition 1. *(Pattern) A pattern, P, is a triple $P = (\langle B_1, \ldots, B_m \rangle, \Theta, \tau)$, where $\langle B_1, \ldots, B_m \rangle$ is a sequence of pairwise disjoint sets of variables v or v^+, $\Theta = \{\theta_1, \ldots, \theta_k\}$ is a set of constraints over variables in $B_1, \ldots, B_m$, and τ is a duration.*

A variable, v, binds a sequence containing a single event, $\langle e_1 \rangle$. A quantified variable, v^+, binds a sequence of one or more events, $\langle e_1, \ldots, e_n \rangle$. Whenever it is clear from the context, we use v to refer to both v and v^+. Set Θ contains constraints over variables that must be satisfied by matching events. Constraints have the form $v.A \phi C$, $v_i.A_k \phi v_j.A_l$, or $prev(v.A) \phi v.A$, where $v.A$ refers to an attribute of a matching event, C is a constant, and $\phi \in \{=, \neq, <, \leq, >, \geq, \}$ is a comparison operator. Finally, τ is the maximal time span within which all matching events must occur.

Query Q1 is formulated as pattern $P = (\langle \{s_1\}, \{s_2\}, \{s_3^+\}, \{s_4\} \rangle, \Theta, 100 \text{ ms})$. The sets, $B_1 = \{s_1\}$, $B_2 = \{s_2\}$, $B_4 = \{s_4\}$, contain one variable each, whereas $B_3 = \{s_3^+\}$ contains one quantified variable. The constraints over the variables are $\Theta = \{s_1.S=\text{'GOOG'}, s_2.S=\text{'GOOG'}, s_3.S=\text{'GOOG'}, s_4.S=\text{'GOOG'}, s_1.P<s_2.P, s_2.P<s_3.P, prev(s_3.P)<s_3.P, s_1.V<s_4.V, s_2.V<s_4.V, s_3.V<s_4.V\}$. Each of the variables s_1, s_2, and s_4 binds a single Google stock trade, and s_3 binds one or more Google stock trades. The constraints over attribute P force the price to increase, and the constraints over V specify that the trade volume of an event that matches s_4 is larger than the trade volume of all other matched events. The maximal time span is 100 ms.

To define the matching of a pattern P over an event stream E, we use a *substitution* $\gamma = \{v_1/\vec{e}_1, \ldots, v_n/\vec{e}_n\}$. Each pair $v_i/\vec{e}_i$ represents a *binding* of a variable v_i to a sequence $\vec{e}_i = \langle e_{i_1}, \ldots, e_{i_k} \rangle$, $|\vec{e}_i| > 0$, of events in E. A substitution contains exactly one binding for each variable in P, and an event can appear in at most one of the bindings. Substitution γ satisfies a constraint of the form $v_i.A\phi C$, if each event $e \in \vec{e}_i$ from binding $v_i/\vec{e}_i$ satisfies $e.A \phi C$. Similarly, γ satisfies a constraint of the form $v_i.A_i \phi v_j.A_j$, if each pair of events, (e_i, e_j), with $e_i \in \vec{e}_i$ and $e_j \in \vec{e}_j$ from $v_i/\vec{e}_i$ and $v_j/\vec{e}_j$ satisfies $e_i.A_i \phi e_j.A_j$. Finally, γ satisfies a constraint of the form $prev(v_i.A) \phi v_i.A$, if each pair of consecutive events, (e_i, e_{i+1}), with $e_i \in \vec{e}_i$ and $e_{i+1} \in \vec{e}_i$ from $v_i/\vec{e}_i$ satisfies $e_i.A \phi e_{i+1}.A$ or $|\vec{e}_i| < 2$.

Definition 2. (Match) *Let $P = (\langle B_1, \ldots, B_m \rangle, \Theta, \tau)$ be a pattern and E be an event stream. A substitution $\gamma = \{v_1/\vec{e}_1, \ldots, v_k/\vec{e}_k\}$ is a* match *of P in E iff*

$$\forall \theta \in \Theta \; (\gamma \text{ satisfies } \theta), \tag{1}$$

$$\forall v_i/\vec{e}_i, v_j/\vec{e}_j \in \gamma \; (v_i \in B_i \wedge v_j \in B_{i+1} \rightarrow e_{i_n}.T < e_{j_1}.T), \tag{2}$$

$$\forall v_i/\vec{e}_i, v_j/\vec{e}_j \in \gamma \; (|e_{i_1}.T - e_{j_n}.T| \leq \tau). \tag{3}$$

Condition 1 requires a match to satisfy all constraints in Θ. Condition 2 ensures that all events in a match that are bound to a variable in B_i must occur before all events that are bound to any variable in B_{i+1}. No order is imposed on events that are bound to variables in the same set. Condition 3 constrains all events in a match to occur within duration τ.

Figure 2(a) illustrates a match of pattern P in stream Stocks. The variables s_1, s_2, and s_4 bind a single event each, whereas s_3 binds two events. All conditions of Def. 2 are satisfied. Condition 1: events e_1, e_4, e_5, e_6, and e_7 are Google stock trades, e.g., $e_1.S = $ 'GOOG'; the price of e_1 is less than the price of e_4, i.e., $e_1.P < e_4.P$; the price of e_4 is less than each of the prices of e_5 and e_6; the price of e_5 is less than the price of e_6, i.e., $prev(e_5.P) < e_6.P$; the volumes of e_1, e_4, e_5, and e_6 are less than the volume of e_7, e.g., $e_1.V < e_7.V$. Condition 2: event e_1 that matches s_1 occurs before e_4 that matches s_2, e_4 occurs before e_5 and e_6 that match s_3^+, and e_5 and e_6 occur before e_7 that matches s_4. Condition 3: the time span between e_1 (first event) and e_7 (last event) is less than 100 ms. The complete list of all possible matches for P is shown in Fig. 2(b).

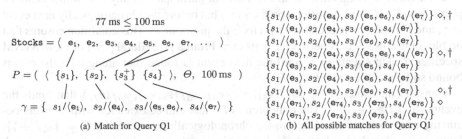

(a) Match for Query Q1

(b) All possible matches for Query Q1

Fig. 2. Examples of Matches for Query Q1

4   Robust Skip-Till-Next-Match

In this section, we introduce RSTNM and compare it to STNM. To facilitate the discussion, we first introduce a few auxiliary concepts.

Consider a pattern $P = (\langle B_1, \ldots, B_m \rangle, \Theta, \tau)$. A *prefix of a pattern* P is a pattern $\widehat{P} = (\langle \widehat{B}_1, \ldots, \widehat{B}_k \rangle, \widehat{\Theta}, \tau)$, $k \leq m$, where $\widehat{B}_j = B_j$ for all $j < k$, $\widehat{B}_k \subseteq B_k$ with $\widehat{B}_k \neq \emptyset$, and $\widehat{\Theta}$ is the set of constraints in Θ that only involve variables from $\widehat{B}_1 \cup \cdots \cup \widehat{B}_k$. For example, pattern $P = (\langle \{s_1\}, \{s_2\}, \{s_3^+\}, \{s_4\} \rangle, \Theta, 100\,\text{ms})$ has four prefixes: $(\langle \{s_1\} \rangle, \widehat{\Theta}_1, 100\,\text{ms})$, $(\langle \{s_1\}, \{s_2\} \rangle, \widehat{\Theta}_2, 100\,\text{ms})$, $(\langle \{s_1\}, \{s_2\}, \{s_3^+\} \rangle, \widehat{\Theta}_3, 100\,\text{ms})$, $(\langle \{s_1\}, \{s_2\}, \{s_3^+\}, \{s_4\} \rangle, \Theta, 100\,\text{ms})$, where $\widehat{\Theta}_1 = \{s_1.S=\text{'GOOG'}\}$, $\widehat{\Theta}_2 = \widehat{\Theta}_1 \cup \{s_2.S=\text{'GOOG'}, s_1.P<s_2.P\}$, $\widehat{\Theta}_3 = \widehat{\Theta}_2 \cup \{s_3.S=\text{'GOOG'}, s_2.P<s_3.P, prev(s_3.P)<s_3.P\}$.

Next, consider a match $\gamma = \{v_1/\vec{e}_1, \ldots, v_k/\vec{e}_k\}$ that satisfies P and binds events $\langle e_1, \ldots, e_i, \ldots, e_n \rangle$. A *prefix of a match* γ is a set of bindings $\widehat{\gamma} = \{v_1/\widehat{\vec{e}}_1, \ldots, v_l/\widehat{\vec{e}}_l\}$, $l \leq k$, that binds events $\langle e_1, \ldots, e_i \rangle$, and where $\widehat{\vec{e}}_j \subseteq \vec{e}_j$ for all $j \leq l$ with $\widehat{\vec{e}}_j \neq \emptyset$. For example, match $\gamma = \{s_1/\langle e_{71} \rangle, s_2/\langle e_{74} \rangle, s_3/\langle e_{75} \rangle, s_4/\langle e_{76} \rangle\}$ has the following prefixes: $\{s_1/\langle e_{71} \rangle\}$, $\{s_1/\langle e_{71} \rangle, s_2/\langle e_{74} \rangle\}$, $\{s_1/\langle e_{71} \rangle, s_2/\langle e_{74} \rangle, s_3/\langle e_{75} \rangle\}$, $\{s_1/\langle e_{71} \rangle, s_2/\langle e_{74} \rangle, s_3/\langle e_{75} \rangle, s_4/\langle e_{76} \rangle\}$.

The union of two sets of bindings is denoted with the symbol $\uplus$ and has the following semantics: $\{v_i/\vec{e}_i, v_j/\vec{e}_j\} \uplus \{v_i/\vec{e}_l, v_k/\vec{e}_k\} = \{v_i/\langle \vec{e}_i \cup \vec{e}_l \rangle, v_j/\vec{e}_j, v_k/\vec{e}_k\}$.

Definition 3. (Skip-till-next-match [2]) *Let* $P = (\langle B_1, \ldots, B_m \rangle, \Theta, \tau)$ *be a pattern, E be an event stream, γ be a match of P in E that binds events* $\langle e_1, \ldots, e_i, e_{i+1}, \ldots, e_n \rangle$. *Furthermore, let $\widehat{\gamma}$ be a prefix of γ that binds events* $\langle e_1, \ldots, e_i \rangle$, γ_p *be a match of a prefix of P in E, and $v/\langle e \rangle$ be a binding. A match γ complies with the skip-till-next-match selection strategy iff*

$$\nexists\, \gamma_p, \widehat{\gamma}, v/\langle e \rangle \; (\gamma_p = \widehat{\gamma} \uplus \{v/\langle e \rangle\} \wedge e_i.T < e.T < e_{i+1}.T).$$

Definition 3 implies two conditions that must hold for a match to comply with STNM. First, since a prefix of a match is also a match γ_p of a prefix of P, no match exists that shares a prefix $\widehat{\gamma}$ with γ and that has an earlier chronologically next event after prefix $\widehat{\gamma}$ than γ. This condition ensures that γ contains the earliest events after the start of the match. Second, since a match γ_p of a prefix of P is not necessarily a prefix of a match, no event e exists that: (1) does not participate in any match that shares a prefix $\widehat{\gamma}$ with γ, (2) occurs after a prefix $\widehat{\gamma}$ in γ but before the chronologically next event in γ, and (3) satisfies together with prefix $\widehat{\gamma}$ the prefix of P. This condition ensures that no blocking noise occurs between events in γ. From a procedural point of view, STNM specifies to skip events in E until the next event is found that along with the events bound so far matches a prefix of P.

Consider the match $\gamma = \{s_1/\langle e_1 \rangle, s_2/\langle e_4 \rangle, s_3/\langle e_5, e_6 \rangle, s_4/\langle e_7 \rangle\}$ that binds the events $\langle e_1, e_4, e_5, e_6, e_7 \rangle$. The only prefix of γ that can be extended with an event occurring after the prefix and before the chronologically next event is $\widehat{\gamma} = \{s_1/\langle e_1 \rangle\}$. However, neither the binding $s_2/\langle e_2 \rangle$ nor $s_2/\langle e_3 \rangle$ extends $\widehat{\gamma}$ in order to become a match γ_p of any prefix of example pattern P (i.e., $\widehat{P} = (\langle \{s_1\}, \{s_2\} \rangle, \widehat{\Theta}, 100\,\text{ms})$). Thus, γ complies with STNM. In contrast, the match $\gamma = \{s_1/\langle e_{71} \rangle, s_2/\langle e_{74} \rangle, s_3/\langle e_{75} \rangle,$

$s_4/\langle e_{76}\rangle\}$ does not comply with STNM, because e_{72} is blocking noise. Event e_{72} does not participate in any match with prefix $\{s_1/\langle e_{71}\rangle\}$, it occurs between e_{71} and e_{74}, and $\gamma_p = \{s_1/\langle e_{71}\rangle, s_2/\langle e_{72}\rangle\}$, satisfies the prefix $(\langle\{s_1\}, \{s_2\}\rangle, \widehat{\Theta}_2, 100\,\text{ms})$ of P. If e_{72} were not present in the stream, match γ would be found. In Fig. 2(b), the matches that comply with STNM are marked with †.

Definition 4. *(Robust Skip-till-next-match)* Let $P = (\langle B_1, \ldots, B_m\rangle, \Theta, \tau)$ be a pattern, E be an event stream, γ be a match of P in E that binds events $\langle e_1, \ldots, e_i, e_{i+1}, \ldots, e_n\rangle$. Furthermore, let $\widehat{\gamma}$ be a prefix of γ that binds events $\langle e_1, \ldots, e_i\rangle$, $\widehat{\gamma}_m$ be a prefix of a match of P in E, and $v/\langle e\rangle$ be a binding. A match γ complies with the robust skip-till-next-match *selection strategy iff*

$$\nexists\, \widehat{\gamma}_m, \widehat{\gamma}, v/\langle e\rangle\ (\widehat{\gamma}_m = \widehat{\gamma} \uplus \{v/\langle e\rangle\} \wedge e_i.T < e.T < e_{i+1}.T).$$

A match complies with RSTNM if no match exists in E that shares a prefix $\widehat{\gamma}$ with γ and that has an earlier chronologically next event after prefix $\widehat{\gamma}$ than γ. In contrast to STNM, a match γ already complies with RSTNM if it contains the earliest events after the start of the match. Noise events cannot block the detection of γ. From a procedural point of view, RSTNM specifies to skip events in E until the next event is found which satisfies P along with the events matched so far and the events that will be matched to complete the match.

For instance, the match $\gamma_1 = \{s_1/\langle e_{71}\rangle, s_2/\langle e_{74}\rangle, s_3/\langle e_{75}\rangle, s_4/\langle e_{76}\rangle\}$ complies with RSTNM since the prefix $\widehat{\gamma} = \{s_1/\langle e_{71}\rangle\}$ extended by $s_2/\langle e_{72}\rangle$ or $s_2/\langle e_{73}\rangle$ does not yield any prefix $\widehat{\gamma}_m$ of any match in Fig. 2(b). Other extensions of prefixes of γ_1 are not possible. In contrast, the match $\gamma_2 = \{s_1/\langle e_{71}\rangle, s_2/\langle e_{74}\rangle, s_3/\langle e_{75}\rangle, s_4/\langle e_{77}\rangle\}$ does not comply with RSTNM, because the prefix $\widehat{\gamma} = \{s_1/\langle e_{71}\rangle, s_2/\langle e_{74}\rangle, s_3/\langle e_{75}\rangle\}$ can be extended by $s_4/\langle e_{76}\rangle$, resulting in the match γ_1, which is a prefix $\widehat{\gamma}_m$ of itself. In Fig. 2(b), the matches that comply with RSTNM are marked with ◇.

RSTNM finds all matches found with STNM plus those that STNM does not detect due to the occurrence of blocking noise in the stream. This follows from two facts. First, a match γ that does not comply with RSTNM cannot comply with STNM, since if there exists a match that shares a prefix with γ and contains an earlier chronologically next event than γ the match disqualifies γ from complying with both, RSTNM and STNM. Second, a match that does not comply with STNM may still comply with RSTNM since the detection of matches that comply with RSTNM cannot be blocked by noise.

5 Automaton-Based Evaluation

In this section, we present an automaton-based solution for the evaluation of event pattern matching with RSTNM. We begin with a basic automaton as used in automaton-based event pattern matching algorithms that use STNM. Then we propose a backtracking mechanism that extends the basic automaton to find all matches complying with RSTNM.

5.1 Basic Automaton

We define a nondeterministic finite state automaton enriched with a match buffer, β, which collects the bindings during the execution of the automaton.

Definition 5. *(Automaton) Let $(\langle B_1, \ldots, B_m \rangle, \Theta, \tau)$ be a pattern with variables $\{v_1, \ldots, v_k\}$. An automaton, N, is a five-tuple $N = (Q, \Delta, q_s, q_f, \tau)$, where $Q = \{q_1, \ldots, q_n\}$, $q_i \subseteq \{v_1, \ldots, v_k\}$ is a finite set of states, $\Delta = \{\delta_1, \ldots, \delta_l\}$ is a finite set of transitions $\delta = (q, v, \Theta_\delta)$, $q_s \in Q$ is the start state, $q_f \in Q$ is the accepting state, and τ is a duration.*

Each state is defined as the subset of the variables in P that have been matched so far. A transition, $\delta = (q, v, \Theta_\delta)$, leads from a source state, q, to a target state, $q \cup \{v\}$, if the input event satisfies the transition condition Θ_δ. The condition Θ_δ contains all constraints from Θ that have the form $v.A \phi C$ or $v_i.A_i \phi v_j.A_j$ and involve only variable v and variables in q. If a transition loops at a state q (i.e., $v^+ \in q$), Θ_δ contains also constraints of the form $prev(v.A) \phi v.A$. If an input event satisfies the condition of multiple transitions, nondeterminism occurs, and the automaton branches into multiple automata. If no transition conditions are satisfied, the input event is skipped. Each transition adds a binding $v/\langle e_i \rangle$ of a variable to an input event to the match buffer β. The execution of an automaton begins in the start state, $q_s = \emptyset$. The accepting state, q_f, marks the acceptance of the bindings in β as a match. Duration τ is the maximal duration of the time interval that can be spanned by the events in β.

Figure 3 shows the automaton for example pattern $(\langle \{s_1\}, \{s_2\}, \{s_3^+\}, \{s_4\} \rangle, \Theta,$ 100 ms) represented as a directed graph. Nodes represent states. Edges represent transitions and are labeled with a variable and a transition condition. The start state is marked with an incoming arrow, the accepting state is doubly circled. To facilitate reading, states are labeled by the concatenation of the corresponding variables, e.g., the node with label $s_1 s_2$ represents state $\{s_1, s_2\}$. The duration τ of the automaton is 100 ms.

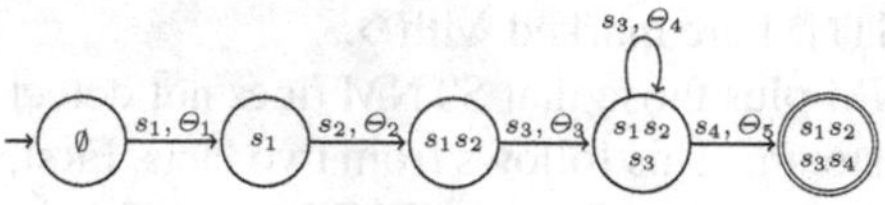

$$\Theta_1 = \{s_1.S = \text{'GOOG'}\}$$
$$\Theta_2 = \{s_2.S = \text{'GOOG'}, s_1.P < s_2.P\}$$
$$\Theta_3 = \{s_3.S = \text{'GOOG'}, s_2.P < s_3.P\}$$
$$\Theta_4 = \{s_3.S = \text{'GOOG'}, prev(s_3.P) < s_3.P\}$$
$$\Theta_5 = \{s_4.S = \text{'GOOG'}, s_1.V < s_4.V, s_2.V < s_4.V, s_3.V < s_4.V\}$$

Fig. 3. Automaton for $(\langle \{s_1\}, \{s_2\}, \{s_3^+\}, \{s_4\} \rangle, \Theta, 100\,\text{ms})$

The automaton in Fig. 3 misses matches that comply with RSTNM. Assume the input events $e_{71}, \ldots, e_{77}$ in Fig. 1. The automaton starts in state $\emptyset$, binds e_{71} to s_1, and changes to state $\{s_1\}$. Then, e_{72} matches s_2 since the price of the Google trade is greater than in e_{71}; the automaton changes to state $\{s_1, s_2\}$. None of the following events e_{73}, e_{74}, e_{75}, e_{76}, and e_{77} satisfies Θ_3, since e_{73} is not a Google trade and the other events do not continue the upward trend of the price. The automaton expires without producing a match. However, $\{s_1/\langle e_{71} \rangle, s_2/\langle e_{74} \rangle, s_3/\langle e_{75} \rangle, s_4/\langle e_{76} \rangle\}$ complies with RSTNM. The reason for the missed match is that s_2 binds noise event e_{72} which indeed represents an increase in the price with respect to the events matched so far. Only later on when the next Google trades are read, it turns out that the partial match cannot be completed and the match is blocked. If blocking noise e_{72} were skipped, the match would be found.

5.2 Automaton with Backtracking

In order to find all matches that comply with RSTNM, we extend the basic automaton with a backtracking mechanism that identifies blocking noise and avoids the blockage of a match. We start by introducing the match window that buffers incoming events.

Given an automaton N with duration τ, a *match window* is a maximal subsequence of the stream E that starts at an event and includes all events that are within distance τ. The match window starting at event e_i is $W = \langle e \in E \mid 0 \le e.T - e_i.T \le \tau \rangle$. For instance, for the automaton in Fig. 3 and the event stream Stocks, the match windows starting at e_1 and e_2 are, respectively, $W_1 = \langle e_1, e_2, e_3, e_4, e_5, e_6, e_7 \rangle$ and $W_2 = \langle e_2, e_3, e_4, e_5, e_6, e_7, e_8 \rangle$.

An automaton starts at the first event in the match window W. If it does not reach the accepting state after reading all events in W, backtracking applies and the last transition is reverted. The automaton returns to the previous state and removes from the match buffer β the event e_i that has been bound by the reverted transition. Event e_i is skipped and the automaton resumes reading at event e_{i+1} in the match window. Hence, backtracking reclassifies e_i from matched to skipped.

To enable backtracking, the automaton needs to keep track of the transitions taken and the events that triggered these transitions. We propose a so-called execution tree to record transitions and positions in the match window. The tree stores dependencies between different automaton instances that branched due to nondeterminism during the execution. Such information allows to stop backtracking before producing matches that do not comply with RSTNM.

Definition 6. *(Execution Tree) Let $N = (Q, \Delta, q_s, q_f, \tau)$ be an automaton and W be a corresponding match window over an event stream. An execution tree, X, is a directed, acyclic graph with nodes V and edges D, where each node has at most one incoming edge. A node represents a pair (δ, c) with $\delta \in \Delta, 1 \le c \le |W|$. Special node $(\circ)$ is the root of X. An edge is a pair $(\delta_i, c_i) \to (\delta_j, c_j)$ with $(\delta_i, c_i), (\delta_j, c_j) \in V$.*

A node, (δ, c), in an execution tree records that the event at position c in the match window W triggered transition δ. The direction of the edges represents the chronology of the transitions taken. Hence, a path from the root to a leaf node in the execution tree represents the sequence of transitions taken so far. If an event triggers multiple transitions (i.e., nondeterminism occurs), X branches into multiple nodes, called *siblings*. Each match window has exactly one corresponding execution tree.

Figure 4(c) shows the execution tree for the automaton in Fig. 3 and the match window $W = \langle e_{71}, e_{72}, e_{73}, e_{74}, e_{75}, e_{76}, e_{77} \rangle$. The execution tree specifies that event $W[1] = e_{71}$ triggered transition $\delta_1 = (\emptyset, s_1, \Theta_1)$, $W[4] = e_{74}$ triggered $\delta_2 = (\{s_1\}, s_2, \Theta_2)$, and $W[5] = e_{75}$ triggered transition $\delta_3 = (\{s_1, s_2\}, s_3, \Theta_3)$. At node $(\delta_3, 5)$, the tree branches into sibling nodes $(\delta_4, 6)$ and $(\delta_5, 6)$, which is due to event $W[6] = e_{76}$ that triggered two transitions, $\delta_4 = (\{s_1, s_2, s_3\}, s_3, \Theta_4)$ and $\delta_5 = (\{s_1, s_2, s_3\}, s_4, \Theta_5)$.

An automaton with backtracking executes as follows. For each match window, W, an execution tree X that contains only the root node $\circ$ is created. The automaton reads the events in W one-by-one, starting at the head $W[1]$. If an input event e at position c in W triggers a transition $\delta = (q, v, \Theta_\delta)$, a binding $v/\langle e \rangle$ is added to match buffer β, the automaton instance changes from state q to state $q \cup \{v\}$, and a new leaf node (δ, c) is added to X. The automaton instance keeps a pointer to the new leaf node. If e triggers multiple transitions, nondeterminism arises. For each transition but one, an automaton instance branches from the original automaton instance. Each instance takes a transition and appends a leaf node to the execution tree X. Since X is shared among all instances, the appended nodes, $(\delta_1, c), \ldots, (\delta_n, c)$, are siblings with distinct transitions, $\delta_i \neq \delta_j$, but equal position c. If e does not trigger any transition and the automaton is not in the start state, it stays in its current state without updating β and X.

When all events in W are processed, automaton instances that reached the accepting state contain a match in the match buffer β. The match is added to the result and the instance terminates. For an automaton instance that is not in the accepting state, backtracking applies. First, transition $\delta = (q, v, \Theta_\delta)$ and position c is retrieved from the leaf node of the execution tree X. Then the automaton instance steps back to state q and removes event $W[c]$ from the binding of variable v in β. The leaf node is removed from X and the instance points to the parent of the removed node. Finally, the instance resumes reading events at $W[c+1]$. If the instance does not lead to a match after reading all events from $W[c+1]$ to the end of W, backtracking applies again, etc.

Uncontrolled backtracking without an additional stop condition would lead to matches that do not comply with RSTNM, as stated in the following lemma.

Lemma 1. *An execution tree X leads to matches that comply with RSTNM iff for all pairs of siblings, (δ_i, c_i), (δ_j, c_j), the following holds: $c_i = c_j$.*

Proof. Assume that the positions c_i and c_j of two sibling nodes are different, and consider two matches, where one corresponds to a path through node (δ_i, c_i) and one to a path through node (δ_j, c_j). The two matches have a common prefix up to their parent node. The subsequent events in the two matches (i.e., the events at position c_i and c_j, respectively) are different, and one event occurs before the other. Without loss of generality, assume that the event at c_i occurs before the event at c_j. The match with the event at c_j would not satisfy Def. 4 of RSTNM because it contains a prefix that, if extended by the event at c_i, is a prefix of the match through tree node (δ_i, c_i).

According to Lemma 1, backtracking must be stopped when it reaches a leaf node, (δ_i, c_i), which has siblings. Backtracking at this point would replace (δ_i, c_i) with (δ_j, c_j), where $c_i < c_j$ (i.e., the event at position c_i is skipped), and the siblings would store different positions which contradicts the lemma. Hence, backtracking stops and node (δ_i, c_i) is removed from the tree. If backtracking reaches the last node of a set of sibling nodes, backtracking is allowed, since all other sibling nodes have already been removed and Lemma 1 is not violated. Another stop condition for backtracking is when the child of the root node is reached. Further backtracking at this point would revert the automaton into the start state and restart the execution at the second event of the match window not considering all events in the stream within a time span of duration τ. To summarize: if backtracking reaches a node in X with siblings or the child of the root node, the automaton instance removes the node from the tree and terminates.

Figure 4 illustrates a few steps of executing the automaton in Fig. 3. Each step shows the triggered transitions δ_i together with the transition graph, the current state q, the match buffer β, and the execution tree X after the transition is taken. For instance, in Fig. 4(a) the automaton instance is in state $\{s_1\}$, and e_{72} triggers the transition $(\{s_1\}, s_2, \Theta_2)$. The instance moves to state $\{s_1, s_2\}$, adds the binding $s_2/\langle e_{72}\rangle$ to β, and appends node $(\delta_2, 2)$ to node $(\delta_1, 1)$ in X. In Fig. 4(b) backtracking applies, since the automaton instance is not in the accepting state after processing the last event e_{77} in W. Node $(\delta_2, 2)$ is removed from X, the binding $s_2/\langle e_{72}\rangle$ is removed from β, and the current state is reset to $\{s_1\}$. The execution resumes at $W[3] = e_{73}$.

In Fig. 4(c), e_{76} is read again and triggers two transitions, i.e., nondeterminism arises. Two automaton instances exist and the execution tree branches. In Fig. 4(d), e_{77} is read and W reaches its end. The automaton instance in state $\{s_1, s_2, s_3, s_4\}$ is accepted and produces match $\{s_1/\langle e_{71}\rangle, s_2/\langle e_{74}\rangle, s_3/\langle e_{75}\rangle, s_4/\langle e_{76}\rangle\}$, whereas the one in state $\{s_1, s_2, s_3\}$ is not accepted. Backtracking cannot be applied to the automaton instance, since the corresponding node $(\delta_4, 6)$ in the execution tree X has a sibling. If backtracking were applied to the instance, node $(\delta_4, 6)$ would be replaced with $(\delta_5, 7)$ which contradicts Lemma 1. The instance would reach the accepting state with match $\{s_1/\langle e_{71}\rangle, s_2/\langle e_{74}\rangle, s_3/\langle e_{75}\rangle, s_4/\langle e_{77}\rangle\}$ which does not conform to RSTNM as explained in Def. 4.

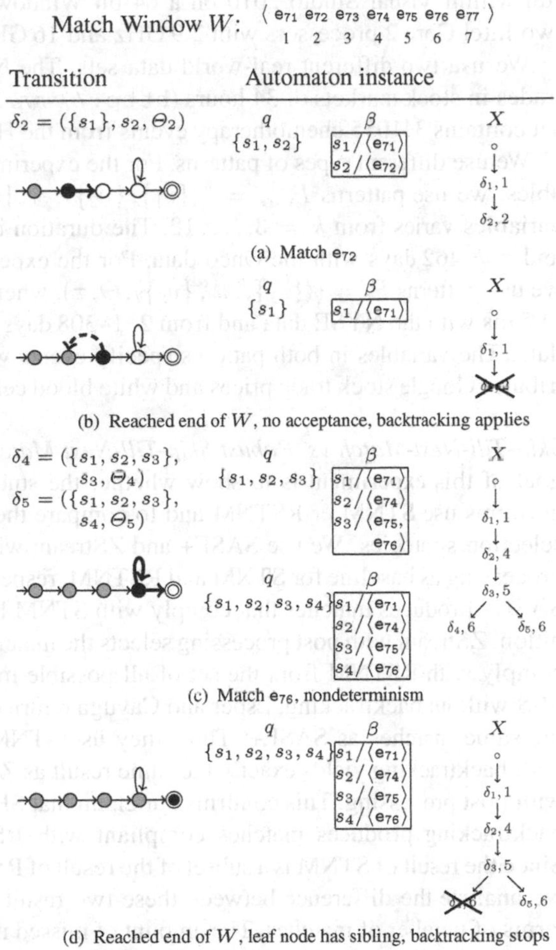

Match Window W: $\langle\ e_{71}\ e_{72}\ e_{73}\ e_{74}\ e_{75}\ e_{76}\ e_{77}\ \rangle$
                   $1\quad 2\quad 3\quad 4\quad 5\quad 6\quad 7$

Transition	Automaton instance		
$\delta_2 = (\{s_1\}, s_2, \Theta_2)$	q $\{s_1, s_2\}$	β $s_1/\langle e_{71}\rangle$ $s_2/\langle e_{72}\rangle$	X

(a) Match e_{72}

q $\{s_1\}$ $\quad$ β $s_1/\langle e_{71}\rangle$ $\quad$ X

(b) Reached end of W, no acceptance, backtracking applies

$\delta_4 = (\{s_1, s_2, s_3\}, s_3, \Theta_4)$,
$\delta_5 = (\{s_1, s_2, s_3\}, s_4, \Theta_5)$

q $\{s_1, s_2, s_3\}$ $\quad$ β $s_1/\langle e_{71}\rangle$ $s_2/\langle e_{74}\rangle$ $s_3/\langle e_{75}, e_{76}\rangle$ $\quad$ X

q $\{s_1, s_2, s_3, s_4\}$ $\quad$ β $s_1/\langle e_{71}\rangle$ $s_2/\langle e_{74}\rangle$ $s_3/\langle e_{75}\rangle$ $s_4/\langle e_{76}\rangle$

(c) Match e_{76}, nondeterminism

q $\{s_1, s_2, s_3, s_4\}$ $\quad$ β $s_1/\langle e_{71}\rangle$ $s_2/\langle e_{74}\rangle$ $s_3/\langle e_{75}\rangle$ $s_4/\langle e_{76}\rangle$ $\quad$ X

(d) Reached end of W, leaf node has sibling, backtracking stops

Fig. 4. Execution of Automaton in Fig. 3

6 Experiments

In this section, we report the results of an empirical evaluation using various event pattern matching algorithms and real-world data. The aim is to compare STNM to RSTNM and to show the advantages of our backtracking mechanism over alternative solutions.

Setup and Data. We implemented in C the automaton-based SES algorithm [3] with and without backtracking as well as the join-tree-based ZStream algorithm [9]. Since ZStream finds all possible matches, we added a post processing step to eliminate matches that do not comply with RSTNM. Furthermore, we used the Esper CEP system 4.9.0 (http://esper.codehaus.org) and SASE+ (http://code.google.com/p/sase-umass), both implemented in Java, as well as Cayuga (http://sourceforge.net/projects/cayuga) and T-REX (http://www.inf.usi.ch/postdoc/margara), both implemented in C++. The experiments were performed on a PC with eight Intel Core i7 processors with 3.4 GHz and 16 GB memory, on which a 64-bit Linux 3.2.0 is installed. Only Cayuga run within Visual Studio 2010 on a 64-bit Windows 2008 R2 Terminal Server with two Intel Core 2 processors with 2.9 GHz and 16 GB memory.

We use two different real-world data sets. The NYSE data set contains 1M share trades in stock markets of 34 hours (http://www.nyxdata.com). The Onco data set contains 341055 chemotherapy events from the Hospital of Meran-Merano.

We use different types of patterns. For the experiments with varying number of variables, we use patterns $P_{vars} = (\langle\{v_1\}, \{v_2\}, \ldots, \{v_k\}\rangle, \Theta, \tau)$, where the number of variables varies from $k = 3, \ldots, 12$. The duration is $\tau = 30\,\text{ms}$ with the NYSE data and $\tau = 462$ days with the Onco data. For the experiments with varying length of τ, we use patterns $P_\tau = (\langle\{v_1\}, \ldots, \{v_8\}\rangle, \Theta, \tau)$, where τ varies from 10–45 ms in steps of 5 ms with the NYSE data and from 231–308 days in steps of 11 days with the Onco data. The variables in both patterns specify events with a downwards trend in one attribute (Google stock trade prices and white blood cell counts per patient, respectively).

Skip-Till-Next-Match vs. Robust Skip-Till-Next-Match. The goal of this experiment is to show whether the studied algorithms use STNM or RSTNM and to compare these two selection strategies. We use SASE+ and ZStream with post processing as baseline for STNM and RSTNM, respectively. SASE+ produces matches that comply with STNM by definition. ZStream with post processing selects the matches that comply with RSTNM from the set of all possible matches. SES without backtracking, Esper and Cayuga return exactly the same matches as SASE+. Thus, they use STNM. SES with backtracking yields exactly the same result as ZStream with post processing. This confirms our claim that SES with backtracking produces matches compliant with RSTNM. Since the result of STNM is a subset of the result of RSTNM, we analyze the difference between these two result sets in

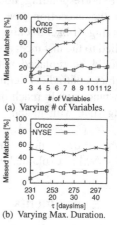

(a) Varying # of Variables.

(b) Varying Max. Duration.

Fig. 5. STNM vs. RSTNM

terms of number of matches. The amount of missed matches with STNM due to blocking noise with respect to RSTNM (y-axis) can be substantial as shown by Figure 5.

SES + Backtracking vs. T-REX. We compare SES with backtracking (SES + bt), which applies RSTNM, to T-REX with the `first-within` selection strategy applied to all variables in the pattern. Though T-REX does not exactly produce matches complying with RSTNM, it is the algorithm closest to RSTNM among all algorithms we analyzed.

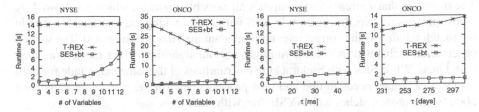

Fig. 6. SES + bt vs. T-REX, Depending on # of Variables and Max. Duration

T-REX differs from SES because it specifies matches starting from the last variable in the pattern, whereas SES specifies matches starting from the first variable in the pattern. Furthermore, T-REX requires a maximal duration for each pair of consecutively matching events rather than a maximal duration for all matching events, as in SES. Due to these differences, it is generally not possible to specify patterns with T-REX and SES that produce identical result sets. To compare the runtime of the two algorithms, we set the maximal duration between two consecutive matching events in T-REX to the maximal duration τ of the pattern divided by the number of variables in the pattern. Of all the settings we tried, this approximation led to the least runtimes for T-REX.

We execute SES with backtracking and T-REX using the patterns P_{vars} and P_τ on the NYSE and the Onco data set and measure the average runtime over three runs. Figure 6 shows the results. The first observation is that SES with backtracking clearly outperforms T-REX for all patterns and data sets. For some patterns SES is by more than an order of magnitude faster than T-REX. The reason is that T-REX first computes a superset of the result set and then selects compliant matches from it. The second observation is that the runtime of SES with backtracking increases with the number of variables and maximal time interval τ. The more variables in the pattern, the more backtracking steps are computed. The larger τ, the more events need to be considered in the match window. The third observation is that the runtime for T-REX is constant for NYSE whereas for Onco the runtime decreases with increasing number of variables k. With increasing k the maximal duration between two consecutive matching events decreases. Partial matches that exceed this maximal duration do not need to be further considered and are discarded. This mechanism affects Onco more than NYSE because Onco contains much less matches than NYSE with respect to the number of discarded partial matches. With increasing maximal duration τ, the runtime of T-REX increases with Onco because the maximal duration between consecutive matching events increases.

SES + Backtracking vs. ZStream + Post Processing. We compare SES with backtracking (SES + bt) to ZStream with post processing (ZStream + pp). Both algorithms produce exactly the same result set. ZStream has shown to outperform automaton-based algorithms at finding all possible matches of a pattern [9].

We execute both algorithms with the patterns P_{vars} and P_τ on the NYSE and the Onco data set and measure the average runtime over three runs. Figure 7 shows the results. The first observation is that, again, SES with backtracking clearly outperforms ZStream with post processing in all experiments. The difference in performance is significant with NYSE, with up to one order of magnitude when varying the number of variables and more than two orders of magnitude when varying the maximum duration.

The reason is that ZStream first computes all possible matches, which is followed by a post processing step. In contrast, in SES with backtracking the first match found in the match window is the correct one; no redundant matches are computed. The second observation is that with NYSE the runtimes of both algorithms differ much more than with Onco. While SES with backtracking has runtimes of the same magnitude for both data sets, ZStream needs to find and post process many (in the worst case more than 5000 times) more matches with NYSE than with Onco.

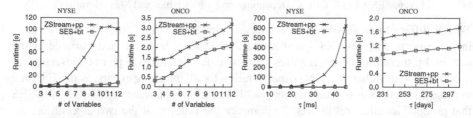

Fig. 7. SES + bt vs. ZStream + pp, Depending on # of Variables and Max. Duration

7   Conclusion

In this paper, we proposed the robust skip-till-next-match selection strategy for event pattern matching. In contrast to the skip-till-next-match selection strategy, robust skip-till-next-match does not allow noise events to block the detection of matches. Robust skip-till-next-match is robust against noise in the event stream. To achieve this, all constraints in the pattern together with a complete match are considered during the matching process. We proposed a backtracking mechanism that extends automaton-based event pattern matching to find all matches complying with robust skip-till-next-match. We conducted extensive experiments using real-world data to show the effectiveness of our approach. The results show that the amount of matches not detected due to blocking noise using skip-till-next-match can be quite substantial with respect to robust skip-till-next-match. In terms of runtime, our approach outperforms alternative solutions. Future work will mainly concentrate on the study of runtime optimizations for the backtracking mechanism.

References

1. Adi, A., Etzion, O.: Amit - the situation manager. The VLDB Journal 13(2), 177–203 (2004)
2. Agrawal, J., Diao, Y., Gyllstrom, D., Immerman, N.: Efficient pattern matching over event streams. In: SIGMOD, pp. 147–160 (2008)
3. Cadonna, B., Gamper, J., Böhlen, M.H.: Sequenced event set pattern matching. In: EDBT, pp. 33–44 (2011)
4. Cugola, G., Margara, A.: Complex event processing with T-REX. J. Syst. Softw. 85(8), 1709–1728 (2012)
5. Demers, A., Gehrke, J., Hong, M., Riedewald, M., White, W.: Towards expressive publish/Subscribe systems. In: Ioannidis, Y., Scholl, M.H., Schmidt, J.W., Matthes, F., Hatzopoulos, M., Böhm, K., Kemper, A., Grust, T., Böhm, C. (eds.) EDBT 2006. LNCS, vol. 3896, pp. 627–644. Springer, Heidelberg (2006)

6. Dindar, N., Güç, B., Lau, P., Ozal, A., Soner, M., Tatbul, N.: Dejavu: declarative pattern matching over live and archived streams of events. In: SIGMOD, pp. 1023–1026 (2009)
7. Harada, L., Hotta, Y.: Order checking in a CPOE using event analyzer. In: CIKM, pp. 549–555 (2005)
8. Liu, M., Rundensteiner, E., Dougherty, D., Gupta, C., Wang, S., Ari, I., Mehta, A.: High-performance nested cep query processing over event streams. In: ICDE, pp. 123–134 (2011)
9. Mei, Y., Madden, S.: Zstream: A cost-based query processor for adaptively detecting composite events. In: SIGMOD, pp. 193–206 (2009)
10. Sadri, R., Zaniolo, C., Zarkesh, A., Adibi, J.: Expressing and optimizing sequence queries in database systems. ACM Trans. Database Syst. 29(2), 282–318 (2004)
11. Schultz-Møller, N.P., Migliavacca, M., Pietzuch, P.: Distributed complex event processing with query rewriting. In: DEBS, pp. 4:1–4:12 (2009)
12. Zemke, F., Witkowski, A., Cherniak, M., Colby, L.: Pattern matching in sequences of rows. Tech. rep. (2007)

CARDAP: A Scalable Energy-Efficient Context Aware Distributed Mobile Data Analytics Platform for the Fog

Prem Prakash Jayaraman[1], João Bártolo Gomes[2], Hai Long Nguyen[2],
Zahraa Said Abdallah[1], Shonali Krishnaswamy[2], and Arkady Zaslavsky[1,*]

[1] CSIRO Computational Informatics, Canberra, Australia
{prem.jayaraman,arkady.zaslavsky}@csiro.au
[2] Institute for Infocomm Research (I2R), A*STAR, Singapore
{bartologjp,nguyenhl,spkrishna}@i2r.a-star.edu.sg
[3] Monash University, Melbourne, Australia
zahraa.said.abdallah@monash.edu

Abstract. Distributed online data analytics has attracted significant research interest in recent years with the advent of Fog and Cloud computing. The popularity of novel distributed applications such as crowdsourcing and crowdsensing have fostered the need for scalable energy-efficient platforms that can enable distributed data analytics. In this paper, we propose CARDAP, a (C)ontext (A)ware (R)eal-time (D)ata (A)nalytics (P)latform. CARDAP is a generic, flexible and extensible, component-based platform that can be deployed in complex distributed mobile analytics applications e.g. sensing activity of citizens in smart cities. CARDAP incorporates a number of energy efficient data delivery strategies using real-time mobile data stream mining for data reduction and thus less data transmission. Extensive experimental evaluations indicate the CARDAP platform can deliver significant benefits in energy efficiency over naive approaches. Lessons learnt and future work conclude the paper.

1 Introduction

Big Data analytics in the *Fog* has created tremendous opportunities to gain new and exciting value from big data. *Fog computing* or briefly *Fog* is a term recently embraced by Cisco Systems [4]. Fog computing extends Cloud Computing paradigm to the edge of the network. A recent vision paper from Intel [1], clearly highlights the value of the data that resides in the Fog and the need for distributed data analytics and techniques that can work in the Fog (closer to the source of data), where some of the *biggest* big data is generated. Further, this approach is also considered as an alternative to alleviate the current big data challenge of processing massive amounts of data in remote Cloud environments.

Fig. 1 presents the big picture of big data in the Fog. The intelligent systems and sensors (Internet of Things), smart city infrastructure, mobile smart phones

---

* Prof. Zaslavsky is a visiting Professor at StPetersburg National Research University of IT, Mechanics and Optics, Russia.

Y. Manolopoulos et al. (Eds.): ADBIS 2014, LNCS 8716, pp. 192–206, 2014.

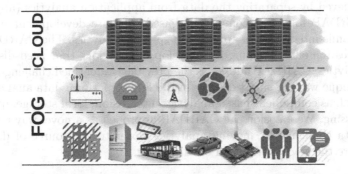

Fig. 1. Smart Devices and Human Entities in the Fog stream data to Cloud

etc., coupled with information from humans are some of the largest volume of streaming complex big data. The data sources as shown in the figure are heterogeneous and distributed. The ubiquitous connectivity available in the Fog (e.g. 2G, 3G, 4G, Internet, Network, WiFi etc) has created new opportunities that can take advantage of these massive data storehouses.

One of the key arguments driving the notion of Fog computing is the emerging wave of Internet deployments, most notably the Internet of Things (IoT) [4]. Although the definition of "Things" has evolved with rapid advancements in technologies, the key notion of enabling "Things" to sense without human intervention has remained the same. IoT is a major generator of live sensor data. The International Data Corporation estimates an installed based of 220 billion "things" by 2020 [2] coupled by the ever increasing growth in mobile smart phone devices. The smart mobile devices can sense the environment and situation around human entities. The enormous increase in data (in petabytes) [1] being generated by the smart devices render the need to move data analytics close to where the data resides/originates. To achieve this, the compute and storage capabilities must be moved to the Fog. To achieve the goal of a distributed analytics infrastructure between the Fog and the Cloud needs addressing of a set of unique challenges including: 1) Smart devices in the Fog need to have a platform to run local analytics in a cost-efficient manner (e.g. energy, visualisation, resources and data transmission); 2) Analytics are very domain dependent e.g. pre-processing of noisy data is different for each domain and 3) Not all application require data in real-time as instant insights are not required. Hence, local data storage and retrieval must be possible.

To this end, in this paper, we propose (C)ontext (A)ware (R)eal-time (D)ata (A)nalytics (P)latform, a context aware distributed mobile data analytics platform for the Fog. The key driving factor behind the development of CARDAP is the ability to efficiently and effectively perform distributed data analytics in Fog/Cloud environments. The proposed CARDAP framework is application domain analytics agnostic i.e. a generic platform that can be extended to suit any application analytics requirement. The CARDAP platform breaks monolithic application silos that are in most cases very difficult and expensive to extend/adapt.

This is achieved by separating the data from application analytics (application logic). CARDAP follows a component based software development model enabling dynamic integration of application-specific analytics. The CARDAP platform reduces the efforts required to develop applications that need distributed mobile analytics. CARDAP addresses the previously identified challenges by providing a unique way for smart devices in the Fog to perform data analytics with features such as component based analytics integration, local storage, query and task processing. We envision the CARDAP approach as a novel way to address the big data challenge by moving analytics closer to the source of data. The following are the key contribution of this paper.

- We propose CARDAP, a context aware distributed mobile data analytics platform. CARDAP enabled efficient data analytics in the Fog by providing a standardised component oriented approach to incorporate the required application-specific analytics. CARDAP also addresses the need for local storage and query processing for application that does not require instant insights.
- We propose a cost model for distributed data analytics in Fog/Cloud environments.
- We conduct experimental evaluations using CARDAP platform to evaluate three distributed data analytics strategies.

The rest of the paper is organised as follows. Section 3 presents the recent work in the area of mobile distributed data analytics platforms for the Fog. Section 2 presents the motivations behind the proposed CARDAP approach. Section 4 provides in-depth details on the proposed CARDAP platform architecture. Section 5 presents a cost model for mobile distributed data analytics. Section 6 provides implementation details of the CARDAP platform followed by discussions on experimental evaluations. Section 7 concludes the paper with remarks on future work.

2   Motivation

A typical example of a IoT big data application in the Fog is a mobile crowdsensing application in smart cities. Mobile crowdsensing popularly called community sensing [9,15] is an autonomous collaborative sensing approach that requires minimal user involvement (e.g. continuous processing of noise level around users' location). Mobile crowdsensing applications takes advantage of a population of individuals to measure large-scale phenomenon that cannot be measured using single individual. In most cases, the population of individuals participating in crowdsensing applications share a common goal.

Let us consider an example scenario *monitoring citizen activity in a smart city*. This example scenario is depicted in Fig. 2. The aim of such an application in a smart city environment is to determine the activity of users in an outdoor park (as highlighted by the red polygon in the Fig. 2). The key requirements to satisfy this scenario includes 1) ability to perform a task on demand, in this

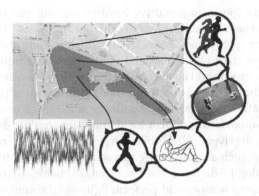

Fig. 2. Scenario: Monitoring Citizen Activity in Smart Cities

example capturing data within a given location from users who are involved in some activity; 2) ability to perform local data analytics, in this example activity recognition on-the-fly [11,10]; 3) ability to store analysed data locally in the Fog that can be queried later by the Cloud (for applications that does not require instantaneous insights) and 4) use an energy-efficient strategy for continuous monitoring and data upload to the Cloud. This scenario can be extended to may such typical smart city example such as monitoring air pollution by mounting sensing component on buses, cars and trams.

As identified by the motivating scenario, there is a need for an efficient mobile data analytics platform that can satisfy the aforementioned requirements to facilitate easy development of IoT applications in the Fog. CSIRO and I2R carry out industry-focussed research and the proposed CARDAP platform has the potential to form the basis for social networking and smart grid applications which have become very popular and have clear commercial advantage.

3   Related Work

Nowadays, with the advent of technology, mobile devices have significantly sensing abilities, computing power, communication and storage resources. They are commonly equipped with various sensors, such as GPS, accelerometer, microphone and camera. Consequently, mobile crowd-sensing (MCS), which analyses crowd behaviour by monitoring large-scale environmental information generated from individual devices, has emerged as a important research topic. Several MCS applications have been successfully developed to discover individual and community trends, such as transportation activities in urban spaces [22], traffic monitoring [23], and collaborative searching [19].

In MCS, mobile devices continuously sense information about the environment and upload the sensed data to the Cloud/remote servers. These processes are obviously energy expensive and may cause battery drain in some cases. Moreover, the raw data generated by physical sensors is usually huge (big data) and can not be used directly as application inputs. Usually, mobile devices need to pre-process the data, perform primitive local processing and only upload inter-mediate results to the backend servers for further analytics. This approach not only helps to reduce energy and network bandwidth consumption, but also avoid overwhelming of raw data at the backend. Therefore, distributed mobile analyt-ics play an important role to the success of many crowd-sensing applications. A taxonomy of distributed mobile data analytics approaches is presented [21].

In distributed data analytics, a key challenge is to design scalable, ease-to-use frameworks that supports to perform local analytics (mobile device) and global analytics (server) effectively [6]. Mobile devices with limited resources can per-form local analytics, such as, converting raw sensed data into application consum-able data (e.g. analog-to-digital converter), remove noise, aggregate/summarise sensed data from many sensors and perform light-weight mining tasks [8]. Mean-while, the aim of global analytics is to discover overall patterns of the environ-ment. Furthermore, the system should understand common information needs of similar procedures in an application or different applications in the same do-mains to avoid duplicate efforts. Another challenge in distributed data analytics is to understand contexts for proper problem-solving in the right circumstances. For example, mobile devices are configured to upload processed data with 3G connection and raw sensed data with wifi connection. Or, in order to save band-width usage, mobile devices only update new data when there are significant changes in the sensed data.

In literature, there are several efforts on designing scalable frameworks for crowd-sensing applications, including high-level abstraction for sensing information [20], task description [16], and a component-based design for quick sensing application deployments [12]. Ye et al. proposed MECA [20] (Mobile Edge Capture and Ana-lytics) middle-ware for crowd-sensing. The framework has three logical layers: data, edge, and phenomena/application. After receiving phenomenon specifications from applications, MECA configures edge nodes and devices for corresponding raw data collection and edge analytics processing. Thereby, users focus on implementing their applications' logic without concern with device interaction.

Ravindranath et al. [16] introduced CITA, a system that eases the develop-ment and running of tasking applications on smart phones. End users create task by writing only on server-side code in form of "condition/action" rules. CITA automatically partitions the code, deals with device coordination, and efficiently executes code on the devices. The framework is currently under developing with their own activity context-aware applications and has not been published yet.

Recently, Jayaraman et al. [12] proposed component-based platform for op-portunistic sensing applications, named MOSDEN. In this approach, each smart-phone occupies a MOSDEN instance in order to run applications with minimal user interaction. Sensors communicate with MOSDEN platform via the concept of plug-in; thereby, a new sensor can be easily added or removed from the sys-tem at running time. A conceptual description of the plugin is in XML format. MOSDEN provides a true zero programming middleware, where users do not need to write program code and supports both push and pull data streaming mechanisms.

Context-aware in mobile crowd-sensing has recently attracted a lot of research work. By understanding the context, mobile devices are not simply data collectors, but they can act dynamically according its sensed data and users' needs. Thereby, context-aware crowd-sensing applications can provide more elegant and meaningful solutions, for example, minimise user intervention and optimise the consumption of resources of mobile devices.

The OPPORTUNITY project has been developed to build a ready-to-use middleware and prototypical implementation for evaluating and testing for applications on opportunistic human activity and context recognition [17]. Some preliminary results have been reported, such as, signal conditioning and feature abstraction, autonomous evowhich can lution and adaptation. Carresra *et al.* [5] proposed an adaptive sampling algorithm for user localisation. The key idea is to trade off the accuracy of the location estimation with the battery consumption by varying the type the localisation methods. The application switch to GPS localisation when there is uncertainty on the user location due to the coarse of network localisation.

Sherchan *et al.* [18] developed Context-Aware Real-time Open Mobile Miner (CAROMM) to facilitate data collection from mobile users for crowdsensing applications. CAROMM aims to reduce energy and bandwidth consumption related to continuous sensing and uploading in crowd-sensing applications. Cooperating with resource-aware clustering, CAROMM send only analysed information from each device when it identifies significant changes in the situation. This approach not only reduces the frequency and amount of transferred data, but also guarantee that no important information will be lost. Based on CAROMM framework, Jayaraman *et al.* [13] later demonstrated another context-aware crowd-sensing application that collect sensory data and activity data from a large number of mobile users. It classifies places into different types of contexts as lively, busy and quiet based on light levels, noise levels, crowd intensity, and user activity levels. The application is able to provide real-time reasoning about different situations/ambience of the locations.

4   CARDAP – Distributed Mobile Data Analytics Platform

In this section, we present the system architecture of CARDAP. The proposed CARDAP system is an outcome of our previous works namely Context-aware open mobile miner (CAROMM) [18] and Mobile Sensor Data Engine (MOSDEN) [12]. We first present a big picture view of the proposed CARDAP concept. We then present architectural details of the CARDAP system.

4.1   A Model for Distributed Mobile Data Analytics

As mentioned in Section 3, a key challenge facing distributed data analytics is to design scalable, ease-to-use frameworks that supports local analytics (mobile device) and global analytics (server) effectively. The focus of this paper is on the local analytics area in the Fog. But to provide a big picture, we first present our model for a distributed mobile data analytics system. The big picture illustrated in Fig. 3 captures our vision. In our model, the smart devices in the Fog can include any Internet connected device such as individual to group of smart mobile devices capturing situational context information from the user and his/her environment, micro-sized processing platforms (e.g. raspberry pi mounted on buses).

We model a request from users/other smart devices/data agencies as a set of tasks $ta_1, ta_2 ... ta_n$. Each task ta_i where $1 < i < n$ has an associated deadline dl_1, $dl_2, ... dl_n$. Each task is also associated with a set of minimum capabilities (C_{min}^{ta}) $c_1, c_2 ... c_m$ that is required to accomplish the task. Examples of capability include specific sensor requirements, data analytic model requirements, etc. $sd_1, sd_2 ...$ sd_k represents the set of mobile smart devices that are in the Fog. Each mobile smart device sd_j where $1 < j < k$ also has the set of associated capabilities C^{sd}. At any given time t, a mobile smart device can perform a task ta_i if and only if task minimum capability $C_{min}^{ta} \subseteq$ mobile device capabilities C^{sd}.

We note, this paper focuses on CARDAP's architecture and experimental evaluations validating its energy-efficiency. The task assignment functionality of the scheduler as depicted Fig 3 based on capabilities and deadlines is outside the scope of this paper.

4.2 System Architecture

As stated earlier, the CARDAP architecture is developed from our previous works namely CAROMM and MOSDEN. CAROMM is a context-aware open mobile miner platform that is underpinned by continuous mobile data stream

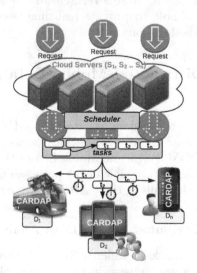

Fig. 3. Distributed Mobile Data Analytics - Big Picture

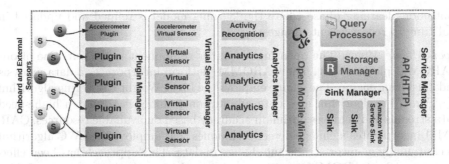

Fig. 4. CARDAP - System Architecture

mining. The mobile data stream mining is used to determine change in contextual information being monitored thus reducing the amount of data being transmitted to the Cloud from the mobile device. The MOSDEN platform is a generic mobile crowdsensing application development platform that breaks the tight coupling between data and application specific processing. MOSDEN framework enables local storage of sensed data. The limitation of CAROMM was its inability to store and query data on-demand and separate analytics and sensing. The limitation of MOSDEN was unavailability of smart processing techniques to reduce bandwidth usage (i.e. data upload). CARDAP was developed to address these limitation of the respective systems. The architecture of CARDAP is presented in Fig. 4.

The five key components of the CARDAP architecture is the data stream capture component, the analytics component, the open mobile miner component, the data sink component and the storage and query processor component.

Data Stream Capture Component: The data stream capture component of CARDAP uses a component-based approach namely plugins and virtual sensors to interface with data sources. The data sources could range from on-board sensors for mobile devices to externally sensors in-case of systems like raspberry pi. The plugin and virtual sensors together enable integration of heterogeneous data sources to the CARDAP platform. The virtual sensor is an abstract representation of a physical/logical sensors in the CARDAP system.

Analytics Component: The analytics component feature allows developers and users to implement application-specific data analytics algorithms. CARDAP, incorporates an mobile activity recognition algorithm namely StreamAR [3]. The StreamAR algorithm is a personalised and adaptive framework for activity recognition that incrementally learns from evolving data stream. The developed framework deals with high speed, multi-dimensional streaming data to learn, model, recognise personalised user's activities. StreamAR system is divided into four phases. A supervised learning phase, where a learning model is built from a set of examples that describe the data domain. An unsupervised learning phase, that employs windowing technique on data stream in order to break down the unlabelled data. A recognition phase handled by an ensemble prediction technique based on a hybrid similarity measures algorithm and finally an incremental and continuous learning phase where the learning model is refined and updated in real time to reflect recent changes.

Open Mobile Miner: The open mobile miner component is the outcome of our CAROMM platform [18]. The CAROMM platform incorporates a data analysis and clustering engine. The engine employs the Light Weight Clustering algorithm [7]. The LWC algorithm uses data adaptation techniques to match high-speed data streams and achieves optimum accuracy based on available resources. CAROMM incorporates a change detect techniques that employs the LWC algorithm to continually monitor significant change in data. This approach has been effectively used by CAROMM for efficient data reduction (reduce number of data transmission) while maintaining a high level of data accuracy [18].

Data Sink Component: The data sink component depicted as Sink in Fig. 4 allows application to push data to any external sink. For e.g. push data to a publish/subscribe bus. This feature is an extension to MOSDEN as MOSDEN only allows local storage and querying of data. Combining the open mobile miner component and the sink, data reduction while transmitting data to a Cloud server can be achieved. The CARDAP platform incorporates sink functionality to upload data to Amazon Web Service (AWS)[1].

Storage and Query Component: The storage and query components of CARDAP perform the functions of storing processed data locally that can be later queried using a RESTful API over HTTP. This feature of CARDAP allows the platform to work independent of a global coordinator (scheduler) thus allowing autonomous task execution. On demand, the global scheduler can perform selective querying of captured and processed data. The query manager uses SQL to resolve incoming requests. The CARDAP platform supports both push and pull approaches to query data from the smart device.

The key feature of the CARDAP platform is its ability to facilitate development of new distributed data analytics applications by wiring the required components using XML configuration files.

5 Cost Model for Mobile Distributed Data Analytics

In this section, we develop cost models for the different distributed mobile data analytic approaches possible in the *CARDAP* . The cost models proposed for the different data collection approaches are:

5.1 Data Transmission Cost Model

The cost of data transmission from n devices to m servers for a considered time period t. The data transmission cost for each device i is defined:

$$Cost_i^{dt} = totaldata(bytes) \qquad (1)$$

---

[1] http://aws.amazon.com/

When multiple devices are considered the cost becomes:

$$Cost_{n*m}^{dt} = \sum_{i=1}^{n} Cost_i^{dt} * m \qquad (2)$$

Let us define the cost of sending all the raw data to the server(s) as $Raw.Cost_{n*m}^{dt}$ and the strategy k (for data collection) data transmission cost as $S_k.Cost_{n*m}^{dt}$. We calculate the bandwidth gain $Gain_{S_k}^{dt}$ of a strategy k in relation to the full raw data sending strategy as:

$$Gain_{S_k}^{dt} = \frac{Raw.Cost_{n*m}^{dt}}{S_k.Cost_{n*m}^{dt}} \qquad (3)$$

5.2   Energy Usage Cost Model

The cost of performing data analytics on the local device, in terms of resource consumption (e.g., energy). The energy drainage (%) on the battery of a single device for a time period t is modelled by $Cost^{eu}$. This value can be assessed for the different strategies used in $CARDAP$ and can be decomposed into $Cost_s^{eu}$ that is the impact that sensing plays on the drain, $Cost_{pr}^{eu}$ that is the energy cost of processing the data and the energy cost of transferring the data $Cost_{dt}^{eu}$.

$$Cost^{eu} = Cost_s^{eu} + Cost_{pr}^{eu} + Cost_{dt}^{eu} \qquad (4)$$

The $Cost_s^{eu} = freq_s * \alpha$ can be described as a function of the sensing frequency $freq_s$, where α is a constant for each device. The $Cost_{pr}^{eu} = CPU_\% * \beta$ component can be described as a function of the $CPU_\%$ usage and β is another constant defined for each device. The $Cost_{dt}^{eu}$ of transferring the data is a function of the number of (bytes) that need to be transferred. For each different strategy k each of the components of the energy usage cost model will be different. For the particular strategy of sending all the raw data $Raw.Cost^{eu}$ the $Cost_{pr}^{eu} \approx 0$, the $Cost_{dt}^{eu}$ is the maximum possible value and the $Cost_{pr}^{eu}$ takes the value of β since it represents the minimum CPU processing value. We evaluate the energy gain $Gain_{S_k}^{eu}$ of a strategy k $S_k.Cost^{eu}$ in relation to the full raw data sending strategy $Raw.Cost^{eu}$ as:

$$Gain_{S_k}^{eu} = \frac{Raw.Cost^{eu}}{S_k.Cost^{eu}} \qquad (5)$$

6   Implementation and Evaluation of CARDAP

6.1   Implementation

In this section, we present implementation details and experimental evaluation outcomes of CARDAP platform. For proof-of-concept implementation purposes, we consider a mobile crowdsensing scenario as described in the motivation section (Section 2). The CARDAP platform has been developed for the Android [2]

---

[2] http://www.android.com/

platform using the Android SDK v4.2.2. For experimentation, we used a Google Nexus 7 tablet (CPU: Quad-core 1.2 GHz Cortex-A9, Memory: 1GB). We implemented interface to sensors as discoverable plugins using the Android interface definition language (AIDL)[3]. This allows CARDAP platform to independently discover interfaces to on-board and external sensors.

The activity recognition engine namely StreamAR was implemented as a data analytics component on CARDAP. To test activity recognition, we wired the accelerometer sensor plugin with the StreamAR component. For experimentation purposes, we used accelerometer dataset from the WISDM lab work on activity recognition [14]. The dataset has 1.1 million data points with activities including walking, jogging, sitting, standing etc.

The LWC algorithm implemented as a part of the CAROMM [18] component in CARDAP detects significant changes in streaming data continuously. Since the activity recognition is state based i.e. either walking or sitting, for evaluation purposes, we use the on-board light sensor to detect significant change in environmental light.

6.2 Evaluation

We evaluate and validate CARDAP's resource and energy efficient performance against the following typical distributed mobile data analytics strategies:

- *Naive approach(baseline/raw data upload)*. All data is collected and sent to the Cloud for further processing (mobile does data collection based on a given time window)
- *Local Analytics (LA)*: Smart mobile device does local analytics and stores data locally which can be queried on demand. For experimentation, we incorporate mobile activity recognition as the local analytics.
- *Local Analytics + Smart data reduction + On-demand sensing (LA-DR-OS)*: Smart mobile device does local analytics, stores data locally and sends data when significant change in the processed data is detected and a pre-defined condition for upload is satisfied. E.g. Record the profile of users who are running within a given location from 5 PM to 8 PM and upload data only when significant change in light value is detected.

We note, the LA and the LA-DR-OS strategies are native features of the proposed CARDAP platform. Depending on application requirements, either of these strategies can be employed for a distributed mobile data analytics application like crowdsensing. The experimental evaluations also validate the performance gain achieved by employing the proposed CARDAP-based strategies namely *Local Analytics - LA* and *Local Analytics + Smart data reduction + On-demand sensing - LA-DR-OS*.

The resource consumption experiment compute the amount of memory and CPU consumed by the CARDAP platform when working under each of the

---

[3] http://developer.android.com/guide/components/aidl.html

aforementioned strategies. The memory consumption is measured in MB and the CPU consumption in jiffy[4]. The energy consumed by each strategy is computed in milli Watts(mW). To compute the energy and resource consumption, we implemented a modified version of the PowerTutor[5] open source android power monitoring application.

The results of the experimental evaluations are presented in Fig. 5a, 5b, 6a and 6b. We compute the gain for strategies *LA* and *LA-DR-OS* over the naive approach using equation (5). An experimental round consisted of replaying the WISDM dataset for each strategy and computing the average CPU, memory and power consumed by CARDAP platform for a time period of 1 hour. The sampling rate for activity recognition was 1 every minute. This resulted in 60 data uploads when experimenting CARDAP using the naive approach.

Fig. 5a and Fig. 5b presents the memory consumption of the CARDAP platform when evaluated under each strategy. It is to be noted that the memory and CPU allocation is controlled by Android operating system. Depending on the process workload, android may allocate more memory and CPU cores to maintain system stability. The CPU consumption of LA-DR-OS is lesser than LA as observed in Fig. 5a. Whereas, the memory consumption is vice-versa i.e. LA approach is lesser than LA-DR-OS as noted in Fig 5b. As stated earlier, since memory and CPU allocation is managed by android, the higher memory consumption trend can be attributed to the overheads involved in managing and maintaining the cluster i.e. continuous data stream clustering to monitor significant change in data streams.

A similar trend is observed with the average power consumption experiment presented in Fig. 6a. LA-DR-OS consumes more power when compared to LA approach due to the following factors 1) overhead to maintain and manage clusters and 2) network consumption due to upload of data when significant change in data is detected and satisfies a pre-defined condition. As indicated by the experimental outcomes, the CARADAP native approach namely LA and LA-DR-OS performs significantly better than the baseline approach of uploading raw data. This observation is supported by the energy gain outcome presented in Fig.6b.

Overall, experimental evaluations clearly validate the resource and energy efficiency of the proposed CARDAP platform irrespective of the strategy. Further, CARDAP's native approaches perform significantly better than the baseline approach making CARDAP strategies an efficient and effective technique to realise the development of energy and resource-efficient distributed mobile data analytics applications (e.g. crowdsensing). The proposed CARDAP platform is the first step in the development of a complete distributed mobile data analytics platform as presented in our big picture in Fig. 3.

---

[4] In computing, a jiffy is the duration of one tick of the system timer interrupt. It is not an absolute time interval unit, since its duration depends on the clock interrupt frequency of the particular hardware platform.

[5] http://powertutor.org/

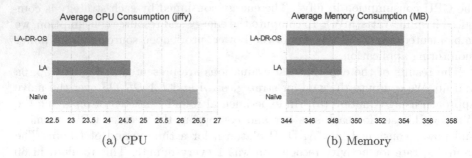

(a) CPU   (b) Memory

Fig. 5. Average CPU and Memory Consumption

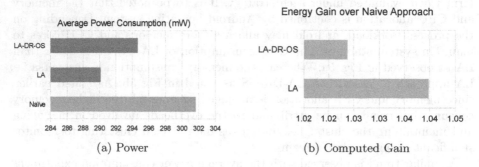

(a) Power   (b) Computed Gain

Fig. 6. Power Consumption and Gain

7   Conclusion

In this paper we have presented CARDAP, a context aware real-time data analytics platform for the Fog. CARDAP is a generic, flexible and extensible, component-based platform capable to deploy complex distributed mobile analytics applications such as on-demand distributed mobile crowdsensing. In addition, we discussed different real-world scenarios where using CARDAP can be significantly beneficial. CARDAP incorporates a number of energy efficient data delivery strategies employing real-time mobile data stream mining for data reduction. Our experimental evaluations indicate that the CARDAP platform can deliver significant benefits in terms of CPU, memory and energy efficiency over baseline approaches. In our future work, we aim to investigate the Cloud part of the proposed distributed data analytics model developing cost-efficient task scheduling and smart device selection approaches.

Acknowledgement. Part of this work has been carried out in the scope of the ICT OpenIoT Project which is co-funded by the European Commission under seventh framework program, contract number FP7-ICT-2011-7-287305-OpenIoT. The authors acknowledge help and support from CSIRO Sensors and Sensor Networks Transformational Capability Platform (SSN TCP).

References

1. Vision paper - distributed data mining and big data (August 2012)
2. The internet of things is poised to change everything, says idc (October 03, 2013)
3. Abdallah, Z., Gaber, M., Srinivasan, B., Krishnaswamy, S.: Streamar: Incremental and active learning with evolving sensory data for activity recognition. In: 2012 IEEE 24th International Conference on Tools with Artificial Intelligence (ICTAI), vol. 1, pp. 1163–1170 (November 2012)
4. Bonomi, F., Milito, R., Zhu, J., Addepalli, S.: Fog computing and its role in the internet of things. In: em Proceedings of the First Edition of the MCC Workshop on Mobile Cloud Computing, MCC 2012, pp. 13–16. ACM, New York (2012)
5. Carreras, I., Miorandi, D., Tamilin, A., Ssebaggala, E.R., Conci, N.: Crowd-sensing: Why context matters. In: 2013 IEEE International Conference on Pervasive Computing and Communications Workshops (PERCOM Workshops), pp. 368–371. IEEE (2013)
6. Gaber, M.M., Gama, J., Krishnaswamy, S., Gomes, J.B., Stahl, F.: Data stream mining in ubiquitous environments: state-of-the-art and current directions. Wiley Interdisciplinary Reviews: Data Mining and Knowledge Discovery 4(2), 116–138 (2014)
7. Gaber, M.M., Krishnaswamy, S., Zaslavsky, A.: Cost-efficient mining techniques for data streams. In: Proceedings of the Second Workshop on Australasian Information Security, Data Mining and Web Intelligence, and Software Internationalisation, ACSW Frontiers 2004, vol. 32, pp. 109–114. Australian Computer Society, Inc., Darlinghurst (2004)
8. Gaber, M.M., Stahl, F., Gomes, J.B.: Pocket Data Mining: Big Data on Small Devices, vol. 2. Springer (2014)
9. K., R., Ye, F., Ganti, H.L.: Mobile crowdsensing: current state and future challenges. IEEE Communications Magazine 49(11), 32–39 (2011)
10. Gomes, J.B., Krishnaswamy, S., Gaber, M.M., Sousa, P.A., Menasalvas, E.: Mars: a personalised mobile activity recognition system. In: 2012 IEEE 13th International Conference on Mobile Data Management (MDM), pp. 316–319. IEEE (2012)
11. Gomes, J.B., Krishnaswamy, S., Gaber, M.M., Sousa, P.A.C., Menasalvas, E.: Mobile activity recognition using ubiquitous data stream mining. In: Cuzzocrea, A., Dayal, U. (eds.) DaWaK 2012. LNCS, vol. 7448, pp. 130–141. Springer, Heidelberg (2012)
12. Jayaraman, P.P., Perera, C., Georgakopoulos, D., Zaslavsky, A.: Efficient opportunistic sensing using mobile collaborative platform mosden. In: 2013 9th International Conference Conference on Collaborative Computing: Networking, Applications and Worksharing (Collaboratecom), pp. 77–86 (October 2013)
13. Jayaraman, P.P., Sinha, A., Sherchan, W., Krishnaswamy, S., Zaslavsky, A., Haghighi, P.D., Loke, S., Do, M.T.: Here-n-now: A framework for context-aware mobile crowdsensing. In: Proc. of the Tenth International Conference on Pervasive Computing (2012)
14. Kwapisz, J.R., Weiss, G.M., Moore, S.A.: Activity recognition using cell phone accelerometers. SIGKDD Explor. Newsl. 12(2), 74–82 (2011)
15. Le, V.-D., Scholten, H., Havinga, P.: Towards opportunistic data dissemination in mobile phone sensor networks. In: Eleventh International Conference on Networks, ICN 2012, pp. 139–146. International Academy, Research and Industry Association (IARIA), France (2012)

16. Ravindranath, L., Thiagarajan, A., Balakrishnan, H., Madden, S.: Code in the air: simplifying sensing and coordination tasks on smartphones. In: Proceedings of the Twelfth Workshop on Mobile Computing Systems & Applications, p. 4. ACM (2012)
17. Roggen, D., Forster, K., Calatroni, A., Holleczek, T., Fang, Y., Troster, G., Lukow-icz, P., Pirkl, G., Bannach, D., Kunze, K.: Opportunity: Towards opportunistic activity and context recognition systems. In: IEEE International Symposium on World of Wireless, Mobile and Multimedia Networks & Workshops, WoWMoM 2009, pp. 1–6. IEEE (2009)
18. Sherchan, W., Jayaraman, P., Krishnaswamy, S., Zaslavsky, A., Loke, S., Sinha, A.: Using on-the-move mining for mobile crowdsensing. In: 2012 IEEE 13th International Conference on Mobile Data Management (MDM), pp. 115–124 (July 2012)
19. Yan, T., Kumar, V., Ganesan, D.: Crowdsearch: exploiting crowds for accurate real-time image search on mobile phones. In: Proceedings of the 8th International Conference on Mobile Systems, Applications, and Services, pp. 77–90. ACM (2010)
20. Ye, F., Ganti, R., Dimaghani, R., Grueneberg, K., Calo, S.: Meca: mobile edge capture and analysis middleware for social sensing applications. In: Proceedings of the 21st International Conference Companion on World Wide Web, pp. 699–702. ACM (2012)
21. Zaslavsky, A., Jayaraman, P.P., Krishnaswamy, S.: Sharelikescrowd: Mobile analytics for participatory sensing and crowd-sourcing applications. In: 2013 IEEE 29th International Conference on Data Engineering Workshops (ICDEW), pp. 128–135. IEEE (2013)
22. Zheng, Y., Zhang, L., Xie, X., Ma, W.-Y.: Mining interesting locations and travel sequences from gps trajectories. In: Proceedings of the 18th International Conference on World Wide Web, pp. 791–800. ACM (2009)
23. Zhou, P., Zheng, Y., Li, M.: How long to wait?: predicting bus arrival time with mobile phone based participatory sensing. In: Proceedings of the 10th International Conference on Mobile Systems, Applications, and Services, pp. 379–392. ACM (2012)

Representing Internal Varying Characteristics of Moving Objects

Ahmed Ibrahim[1], Ulanbek Turdukulov[2], and Menno-Jan Kraak[1]

[1] Faculty of Geoinformation Science and Earth Observation (ITC),
University of Twente, The Netherlands
[2] Western Australian School of Mines, Curtin University, Australia

Abstract. Recent data acquisition tools have resulted in huge amounts of data that have spatial and temporal components. The movement represents an important category of such data. Some phenomena may have attributes that vary continuously over space, such as wildfires and storms. Nevertheless, for simplification purpose, most applications represent such phenomena as objects by neglecting their internal continuous structure. Moreover, little consideration has been given to such characteristics in moving objects database. At this end, this paper presents a data model for managing raster data and internal heterogenous attributes in moving objects. The data model utilizes the abstract data types. We add two abstractions (moving raster, and combined type) to describe the change of the raster data and internal varying characteristics of the moving objects along with specific operations that permit to analyse them. Query examples are provided to demonstrate the application of these operations.

Keywords: Spatiotemporal data model, moving objects, internal varying characteristics, moving objects databases.

1 Introduction

Recent data acquisition tools have resulted in huge amounts of data that have spatial and temporal components. The movement represents an important category of such data. Moving objects can be human, animals, cars, and maybe also hurricanes, wildfire, or storm. Movement data are collected as discrete recordings of position and timestamp. In some applications, these moving objects may need to be coupled with other continuous dynamic phenomena such as temperature and wind; for example, bird movement might be influenced by temperature, or car movement can be affected by wind. Moreover, some moving objects could be inhomogeneous; they might have internally varying characteristics conceptualized as a field. These fields are usually represented as rasters that change over time, such as storms, wildfires, and hurricanes [1]. These values are usually clipped from a larger raster coverages. For example, the wind speed continuously varies within the wind storm extent and this variation changes during the movement and is clipped from larger raster coverage representing the wind

Y. Manolopoulos et al. (Eds.): ADBIS 2014, LNCS 8716, pp. 207–218, 2014.

speed. Nevertheless, for simplification purpose, most applications represent such phenomena movement as objects by neglecting their continuous characteristics. Gaining insight into these complex phenomena requires answers to queries that include the object movement and the raster representation. Consequently, there is a need to represent the movement of objects, the dynamics of raster data, as well as the internal structure of the moving objects.

Today, moving objects databases provides a comprehensive framework using a set of data types and operations to handle the continuous movement. The term moving in this technology does not only refer to the position change, but to the change in general. This database technology allows the user to model, store, retrieve, and query the objects change over time. Two main spatial abstractions, moving point and moving region, are used to represent the change of the location and shape/extent, respectively. They can represent the changes of homogenous attributes. However, this technology does not provide the enough support for either the raster representations or the internal heterogeneity of the moving objects. Therefore, it is needed to add new abstractions to provide such support.

This work focuses on the design and implementation of such abstractions in moving object database. The proposed model utilizes the abstract data types, adopted the sliced representation defined by Forlizzi et al. [2], and developed in PostGIS. First, we present a general moving data type to represent different types of temporal data. Our moving data types are based on observations with evolution function to describe the movement to the next observation. The spatial observations store a geometry or raster to represent the location/shape of an object or a continuous field. Second, we provide two data types: Moving Raster (or *mraster* in short) to model the change in raster data, then combined type (or *CType* in short) to represent the internal structure of moving objects. *CType* combines a moving geometry and moving raster by referencing them. The moving geometry defines the temporal and spatial extent of the moving object and the reference to the moving raster defines the internal structure of the objects within that extent. Finally, we extend the existing query language with new operations that are necessary in managing the new data types.

The remainder of this paper is organized as follows: in the Section 2, we provide the fundamentals of phenomena representation and the need to represent the internal structure of the moving objects. Section 3 summarizes the related work. Section 4 describes the proposed data model. It presents the new moving data types and some operations and demonstrates their use by expressing some query examples. Finally, we present our conclusions and guidelines for future work.

2   Phenomena Representation

Humans perceive the geographic phenomena in two main conceptual views either as object-based (discrete objects that occupy the real world) or field-based (continuously varying fields) [1, 3–5]. Field-based model conceptualizes the space as a set of locations related to each other and a set of attributes where each location is mapped to a value for each attribute [6]. Field-based model is used to describe the continuous variation across the space such as elevation, and temperature. This model is usually represented in tessalation or raster form. Object model conceptualizes the real world as a set of discrete entities called objects as schools, towns, cities, seas, or etc., with a set of attributes including the location are associated with each object such as for building, building name, number of people, owner. The objects are represented in vector form as point, line, and polygon. Perceiving the phenomena as either object-based or field-based is not sufficient to represent some phenomena such as wildfire and hurricane. These phenomena contain object-like and field-like characteristics and called complex phenomena[5] or multi-representations phenomena [7], or evolving moving objects [1]. These phenomena need to integrate both of object-based and field-based view into one model.

Goodchild et al. [1] defined three dimensions to describe the object changes that occur over time: movement, geometry, and internal structure. Each dimension differentiates between static and dynamic object behaviours. Movement dimension represents the object location and distinguishes between stationary and moving objects. Geometry represents the object shape and distinguishes between the rigid and elastic objects. Internal structure distinguishes between the uniform and evolving objects. Evolving objects describe the objects with internal heterogeneity. In other words, they have attributes that varies from one location to another within the spatial extent. According to these dimensions, two types of evolving moving objects can be distinguished: (1) Evolving Moving Rigid and (2) Evolving Moving Elastic. They have conceived as discrete objects that move, but also have internal variables within their spatial extent, which changes over time too (Fig. 1). For example, storm moves and its shape and severity changes over time. It can be represented as a moving object to define its boundary, but this representation does not describe the wind speed change along the storm lifetime. Given such objects, it is easy to realize the need to design a database

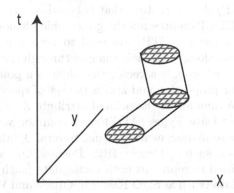

Fig. 1. Moving object with internal characteristics

model combining their continuous movement and varying internal structure. We consider the following motivating queries to illustrate this need throughout the paper:

- Query1: What are the regions with the maximum wind speed within a given storm?
- Query2: When did the maximum, minimum wind speed of a given storm enter/leave a given city?
- Query3: What is the minimum and maximum wind speed of a given storm within a given city?
- Query4: What is the maximum wind speed during each car movement within a given storm?

While the abstractions defined in moving objects databases are capable of representing the geometry movement, they are not suitable for the above queries. Indeed, the wind speed is represented in raster that associates a value to each location within the spatial extent of the storm and are part of a larger region of interest. Therefore, besides the representation of moving geometries, we need to represent the dynamics in raster data within moving object database technology. Then the integration of both representations to reflect the relationship between the internal structure and movement. This can be done by extending the existing model with new data types and operations for the continuous data.

3   Related Work

Different spatiotemporal data models have been proposed to represent the continuous characteristics of the objects. Three-domain model [8] consists of semantics, time, and space domains to represent the spatio-temporality of wildfire. Yuan [9] presented the logical structure for this model as semantic objects, temporal objects, spatial objects, and domain links as relational tables. EDGIS [10] abstracts the real world as set of atoms that can be used to represent both object-like and field-like characteristics of geographic phenomena. It uses the space-time point (STP) data structure that is based on geo-atom primitives of Goodchild et al. [1]. EDGIS represents the geographic phenomena in three concepts: STPs, features, themes. STPs are used to represent both discrete and continuous phenomena, along with their changes through time. Each STP is defined as $<x; Z; z(x)>$ where x is a vector that defines a point in space-time; Z is a set of attributes or properties; and $z(x)$ is the set of specific attribute values at that point in space-time for the associated attribute Z.

Nowadays, Abstract Data Types (ADT) are a common way to represent the spatial data in database management systems. Several database providers support spatial data types, such as Oracle, DB2, PostgreSQL, and Microsoft SQL Server. They can be used to represent both vector data (such as point, line, and polygon) and raster data (such as SDO.Raster in Oracle and WKTraster in Post-GIS). Erwig et al. [11] and Güting et al. [12] proposed a data model for moving objects that is able to define the continuous movement using a set of abstract data types and a collection of operations. They introduced a type constructor moving that transforms static types into moving types. The sliced representation decomposes movement into a set of slices, where the movement within each slice can be described by a simple function. A set of operations is defined on these data types to manage the movement data. Using the temporal lifting process on the operation of static types, these operations can be applied over the moving data types. Since then, ADTs have received a growing attention for representing the movement in databases. Kim and Kiyoki [13] presented a moving data type to model the continuous fields of moving objects by decomposing the coverage into geometric features and dealing with each part as moving geometry with value.

4 Proposed Data Model

This paper focuses on the representation of evolving moving objects, i.e., moving objects that have internal variables changing continuously over time. The aim is to propose a data model to represent those objects. To this end, we introduce a new data type (moving raster) to model raster data changing over time, then a new data type to combine the movement and raster data. The idea is to apply the movement concept of moving geometry for raster data. As a consequence, we can provide the continuous movement (change) in field data. The proposed model, developed in PostGIS, adopted the discrete model defined by Forlizzi et al. [2] that decomposes movement into a finite set of units as a pair of a period and a function that maps the time to a value. Indeed, the movement data are usually collected as a set of observations at discrete times $(t_1, .., t_n)$. Therefore, the movement can be represented by storing the observations at each of these times, together with an interpolation method to be used for the period $[t_i, t_{i+1}]$.

We informally define an observation as the state of thematic or spatial attribute of the phenomenon under study at a time. The observations can be viewed from different perspectives, depending on the type of attributes under interest. The observation can be thematic, geometric, or raster. The thematic observations define the thematic (non spatial) attributes, where each can be represented as base types such as a real, integer, bool, or string. Geometric observations are used to describe spatial attributes represented as point, line, or polygon. The raster observations represent regular tessellation-based data, where each value is represented as a raster. For example, the raster observations can represent the wind speed variation, whereas in the car movement, the geometric observations represent the positions of the car and base observations represent the number of occupants.

Definition 1. *An observation is a pair of value (V) and instant (T).*

Observation $:= \{ (V, T) \mid V \in Base \cup Geometry \cup Raster \wedge T \in Instant \}$

Where V represents the state of the attribute and T is the observation time. *Base* types and *Instant* are mapped into the corresponding data types defined in Postgres. *Geometry, Raster* are mapped into the corresponding data types defined in PostGIS. We consider that the observation is a unified unit to represent movement.

Definition 2. *A moving type is an ordered collection (n) of observations with an evolution function (F) to describe the change over the observation to the next observation.*

$$Moving := \{ ([O], F) \mid O \in Observation \wedge$$
$$F \in \{LINEAR, CONSTANT, NEARESTNEIGHBOR\},$$
$$i. \quad (O_i.T < O_{i+1}.T) \wedge$$
$$ii. \quad (n \geq 2) \}$$

The first condition does not allow the temporal overlay. The second condition ensures that at least two known observations exist in the moving type. Three evolution functions are available: Linear, Constant, and NearestNeighbor. Different moving types can be defined based on the observation type, where the type is described by its modifier. The evolution function and the spatial reference identifier (SRID) can be given by the modifier of the moving data type.

4.1   Raster and Movement

mraster models the evolution of regular tessellation-based data. We define *mraster* as an ordered collection of raster observations with an evolution function. This function describes the change from the value of each pixel in an observation to the value of the corresponding pixel in the consequent observation. This means that we can retrieve the raster coverage at any instant in the temporal extent of *mraster* by estimating the value of each pixel within the coverage at this time using the evolution function. In this work, we assume that the coverage width, height, and pixel size do not change over the time.

Three modes are available for representing the change in a raster coverage using this abstraction: two observations per row (whole table describes change in one raster coverage), all observations within one row (one row represents the whole change of this raster coverage), or hybrid (set of observations per row). Each mode has its advantages and disadvantages, and the selection of mode is based on different criteria, particularly the raster size, lifetime, and computer memory. The third mode is selected in the current prototype.

Several operations are needed to manipulate the moving raster. Those include the extension of static raster operations supported by PostGIS, such as Resampling and Union. The new versions apply the procedure to every observation in the moving raster. In addition, we defined a set of new operations to: (1) deal with temporal data, such as *ST_TemporalAggregate* to compute a temporal raster at a given granularity; (2) deal with integration with moving geometries, such as *ST_Integrate* to aggregate the raster cells that intersect the moving geometries; (3) restrict the raster data to specific parameters such as *ST_AtMax* and *ST_AtMin*; and (4) summarize the pixel values, such as *ST_Max* and *ST_Min*.

4.2   Representation of Evolving Moving Objects

Evolving moving objects combines two types of attributes: (1) discrete attributes that describe the object-based characteristics such as position and spatial extent; (2) continuous attributes describe the internal varying characteristics within their spatial extent such as wind speed of the storm. However, the latter is not a mere separated coverage. This may be the case for some phenomena, but more often, these attributes are part of larger region of interest. The proposed approach uses the above desribed moving types to represent these both attributes. The object-based attributes are represented as moving base and moving geometry. The main raster coverage is represented as moving raster in different table. Consequently, a new moving data type *CType* is defined to store a reference to the main moving raster table as an internal structure and moving geometry as spatial extent. To formally define *CType*, we need to define two auxiliary types:

Definition 3. *Let mrasterId be uniquely identified mraster defined as:*

$$mrasterId := \{ \, (\, Id, mrast \,) \, | \, Id \in int \wedge mrast \in mraster \, \}$$

Definition 4. *let mrasterIdSet be a set of identified mraster defined as:*

$$mrasterIdSet := \{ \, [mrastId] \, | \, mrastId \in mrasterId, \\ (\, \forall i \neq j \Rightarrow mrastId_i \neq mrastId_j \,) \}$$

Definition 5. *CType is a a pair of moving polygon and identified mraster set as:*

$$CType := \{ \, (\, mgeom, mrastset) \, | \, mgeom \in mpolygon \wedge \\ mrastset \in mrasterIdSet, (mrasterset.mrast \cap mgeom \,) \}$$

The condition ensures that the moving raster spatiotemporaly intersect the moving geometry within *CType*. The continuous spatial component of this type is identified using the references to the moving raster by spatiatemporal overlay with the moving geometry data. Fig. 2 shows how *CType* is perceived and represented. *CType* attempts to represent the internal structure of a moving object. Therefore, the operations on this type contain: movement operations and raster operations. Movement operations presented in Lema et al. [14] are overloaded to accept moving raster and *CType* as one of their inputs, such as diftime, at, atperiod. In addition, some functions, defined for moving raster, are extended to deal with the raster part referenced by this type, such as *ST_Integrate*, *ST_ATMIN*, and *ST_ATMAX*. The spatial part of those are implemented on the moving polygon that represents the spatial extent within *CType*. In the next paragraphs, we present the algorithm to return the snapshot of *CType* at a given time.

ST_Snap. This operation maps the time to the raster part in *CType*. It returns a raster type that represents the projection of the moving raster component at a given time instant or null if it does not exist. This operation is done in three main

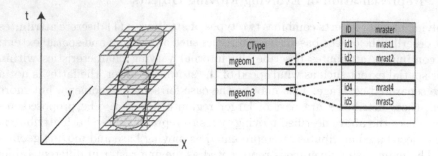

Fig. 2. *CType* representation

steps: retrieve the spatial extent at a given time, spatially overlay the resulting extent with the raster coverage; and compute the pixel values using the defined evolution function. This operation first checks for the existence of the given time during the temporal extent of the combined type. If true, it retrieves the spatial extent using *St_AtInstant* function of the moving geometry component within *CType*. Then, it finds the observation index of the given time instant. If it is equal to the observation time, it simply returns the result of a spatial overlay of spatial extent with the raster data of this observation. If not, it creates an empty raster with an initial value for the pixels that spatially overlay the spatial extent, then it invokes another sub-algorithm to calculate the values at this time, based on the evolution function defined for the moving raster component. The result of this operation is a new raster with values for the pixels in the spatial extent and nodata for others.

```
Algorithm: ST_Snap
Input: ctype: Combined Type, gt:Instant.
Output: raster at this time or null if unknown
Begin
If (gt is not between ctype.TemporalExtent) then
Return NULL;
//use the St_AtInstant function for moving geomtery to
 get the spatial extent t this time.
Geom := St_AtInstant(ctype.Mgeomtery, gt)

//Get the Observation index
// to check if the index is equal to observation Instant
InstEqualInd := False
Index := GetInstantIndex(ctype, gt, InstEqualInd)

// if the given instant is equal to one of the
observations time, then Overlay with the observation
snapshot, and return the result
If ( InstEqualInd ) then
```

```
Begin
Temp_rast := ctype.Mraster.Observations[Index].Value
Return GetpixelOverlay(Temp_rast, Geom)
EndIf
// Get Empty raster with an initial value for pixels that
overlay the geometry and nodata for others.

Rast := IntialpixelOverlay(ctype.Mraster, Geom, 0)
//Interpolate to fill the value of the required pixels
Obs0 := ctype.mraster.observations[i-1]
Obs1 := ctype.mraster.observations[i]
Return raster_Interpolate(rast, Obs0, Obs1, gt,
   ctype.mraster.Evolfunction)
End


Algorithm: raster_ Interpolate
Input: raster: Raster, Observation0: Raster_Observation,
Observation1: Raster_Observation, gt: Instant, efunc:
Evolution Function
Output: raster at this time or null if unknown
Begin
Tstart   := Observation0.ts
Tend   := Observation1.ts
Rast0   := Observation0.value
Rast1   := Observation1.value

//iterate for the pixel that have initial value and
compute the new value.
Foreach (IntializedPixel pixel in     raster)
New_val := linear_interpolate (val(Rast0,pixel),
val(Rast1,pixel),  Tstart,  Tend, t)
Setvalue (raster , pixel, New_val);
Loop;
Return raster;
End
```

4.3 Query Samples

To demonstrate the capabilities of the data types defined above, we present answers to a set of queries related to a storm movement analysis. These data types are integrated in a DBMS data model and the operations are invoked through SQL queries. We will present the query examples using a relational schema with three tables, that contain geometry information on storms, cities, and car trips, and one table stores information on wind as moving raster data as follows:

```
City: (CityID :Int, CityName:varchar, CityGeom: Geometry(Polygon))
Car_Trip: (TripID:Int, CarID:Int, OwnerID:Int,
            Trip:Moving(Point,4236,Linear))
Wind: (RID:Int, Winddata: Moving(Raster,4236,Linear))
Storm: (StormID:Int, Geom: Moving(Point,4236, polygon),
        Intensity: Ctype)
```

City relation contains a set of polygon geometries describing the territories of several cities. Car_Trip relation stores the cars movement within cities where Trip attribute is a moving point representing the trajectory within the city. Wind relation stores the change in the wind of over specific region and as moving moving raster. Storm relation stores the storm events that pass the cities where Geom is a moving polygon representing its shape and Intensity is a *CType* representing the internal varying wind speed within the storm by referring to Wind relation. The Intensity attribute has a moving polygon as Geom attribute to represent the geometric component of *CType*.

– Query1: What are the regions with the maximum wind speed within a given storm?

```
SELECT (ST_DumpAsPolygons ( ST_ATMax ( Intensity ) ) ).geom
 AS Areas
FROM Storm
WHERE stormID=sid;
```

– Query2: When did the maximum wind speed of a given storm enter and leave a given city?

```
SELECT ST_Deftime ( WindInCity.MaxWind ).TStart
AS Start Time, ST_Deftime (WindInCity.MaxWind).End
AS Leave Time
FROM
(SELECT ST_ATMax (ST_Clip ( intensity, CityGeom) ) AS
 MaxWind  FROM Storm, City WHERE StormID = id and
 ST_Intersects (Intensity, CityGeometry) ) AS WindInCity
```

– Query3: What is the maximum wind speed during each car movement within storm id?

```
SELECT ST_Integrate(Intensity, Trip,Max)
FROM
FROM Storm, Car_Trip
WHERE StormID = id
```

– Query 4: What are the minimum and maximum wind speeds of storms that pass a given city?

```
SELECT ST_Max(Intensity), ST_Min (Intensity),
FROM
FROM Storm, City
WHERE CityID = cid And ST_Passes(Intensity, CityGeom)
```

5   Conclusion and Future Work

Early works on moving objects databases have focused on the definition of data models and query languages for dealing with objects where their geometry change over time. However, those geomteries are usually related to other geographic phenomena represented in raster form. In addition, some phenomena have an internal varying structure that changes during the movement. These objects such as storms, are often modelled as moving polygon, neglecting their internal structure. The main focus of this work was to design and implement a data model for representing the change in raster data of such moving objects. The proposed data model is based on abstract data types and implemented on the top of PostgreSQL DBMS and PostGIS.

The model follows the basic movement concepts used in sliced representations and introduces new data types that allow representing the change in raster data as well as the internal structure of moving objects. Observation is used to build the moving types as a collection of observations and an evolution function. This function describes the change to the next observation in the collection. The value of raster observations is stored as Raster type defined by PostGIS. Moving raster is defined as an ordered collection of raster observations with an evolution function to describe the change of pixel values to the next observation. Several operations are defined to manipulate the moving raster, dealing with temporal data, integration with moving geometries, restricting the raster data to specific parameters, and summarizing the pixel values.

CType is defined to represent the internal varying attributes in the moving objects. It stores the references to the a set of moving raster records as the internal structure and moving geometry as the spatial extent. The temporal and spatial data are identified using the moving geometries, and the raster data are identified using the spatiotemporal overlay with the referenced moving rasters. Several operations for this type need to deal with both its raster and movement components. These operations are defined by overloading those of moving geometry and moving raster. Storing only the reference to moving raster records creates some challenges related to the performance because they need to implement the overlay when this part is acquired. On the other hand, this eliminates storing the raster data many times.

The future research will include the following three phases: (1) implement an interactive tool for querying, mining and visualizing and this type of data; (2) test the current data types with different case studies to meet the needs of different query types; and (3) evaluate the performance of such data types, since they are currently used for small raster datasets.

References

1. Goodchild, M.F., Yuan, M., Cova, T.J.: Towards a general theory of geographic representation in GIS. International Journal of Geographical Information Science 21(3), 239–260 (2007)

2. Forlizzi, L., Güting, R.H., Nardelli, E., Schneider, M.: A data model and data structures for moving objects databases. SIGMOD Rec. 29(2), 319–330 (2000)
3. Cova, T.J., Goodchild, M.F.: Extending geographical representation to include fields of spatial objects. International Journal of Geographical Information Science 16(6), 509–532 (2002)
4. McIntosh, J., Yuan, M.: A framework to enhance semantic flexibility for analysis of distributed phenomena. International Journal of Geographical Information Science 19(10), 999–1018 (2005)
5. Yuan, M.: Representing Complex Geographic Phenomena in GIS. Cartography and Geographic Information Science 28(2), 83–96 (2001)
6. Galton, A.: Space, Time, and the Representation of Geographical Reality. Topoi 20(2), 173–187 (2001)
7. Galton, A.: Fields and Objects in Space, Time, and Space-time. Spatial Cognition & Computation: An Interdisciplinary Journal 4(1), 39–68 (2004)
8. Yuan, M.: Wildfire conceptual modeling for building GIS space-time models. In: GIS/LIS, pp. 860–869 (1994)
9. Yuan, M.: Use of a Three-Domain Repesentation to Enhance GIS Support for Complex Spatiotemporal Queries. Transactions in GIS 3(2), 137–159 (1999)
10. Pultar, E., Cova, T.J., Yuan, M., Goodchild, M.F.: EDGIS: a dynamic GIS based on space time points. International Journal of Geographical Information Science 24(3), 329–346 (2010)
11. Erwig, M., Güting, R.H., Schneider, M., Vazirgiannis, M.: Spatio-temporal data types: An approach to modeling and querying moving objects in databases. GeoInformatica 3(3), 269–296 (1999)
12. Güting, R.H., Bhlen, M.H., Erwig, M., Jensen, C.S., Lorentzos, N.A., Schneider, M., Vazirgiannis, M.: A foundation for representing and querying moving objects. ACM Transactions on Database Systems 25(1), 1–42 (2000)
13. Kim, K.-S., Kiyoki, Y.: An Object-Field Perspective Data Model for Moving Geographic Phenomena. In: Yoshikawa, M., Meng, X., Yumoto, T., Ma, Q., Sun, L., Watanabe, C. (eds.) DASFAA 2010. LNCS, vol. 6193, pp. 410–421. Springer, Heidelberg (2010)
14. Lema, C., Antonio, J., Forlizzi, L., Güting, R.H., Nardelli, E., Schneider, M.: Algorithms for Moving Objects Databases. The Computer Journal 46(6), 680–712 (2003)

User Identification within a Shared Account: Improving IP-TV Recommender Performance

Zhijin Wang, Yan Yang, Liang He, and Junzhong Gu

Department of Computer Science and Technology,
East China Normal University,
Shanghai, China
zhijin@ecnu.cn, {yangyan,lhe,jzgu}@cs.ecnu.edu.cn

Abstract. Multiple users share a common account in Internet Protocol Television (IP-TV) services. Can such shared accounts be identified solely on the basis of logs recorded by set top boxes (STBs)? Once a shared account is identified, can the different users sharing it be identified as well? We suppose different users within a shared account not only have different preferences for TV programs, but also get used to consuming services in different periods (e.g., after dinner or at weekend). We propose an algorithm to decompose users in composite accounts based on mining different preferences over different periods from consumption logs. In our experiments, the proposed algorithm outperforms traditional user-based collaborative filtering method 3-8 times when leveraging the decomposed users for personalized recommendation.

Keywords: User identification, Shared account, IP-TV recommendation, Experimentation.

1 Introduction

Internet Protocol Television (IP-TV) [10,16] services have been widely consumed in our daily life. It's a common phenomenon that families or roommates share television programs after they get back home or dormitory from work. With the development of IP-TV technologies, more and more multimedia resources (e.g., channels, programs and videos) are integrated into the services. In order to retrieve the preferred programs efficiently, recommender systems [1] are introduced into the services for IP-TV recommendations.

User experience can be improved when recommender systems are introduced into the IP-TV services. However, user identification is one of the most challengeable problems which degrades recommender performance [4]. The use of a single account by multiple users poses a challenge in providing accurate personalized recommendations. Informally, the recommendations provided to a shared account, comprising the ratings of two dissimilar users, may not match the interests of either of these users [21].

We aim to improve recommender performance by addressing the challenge of user identification in IP-TV services. According to a log recorded by a STB, a log contains: account id, program id, start time, end time, and genre(s).

Y. Manolopoulos et al. (Eds.): ADBIS 2014, LNCS 8716, pp. 219–233, 2014.

The user information is unavailable since the services are indistinctly shared by the users in a shared account. What's more, the interaction between user and set top box (STB) is very weak. Therefore, none of individual user informationcan be directly used for recommendation.

We suppose users within a shared account not only have different preferences for programs, but also get used to consuming services in different periods (e.g., after dinner or at weekend). There are two questions: (1) How to capture user preference over period accurately? (2) How to identify users based on the captured user preference over period?

To address these questions, we define: (1) a period which consists of non-overlapping sub-period(s); (2) a user who consumes services during these sub-periods. Based on these definitions, we propose an algorithm to carry out identification task.

To summarize, our contributions are as follows:

1. To the best of our knowledge, we are the first to study user identification as a problem of identifying preference and consumption time within a shared account.
2. We consider that user consuming behavior is periodic in IP-TV services, and a continuous period consists of non-overlapping sub-periods. Hence, the account preference over period can be captured by leveraging the designed implicit rating technique in the $\{account \times item \times time\}$ 3-dimensional space [2].
3. Based on the captured preferences, we design a user identification algorithm to identify users which mainly includes virtual user split stage and virtual user merge stage.
4. Finally, we demonstrate how the methods above can be applied to improve recommendation. We also study the effects of user splitting and user merging on recommendation.

The rest of the paper is organized as follows. Section 2 provides a brief review of related work on user identification and IP-TV recommendation. Section 3 describes our proposed approach to carry out identification task. Section 4 shows the settings in our experiments. Section 5 presents the experimental results. Section 6 analyses the users identification and recommender performance. Finally, we conclude in Section 7.

Table 1. Symbols

Symbol	Description
A, I, U	account set, item set, identified user set
P	the continuous period
p_k	the sub-period within P
v_{ak}	the virtual user within account a in sub-period k
s_h	several sub-periods within P
u_{ah}	the h-th identified user within account a
$u_{a\cdot}$	the identified users within account a
$|A|$	number of all accounts
$|I|$	number of all items
$|U|$	number of all identified users
$|P|$	number of all sub-periods
G	similarity graph of virtual users
d_{ait}	the duration of item i consumed by account a in sub-period t
r_{ait}	the implicit rating of account a to item i in sub-period t
S_{aij}	the preference similarity between virtual user i and j within account a
ρ	the threshold of preference similarity between two virtual users

Table 1 gives the main symbols used in this paper.

2   Related Work

In this section, we introduce related work on user behavior in IP-TV services at first, and then describe the current research on user identification for recommendation which related to our proposed algorithms.

2.1   User Behavior in IP-TV Services

User behavior data can be categorized into implicit feedback [14], explicit feedback [5], and their combination [15]. Unlike searching webs with strong intentions [3,7] (e.g., click links, type text or speech input in search bars), user behavior in IP-TV services is often implicit or unconscious.

One of the most common used implicit rating technique in IP-TV services is binary rating $r_{ui} \in \{0, 1\}$(e.g., [8,20]). But it can't capture the difference of duration within a shared account, hence, the percentage of play time the account has watched is exploited to capture account preference which we will further discuss in 3.2.

In the aspect of temporal features, user consuming behavior can be divided into long-term & short-term behavior [6] and periodic [12] behavior. We suppose that user consuming behavior (temporal feature) in IP-TV services is periodic, and the preferences for programs may change as time goes by.

2.2   User Identification for Recommendation

It's still a challenge to identify users of a STB since the STB is typically used by multiple users, e.g. family members or roommates. The issue of user identification within a shared account has received attention only recently.

Zhang et al. studied the user identification as a subspace clustering problem at [21], a composite account was regarded as a union of linear subspaces and used subspace clustering for carrying out identification task. They applied EM, GPCA and other clustering algorithms to identify users who belong to the same account on CAMRa2011 [18] dataset and Netflix [5] dataset. Their target was to label accounts to users. In contrast, we are trying to decompose users within a shared account.

Said et al. at [19] regarded time as a type of contextual information, and used it to split account into sub-profiles which is a method of recommending movies to specified users. A single user profile was split into several, possibly overlapping, contextual sub-profile (home and cinema sub-profiles as presented) in contextual pre-filtering stage. The split sub-profiles were integrated into recommenders in contextual post-filtering stage. However, they didn't consider incorporating overlapping contextual sub-profile into one that often decreases recommendation performance somehow which we will discuss in section 5.

3 Our Proposed Approach

In this section, we illustrate our proposed approach. The approach contains problem definition, notations and user identification algorithm named VUI.

3.1 Problem Definition and Notations

To start with, let us consider a common scene that multiple users share a common account in IP-TV services. As figure 1 shows, an account corresponding to a STB shared by 3 kinds of family members: senior, younger and kids. The senior get used to demanding history series in the morning or afternoon, kids would like to play the sort of cartoon programs after school or dinner, and younger might prefer films after kids go to the bed.

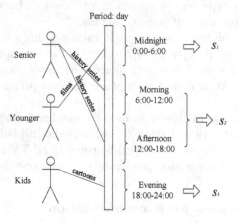

Fig. 1. An example of users sharing an account

From the common scene above we can find: (1) the consuming behavior is periodic; (2) different users get used to consuming the services in different part(s) of a period; (3) different users often have different preferences for programs (genres) provided by the services. Based on this common phenomenon, the problem of identifying users sharing a common STB can be regarded as distinguishing preference over period.

We introduce period P to describe periodic behavior. The period is defined as:

$$P := \bigcup_{k=1}^{|P|} p_k, \emptyset := \bigcap_{k=1}^{|P|} p_k \tag{1}$$

this definition means that a continuous period P (e.g., week) consists of several, non-overlapping sub-periods p_k (e.g., day of week). A user consumes services in more than one sub-periods, therefore the user is defined as:

$$u_{ah} := \{a, s_h | a \in A \cap s_h \subseteq P \cap s_h \neq \emptyset\} \tag{2}$$

where u_{ah} is the h-th user within account a who consumes services at s_h (e.g., Saturday and Sunday) in period P, A denotes all accounts in the system. As figure 1 provides, the senior consume services at s_1, the younger consume services at s_2, and kids s_3. Hence, all users U in system can be defined as:

$$U := \bigcup_{a \in A} \bigcup_{s_h \subseteq P} u_{ah} \tag{3}$$

Our goal is to identify U, and providing accurate recommendations for A by means of U. We are trying to reach the goal by addressing the two questions: (1) How to capture the preferences of user u_{ah} (h-th user within account a)? (2) How to determine the consuming time s_h of h-th user within each account?

In order to capture the preferences of user u_{ah}, we introduce virtual user v_{ak} to present activities of an account in a sub-period. The virtual user is defined as:

$$v_{ak} := \{a, p_k | a \in A \cap p_k \in P \cap p_k \neq \emptyset\} \tag{4}$$

where v_{ak} means the activities of account a in sub-period k, and an identified user is a composite of virtual user(s). Therefore, the user preference can be composed by the preferences of corresponding virtual users. The problem of determining the consuming time s_h of h-th user within account a is equivalent to assigning virtual users to identified users within an account.

We suppose that users in reality have different preference for both programs (or genres) and periods. Hence, the users in an account can be identified by combinations of virtual users. In order to study how the combinations affect identified users and recommendation performance, we introduce the similarity graph G, which uses vertexes to denote virtual users, and uses edges to denote the similarity between virtual users. The similarity graph G is defined as:

$$G := G(v_{a\cdot}, s_{a\cdot\cdot}) \tag{5}$$

where $v_{a\cdot}$ presents all virtual users within account a, $s_{a\cdot\cdot}$ presents similarities among all virtual users.

3.2   Algorithm for User Identification

In this section, we illustrate the algorithm for identifying users within a shared account. Detail steps of Virtual user based User Identification algorithm (VUI) are given in Algorithm 1.

Implicit Rating Technique in 3d Space. We adopt time concerned 3-dimensional space $\{account \times item \times time\}$ to present account preference over period (or sub-periods). As figure 2 shows, the coordinates denote accounts A, items I and period P, the symbols (e.g., triangle, square, and star) in it means the corresponding programs consumed over period. And items present programs or genres. We formulate the implicit rating as follows:

$$r_{ait} = \frac{\exp(d_{ait})}{\sum_{t \in P} \exp(d_{ait})} \tag{6}$$

Algorithm 1. Pseudo code of VUI to identify users within shared accounts.

Input:
  A, All accounts; I, All items; P, All sub-periods;
  D, Duration of accounts to items.
Output:
  U, All identified users.
 1: // Implicit ratings R
 2: **for** each account a, item i, sub-period t in A, I, P **do**
 3: $r_{ait} \leftarrow$ calculate implicit rating by means of d_{ait}.
 4: **end for**
 5: **for** each account a in A **do**
 6: // Split
 7: VirtualUsers V;
 8: **for** each sub-period p_k in P **do**
 9:   $V.append(v_{ak})$;
10: **end for**
11: SimilarityGraph G;
12: **for** each virtual user i in V **do**
13:   **for** each virtual user j in V **do**
14:    $S_{aij} \leftarrow$ calculate preference similarity between i and j by means of implicit rating R;
15:    $s_{aij} \leftarrow S_{aij}, \rho$ // Threshold ρ
16:    **if** $s_{aij} == 1$ **then**
17:     $G.insert(vertex(i), vertex(j), s_{aij})$;
18:    **end if**
19:   **end for**
20: **end for**
21: // Merge
22: **for** each vertex v in $G.vertexes$ **do**
23:   **if** $v.visited == false$ **then**
24:    $v.visited = true$;
25:    $u_{ah}.add(v)$; // Add v to h-th user;
26:    Get list of vertexes that connect to v as L;
27:    Visit each vertex in L and add to u_{ah} recursively;
28:    $h = h + 1$;
29:   **end if**
30: **end for**
31: $U.add(u_{a.})$; // Add identified users within account a
32: **end for**
33: **return** U;

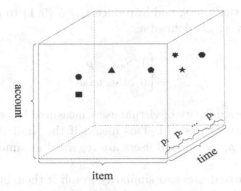

Fig. 2. $\{A \times I \times P\}$ 3d space

where d_{ait} is the duration of item i consumed by account a in sub-period t. Note that an account may demand an item more than one time, therefore, we do not choose the binary implicit rating techniques [8,20] or percentage of a program watched to the length of it [17]. The reasons are: (1) Accuracy. The binary implicit rating technique can describe a user has scanned a video, but it can't describe the degree of how the program is preferred. (2) Without length of programs. The provided dataset does not have the length of a file in terms of showing time of programs.

Split of Virtual Users. The split of virtual users is the process of determining sub-periods. We try two different methods to determine sub-periods. One is (1) empirical split method. In this method, the sub-periods are assigned by experience. The other is (2) average split method. The method split P in k equal length sub-periods. The split of virtual users is described at line 7-10, Algorithm 1.

Similarity Measurement between Virtual Users. Once virtual users in an account are obtained, the similarities among them can be measured, we adopt cosine method to measure the similarity between two virtual users since it's widely used. The similarity between virtual user v_{ai} and virtual user v_{aj} is defined as:

$$S_{aij} = \cos(v_{ai}, v_{aj})$$
$$= \frac{\sum_{k \in I(v_{ai}) \cap I(v_{aj})} r_{aki} \cdot r_{akj}}{\sqrt{\sum_{k \in I(v_{ai})} r_{aki} \cdot \sum_{k \in I(v_{aj})} r_{akj}}} \qquad (7)$$

where r_{aki} denotes items k consumed by account a in sub-period i, and $I(v_{ai})$ denotes items set consumed by virtual user v_{ai}. The similarity measurement is used at line 14, Algorithm 1.

Similarity Threshold ρ. In order to control the process of combinations of virtual users in the similarity graph, we use parameter $\rho \in [0, 1]$ to threshold the similarities of virtual users, and introduce $s_{aij} \in \{0, 1\}$ to present similarity. The binary similarity s_{aij} is defined as:

$$s_{aij} = \begin{cases} 1, S_{aij} \geqslant \rho \\ 0, otherwise \end{cases} \tag{8}$$

where S_{aij} means the similarity of virtual users measured by cosine method, and ρ denotes the similarity threshold. This means if the similarity of two virtual users is greater than ρ, the virtual users are regarded as similar, otherwise not similar.

We add similar virtual users to similarity graph if their binary similarity is equal to 1. The thresholding process is described at line 15-18, Algorithm 1.

Merging of Similar Virtual Users. According to steps above, the similarity graph can be obtained. Once a similarity graph is generated, the users can be identified by merging similar virtual user.

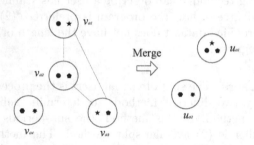

Fig. 3. An example of merging virtual users

As figure 3 shows, an example of merging similar users on similarity graph, the connected vertexes are merged to one as an identified user. In a word, we adopt deep-first-search (DFS) algorithm to carry out the merging task. An alternative way to carry out the task is bread-first-search (BFS) algorithm. The merge operation is described at line 22-31, Algorithm 1.

4   Experimental Setup

In this section, we illustrate dataset collection, evaluation metrics and algorithms for recommendation. We evaluate algorithm VUI on the dataset collected from the content provider SMG[1] in Shanghai, China. It should be noted that we focus on the evaluation of recommender performance by means of identified users, rather than the accuracy evaluation of algorithm VUI.

---

[1] http://www.smg.cn/

4.1   Dataset Collection

The logs in the services from SMG are during the period between March 1, 2011 till March 31, 2011. A log describes an account consumed a movie as well as genre, also the start time and end time of the services. We filter out logs of play time (calculated by start time and end time) less than 10 minutes. It contains 376,038 records, 5,933 videos categorized into 66 genres consumed by 14,856 accounts after being filtered. The records before March 25, 2011 are used for training, and the rest are as test set.

In order to avoid problems related to cold start, for both accounts and items, we decide that accounts in the evaluation sets have to consume at least 100 programs. We evaluate our results on a subset of 100 randomly selected accounts due to the long running time of the experiments when the full dataset is used.

4.2   Evaluation Metrics

We use Precision and Recall metrics to measure the performance of all the mentioned algorithm, since they often attract lots of attention in a running system and are well known. The Precision metric is defined as:

$$Precision@N = \frac{\sum_{u \in U} |R(u, N) \cap T(u)|}{\sum_{u \in U} |T(u)|} \tag{9}$$

where N denotes the length of a recommendation list, $R(u, N)$ denotes the recommendation list to user u with length N, $T(u)$ means items has been consumed by identified user u in test set. The Recall metric is defined as:

$$Recall@N = \frac{\sum_{u \in U} |R(u, N) \cap T(u)|}{\sum_{u \in U} |R(u, N)|} \tag{10}$$

From these definitions, we can see that a larger $Precision@N$ or $Recall@N$ indicates a better performance.

4.3   Recommendation Algorithms

We adopt, one of the most famous collaborative filtering methods, K-Nearest Neighbor (KNN) method to provide recommendations, since it performs very well in practice (e.g., [9,11,13]), and we can also learn the benefit from identified users by comparing with recommendations without identification.

The Cosine method is used to measure the similarity among accounts in algorithm KNN. For convenience, we name recommendations for accounts as AccountKNN, and recommendations for identified users as VUI-KNN, respectively.

The Contextual User Profile (CUP) method [19] is implemented and used to compare with VUI-KNN, since (a) the method regards an account consists of two contextual user profiles (home and cinema) by consuming time, which is similar to identify users but not, and (b) the authors also use KNN method to provide recommendations. We name recommendations according to contextual user profiles as CUPs.

5   Experimental Results

In this section, we conduct several experiments to compare different parameters
of VUI and different methods. Our experiments are intended to address the
following questions:

- How the parameters ($|P|$ and ρ) affect recommendations? In other words,
  how the assignment of sub-periods and combinations of virtual users affect
  recommendations?
- How the split methods affect recommendations?
- Can the KNN method take the advantage of VUI? Can the performance of
  VUI outperform that of CUPs?

5.1   Effects on Parameters $|P|$ and ρ

To study how the assignment of sub-periods and the process of combinations in
similarity graph affect user identification and recommendations, we measure the
performance in terms of precision and recall as $|P|$ or ρ change while holding
other parameter. Here, we use average split method to assign equal length sub-
periods. Note that, when $|P| = 1$ or $\rho = 0$, the VUI regards an account as a
user, the AccountKNN algorithm is obtained.

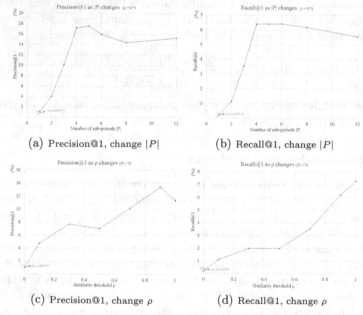

(a) Precision@1, change $|P|$ (b) Recall@1, change $|P|$

(c) Precision@1, change ρ (d) Recall@1, change ρ

Fig. 4. The effect of $|P|$ and ρ on results

We fix ρ at 0.7 and change $|P|$ to measure precision and recall when mak-
ing only one genre recommendation. As shown in figure 4(a) and 4(b), (1) the

precision and recall value is significantly improved, when comparing with AccountKNN ($|P| = 1$, Precision: **1%**, Recall: ¡**1%**); (2) the two optimal values are obtained at $|P| = 4$ and $|P| = 5$, and the precision value is slight over **17%**; (3) the performance starts degrading when $|P| > 5$.

To study effects on similarity threshold ρ, we hold $|P|$ at 3 (according to experience), and measure the performance of making one genre recommendations. Figure 4(a) and 4(b) reveals, (1) the recall value is still increasing as ρ increases; (2) the optimal precision is **13%**, ρ corresponding to 0.9;

5.2   Empirical Split versus Average Split

To study how split methods affect recommendations, we compare the designed sub-periods with the equal length sub-periods in terms of recommender performance.

According to the conducted experiments above, the optimal values are obtained at $|P| = 4$ and $|P| = 5$ (a slight better). We set $|P| = 4$ to compare the split methods, since it's more easier to empirically split up sub-periods than $|P| = 5$. The split sub-periods are shown in table 2, the differences are the end edge of afternoon and evening.

Table 2. The split up sub-periods ($|P| = 4$)

	Midnight	Morning	Afternoon	Evening
Average	0:00-6:00	6:00-12:00	12:00-18:00	18:00-24:00
Empirical	**23:50-6:00**	6:00-12:00	12:00-**19:00**	**19:00-23:50**

Recommendations are provided according to VUI-KNN, we compare the results on top-N recommendation as the length of recommendation list N changes.

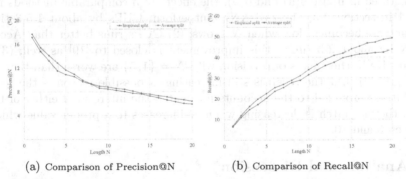

(a) Comparison of Precision@N   (b) Comparison of Recall@N

Fig. 5. Comparison of empirical split method and average split method

As figure 5(a) and 5(b) states, the empirical split method gains a slight improvement when comparing with the average split method, but the improvement is not stable. A possible reason for the improvement is that users are off work after 18:00, they need to spend time on the way and can't receive programs immediately. The benefit of average split method is its simpleness and can be applied automatically. Actually, we used the average split method to carry out the split mission when $|P|$ is greater than 4.

5.3   Comparing with CUPs

The CUPs is configured as: (1) Each account is regarded as two context user profiles (Morning and Afternoon) by means of start time, context users are recognized as 'Morning' user if they consume items before 12:00, and 'Afternoon' user if after 12:00; (2) Recommendations are obtained by KNN method as well as our proposed VUI-KNN. Note that, we implement CUPs on SMG, instead of Moviepilot[2] dataset.

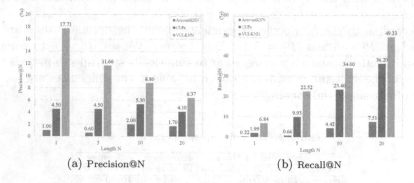

(a) Precision@N   (b) Recall@N

Fig. 6. Comparison of methods in terms of *Precison@N* and *Recall@N* with $N = \{1, 5, 10, 20\}$

As stated in figure 6(a) and 6(b), the effects of 3 comparable methods mentioned in section 4.3, (1) VUI-KNN outperform CUPs by about 1.5-3 times, the increase becomes slow when N grows; (2) CUPs runs better than AccountKNN by about 2-3 times, this improvement is closed to [19] as well; (3) for AccountKNN, the effects of precision with $N = \{1, 5\}$ are worse than that with $N = \{10, 20\}$, but the recall is still increasing, a possible reason is the recommendations provided to the accounts not match the interests of either of these users, the mismatch is decreasing when N increases to a proper value which is between 5 and 10.

6   Analysis and Discussion

In this section, we analyze how the parameters affect the number of identified users, and also how the identified users affects recommender performance. According to these analyses, we discuss about the limitations of the proposed algorithm VUI.

---

[2] http://www.moviepilot.de

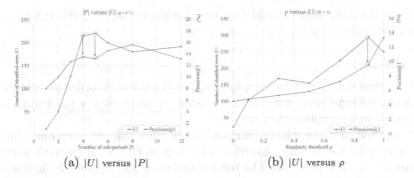

(a) $|U|$ versus $|P|$ (b) $|U|$ versus ρ

Fig. 7. Number of identified users $|U|$ and Precision@1 as parameters change

6.1 Identification and Performance Analysis

We try to discover the relationship between the number of identified users and the recommender performance. The experimental results are plotted in figure 7, and the performance is in terms of precision.

As figure 7(a) reveals, (1) the precision value increases when users are identified; (2) the two optimal precision values are found at $|U| = \{165, 169\}$; (3) the precision starts decreasing when $|U| > 165$. The optimal precision value is found at $|U| = 294$ in figure 7(b). According to figure 7(a) and 7(b), the optimal precision value is between $|U| = \{165, 169\}$.

The following summarizes the key conclusions we observe from the results: (1) The recommender performance is improved when users within accounts are identified for personalized recommendations. A reason for the improvement is that recommendations based on identified users alleviate the problem of recommending given item to wrong users within a shared account. (2) There exists a pair of P and ρ leading to the best performance of recommendations. In other words, less or more identified users ($|U|$ is too small or too large) will degrade the performance.

1. Less users identified, when P is split into few sub-periods and ρ is set very close to 1, which may regard two (or more than two) real users as one. Hence, a possible reason for the performance degrading is recommending items to who unlike them.

2. More users identified, when P is split into too many sub-periods and ρ is set very close to 0, which may regard a real user as two (or more than two) identified users, and the preferences of the real user are divided into several parts by the identified users. Hence, the opportunity of recommending right items to the real user may decrease since the KNN recommends items other users also preferred.

The best pair of P and ρ found in the pervious section reflects user consuming behavior in IP-TV services. when $|P|$ is set to 4, P consists of the four sub-periods: Midnight (0:00-6:00), Morning (6:00-12:00), Afternoon (12:00-18:00) and Evening (18:00-24:00), and $\rho = 0.7$, we get the best performance.

It also means that, a user has his/her own preference in sub-periods when consuming the services.

6.2 Discussion

Multiple users share a common account in IP-TV services, in order to recommend right items to right users within these shared accounts, we try to identify users for personalized recommendations.

We suppose that users have different tastes, thus different recommendations are required for them. The algorithm VUI is proposed to distinguish users by time and preference patterns. The recommendation performance is significantly improved by the identified users.

The performance of VUI is affected or decided by the parameters. We learn the parameters by cross-validation method in our experiments. But the parameters (e.g., P and ρ) can't be obtained automatically. The potential work is to learn the parameters, which can be regarded as learning user consuming behavior in terms of time and preference.

7 Conclusion and Future Work

In this paper, we define the problem of user identification as mining different preferences over different periods from consumption logs. According to this definition, an algorithm for user identification is proposed to predict users within a shared account in IP-TV services. The process of user identification consists of two phases. The first is to partition a day and identify behavior specific to different periods. Secondly, periods for which discovered usage patterns are similar are regarded as associated with the same actual user. The association process is carried out by leveraging DFS algorithm in a similarity graph.

The predicted users are able to improve recommender performance in terms of precision and recall. The optimal precision value and recall value are obtained when $|P| = 4$ and $\rho = 0.7$, $|P| = 4$ also reflects the kinds of user consuming preferences in terms of periods, and the number of identified users corresponding to the optimal performance can also be found by the cross-validation method in the conducted experiments.

The evaluation of such methods is a potential future direction of this work.

Acknowledgments. We would like to thank SMG for sharing the IP-TV consumption logs, and the anonymous reviewers for their valuable comments and suggestions to improve the quality of this paper. This work is supported by the Shanghai Science and Technology Commission Foundation (No.12dz1500205 and No.13430710100).

References

1. Adomavicius, G., Tuzhilin, A.: Toward the next generation of recommender systems: A survey of the state-of-the-art and possible extensions. IEEE Trans. Knowl. Data Eng. 17(6), 734–749 (2005)

2. Adomavicius, G., Tuzhilin, A.: Context-aware recommender systems. In: Recommender Systems Handbook, pp. 217–253 (2011)
3. Ageev, M., Lagun, D., Agichtein, E.: Improving search result summaries by using searcher behavior data. In: SIGIR 2013, pp. 13–22 (2013)
4. Bambini, R., Cremonesi, P., Turrin, R.: A recommender system for an iptv service provider: a real large-scale production environment. In: Recommender Systems Handbook, pp. 299–331 (2011)
5. Bell, R.M., Koren, Y.: Scalable collaborative filtering with jointly derived neighborhood interpolation weights. In: ICMD 2007, pp. 43–52 (2007)
6. Bennett, P.N., White, R.W., Chu, W., Dumais, S.T., Bailey, P., Borisyuk, F., Cui, X.: Modeling the impact of short- and long-term behavior on search personalization. In: SIGIR 2012, pp. 185–194 (2012)
7. Grasch, P., Felfernig, A., Reinfrank, F.: Recomment: towards critiquing-based recommendation with speech interaction. In: Recsys 2013, pp. 157–164 (2013)
8. Hu, Y., Koren, Y., Volinsky, C.: Collaborative filtering for implicit feedback datasets. In: ICDM 2008, pp. 263–272 (2008)
9. Katz, G., Ofek, N., Shapira, B., Rokach, L., Shani, G.: Using wikipedia to boost collaborative filtering techniques. In: Recsys 2011, pp. 285–288 (2011)
10. Kim, E., Pyo, S., Park, E., Kim, M.: An automatic recommendation scheme of tv program contents for (ip)tv personalization. TBC 57(3), 674–684 (2011)
11. Koren, Y.: Collaborative filtering with temporal dynamics. In: KDD 2009, pp. 447–456 (2009)
12. Li, Z., Wang, J., Han, J.: Mining event periodicity from incomplete observations. In: KDD 2012, pp. 444–452 (2012)
13. Liu, N.N., Zhao, M., Xiang, E.W., Yang, Q.: Online evolutionary collaborative filtering. In: Recsys 2010, pp. 95–102 (2010)
14. Ma, H.: An experimental study on implicit social recommendation. In: SIGIR 2013, pp. 73–82 (2013)
15. Pero, Š., Horváth, T.: Opinion-driven matrix factorization for rating prediction. In: Carberry, S., Weibelzahl, S., Micarelli, A., Semeraro, G. (eds.) UMAP 2013. LNCS, vol. 7899, pp. 1–13. Springer, Heidelberg (2013)
16. Pyo, S., Kim, E., Kim, M.: Automatic and personalized recommendation of tv program contents using sequential pattern mining for smart tv user interaction. Multimedia Syst. 19(6), 527–542 (2013)
17. Ricci, F., Rokach, L., Shapira, B., Kantor, P.B.: Recommender Systems Handbook. Springer (2011)
18. Said, A., Berkovsky, S., Luca, E.W.D., Hermanns, J.: Challenge on context-aware movie recommendation: Camra2011. In: Recsys 2011, pp. 385–386 (2011)
19. Said, A., Luca, E.W.D., Albayrak, S.: Inferring contextual user profiles - improving recommender performance. In: Proceedings of the 3rd Workshop on Context-Aware Recommender Systems. IEEE (2011)
20. Xu, M., Berkovsky, S., Ardon, S., Triukose, S., Mahanti, A., Koprinska, I.: Catch-up tv recommendations: show old favourites and find new ones. In: Recsys 2013, pp. 285–294 (2013)
21. Zhang, A., Fawaz, N., Ioannidis, S., Montanari, A.: Guess who rated this movie: Identifying users through subspace clustering. In: UAI 2012, pp. 944–953 (2012)

P-TRIAR: Personalization Based on TRIadic Association Rules

Selmane Sid Ali, Omar Boussaid, and Fadila Bentayeb

Laboratoire ERIC, Université Lyon 2, Bron, France

Abstract. This article describes a new personalization process on decisional queries through a new approach of triadic association rules mining.
This process uses the query log files of users and models them in new way
by taking into account their triadic aspect. To validate our approach, we
developed a personalization software prototype *P-TRIAR* (Personalization based on TRIadic Association Rules) which extracts two types of
rules from query log files. The first one will serve to query recommendation by taking into account the collaborative aspect of users during their
decisional analysis. The second type of rules will enrich user queries.
The approach is tested on a real data warehouse to show the compactness of triadic association rules and the refined personalization which we
propose.

1 Introduction

OLAP [1] systems users formulate decisional queries to meet their needs of specific analysis for decision support. OLAP tools are known to be intuitive as their
end users are not necessarily computer scientists. However, the large volume of
data and the complexity of analytical queries which involve a lot of aggregations
make this task of analysis more difficult to users. So it seems necessary to provide
them solutions best suited to their way of thinking through methods of recommendation and enrichment of their analytical queries. These methods are called
personalization. In this paper, we propose a new personalization process of analytical queries. We are particularly interested in collaborative recommendation
and enrichment of decisional queries based on log files.

The personalization works which exploit query log files use in most cases
frequent itemsets [11] and association rules [20]. However, the large number of
frequent itemsets and association rules obtained makes the task of personalization more difficult. Contrary to these approaches, the work we propose is based
on another type of more compact rules called triadic association rules. These
rules convey a richer semantic than conventional rules as they are formed in
addition to the premise and the conclusion of a condition which enrich the rule.
Our personalization process consists of five steps:

1. Modeling data of OLAP servers query log files by a triadic context. This
   triadic context will consist of the set of users, the set of queries, the set of

---

[1] *On-line Analytical Processing abr. OLAP.*

Y. Manolopoulos et al. (Eds.): ADBIS 2014, LNCS 8716, pp. 234–247, 2014.

attributes (descriptors and measures) in the SELECT clause and a ternary relation between these three sets.

2. The mapping of a triadic (tridimensional) context into a dyadic (bidimensionnel) one will be done by flattening the set of users over the set of attributes.
3. The computation of dyadic association rules *(premise → conclusion)*.
4. The generation of triadic association rules $(premise \rightarrow conclusion)_{(condition)}$ through a *factorization* of dyadic ones.
5. The exploitation of these triadic association rules for personalization. To validate our approach we developed a personalization software prototype *P-TRIAR* (Personalization based on TRIadic Association Rules) to extract two types of rules from query log files. The first one will serve query recommendation by taking into account the collaborative aspect of users during their analysis. This recommendation will be carried out by the user communities discovered across multiple links between them. The second type of rules aims to enrich user queries by recommending attributes to add to his query.

The rest of the paper is organized as follows. Section 2 presents the modeling of log data with Formal Concept Analysis (FCA) and Triadic Concept Analysis (TCA) while recalling their basic concepts. Section 3 describes the proposed approach and algorithms for producing triadic association rules. Then in Section 4 we detail *P-TRIAR* and our process of personalization. Section 5 sheds light on works in personalization and association rules mining from multidimensional data. Experiments are performed in Section 6 to illustrate the compactness of triadic rules compared to dyadic rules and their contribution to personalization. Finally, some perspectives for future work are presented in Section 7.

2 Modeling Data Log Based on Formal Concept Analysis

In this section, we develop the data modeling process based on FCA. These data are implicitly collected from query log files of *OLAP* servers. We are interested in this work, especially on three data contained in a *SQL server* query log files[2], namely *MSOLAP_User* which identifies users, *Dataset* which contains the query and their attributes and *StartTime* which indicates the date and launch time of the query. This last is used to determine the date from which the log file could be exploited. These three data are easily accessible in data warehouses query log unlike data on user profiles, which are often hidden because of their private aspect. Our following definitions of a triadic context and its equivalent are based on those introduced in [13].

Definition 2.1. (Triadic context) a *triadic context* is a quadruplet of the form $\mathbb{K} := (R, U, A, Y)$ where:

- R, U, A respectively define que**R**ies, **U**sers and **A**ttributes (descriptors and measures) of the query SELECT clause.

---

[2] http://technet.microsoft.com/en-US/library/cc917676.aspx

Table 1. (a) Triadic context $\mathbb{K} := (R, U, A, Y)$, formed from $R = \{R_1, R_2, R_3, R_4, R_5\}$ (queries), $U = \{U_1, U_2, U_3, U_4\}$ (users) and $A = \{a_1, a_2, a_3, a_4, a_5\}$ (attributes). (b) Equivalent dyadic context $\mathbb{K}^{(1)}$ obtained from $\mathbb{K}$.

(a)

$\mathbb{K}$	U_1	U_2	U_3	U_4
R_1	$a_1a_2a_4$	$a_1a_2a_4a_5$	a_1a_3	a_1a_5
R_2	$a_1a_4a_5$	$a_2a_3a_4$	$a_1a_2a_4a_5$	a_4a_5
R_3	$a_1a_2a_4$	a_4a_5	a_1a_2	a_1a_5
R_4	$a_1a_2a_4a_5$	a_2a_4	a_1a_2	a_4a_5
R_5	$a_1a_4a_5$	$a_1a_4a_5$	$a_1a_2a_4a_5$	a_1a_5

(b)

$\mathbb{K}^{(1)}$	U_1 $a_1a_2a_3a_4a_5$					U_2 $a_1a_2a_3a_4a_5$					U_3 $a_1a_2a_3a_4a_5$					U_4 $a_1a_2a_3a_4a_5$				
R_1	1	1		1		1	1		1	1	1		1			1				1
R_2	1			1	1		1	1	1		1	1		1	1				1	1
R_3	1	1		1					1	1	1	1				1				1
R_4	1	1		1	1		1		1		1	1							1	1
R_5	1			1	1	1			1	1	1	1		1	1	1				1

– $Y \subseteq R \times U \times A$ represents a ternary relation where each $y \subseteq Y$ represents a triple: $y = \{(r, u, a) | r \in R, u \in U, a \in A\}$. In other words, query $\boldsymbol{q}$ is launched by user $\boldsymbol{u}$ and which involves the attribute $\boldsymbol{a}$.

We illustrate through an example (Table 1.a), the transition from log data to a triadic context. Each user in $U =(U_1, U_2,..., U_4)$ performs analysis by launching a sequence of queries denoted $R =(R_1, R_2,..., R_5)$ where each query is composed of a set $A =(a_1, a_2,...,a_5)$ of attributes from different facts and dimensions of the data warehouse.

For example, the value $a_1a_2a_4$ located at the intersection of the first column and the first row means that the user U_1 launched the query R_1 composed of the attributes a_1, a_2 and a_4.

Definition 2.2. (Dyadic context) In Formal Concept Analysis a *dyadic formal context* is a triplet $\mathbb{K}^{(1)} := (G, M, I)$ where G is a set of objects, M a set of proprieties and I a binary relation between G and M. Our equivalent dyadic context is formed by flattening of the triadic context we defined (see Definition 2.1). The objects in G are the queries in R and proprieties in M are pairs (a_j, a_k) in the projection of set users into set of attributes $U \times A$. The table 1.b represents the dyadic context $\mathbb{K}^{(1)}$ obtained from the triadic context $\mathbb{K}$, thus: $\mathbb{K}^{(1)} := (R, U \times A, Y^{(1)})$ with ($(a_i, (a_j, a_k)) \in Y^{(1)} \iff (a_i, a_j, a_k) \in Y$). The value 1 of the first row and the first column means that the user U_1 launches the query R_1 which implies the attribute a_1. In what follows, the pair $(a_j, a_k) \in U \times A$ will be noted in a simplified manner by a_j-a_k.

Definition 2.3: (Derivation) For $X \subseteq G$ and $Z \subseteq M$, two subsets $X' \subseteq M$ and $Z' \subseteq G$ are respectively defined as a set of proprieties common to the objects in X and a set of proprieties which share all attributes in Z. Formally, the derivation denoted $'$ is defined as follows:

$$X' := \{a \in M \mid o I a \; \forall o \in X\} \quad \text{and} \quad Z' := \{o \in G \mid o I a \; \forall a \in Z\}.$$

This proposal defines a pair of correspondence $(','$) between the set of parts of G and the set of parts of M representing a Galois correspondence. The closure operators in G and M are denoted by $''$. For example, the closure of $U_2 - a_4$ is given by:

$(U_2 - a_4)'' = ((U_2 - a_4)')' = \{R_1, R_2, R_3, R_4, R_5\}' = \{U_1 - a_1, U_1 - a_4, U_2 - a_4, U_3 - a_1, U_4 - a_5\}$.

Definition 2.4: (formal concept) A formal concept (cf) is a pair (X, Z) with $X \subseteq G$, $Z \subseteq M$, $X = Z'$ and $Z = X'$. The set X is called *extension* of cf while Z is its *intention*. A formal concept (dyadic) corresponds to a maximum rectangle in a dyadic context.

Example. As $\{R_1, R_2, R_3, R_4, R_5\}' = \{U_1 - a_1, U_1 - a_4, U_2 - a_4, U_3 - a_1, U_4 - a_5\}$ and $\{U_1 - a_1, U_1 - a_4, U_2 - a_4, U_3 - a_1, U_4 - a_5\}' = \{R_1, R_2, R_3, R_4, R_5\}$, then the pair$(\{R_1, R_2, R_3, R_4, R_5\}, \{U_1 - a_1, U_1 - a_4, U_2 - a_4, U_3 - a_1, U_4 - a_5\})$ form a formal concept.

Definition 2.5: (Dyadic association rule) Let (G, M, I) a formal dyadic context. An association rule (R) [2] has the following format $R : B \rightarrow C\ (s, c)$ where $B, C \subseteq M$ with $B \cap C = \emptyset$. The support s of a rule R is calculated by the formula: $Supp(R) = \frac{|B' \cap C'|}{|G|}$. The confidence c is given by: $Conf(R) = \frac{|B' \cap C'|}{|B'|}$. We speak of implication when the confidence of the association rule is equal to 1.

In the following section, we will show how to exploit the dyadic association rules for generating triadic ones. The dyadic rules are produced from our context $(\mathbb{K}^{(1)})$ by applying *Pasquier* algorithms [16].

3 Triadic Association Rules Extraction

3.1 Definitions

It is apparent from the literature study so far that [4] is the first to study the implications extraction problem in triadic contexts. A *triadic implication* has the following form: $(A \rightarrow D)_C$. This implication is true if *"whenever A is true under all conditions in C, then D is also true under all conditions"*. Afterwards, [7] have extended the work of *Biedermann* and defined three types of implications: attribute - condition implications, conditional attribute implications, attributional condition implications. [14] extended these definitions to association rules and proposed three types: *Attributes-Conditions Association Rules* (**A-CARs**); *Conditional Attribute Association Rules* (**CAARs**); *Attributional Condition Association Rules* (**ACARs**). In what follows, we consider our example (Table 1) of a triadic context $\mathbb{K} := (R, U, A, Y)$ and its equivalent dyadic one $\mathbb{K}^{(1)} := (R, U \times A, Y^{(1)})$ to define the different types of association rules.

Definition 3.1.1: An *Attribute-Condition Association Rule* (**A-CAR**) is a dyadic association rule in the form $A \rightarrow D\ (s, c)$, where A and D are subsets of $U \times A$, s and c represent respectively the support and confidence. These dyadic association rules are extracted from the dyadic context $\mathbb{K}^{(1)}$.

Example: $U_2 - a_1 \rightarrow U_2 - a_5, U_2 - a_4, U_3 - a_1, U_1 - a_1, U_1 - a_4, U_4 - a_1, U_4 - a_5$ $(0.4, 1)$ is an *A-CAR* with support equal to 40% and a confidence equal to 100%.

Definition 3.1.2. A *Conditional Attribute Association Rules* according to *Biedermann* formalism (**BCAAR**) is a triadic association rule with the following notation: $(A \to D)_C$ (s, c), where A and D are subsets of U, and C a subset of A and means that A implies D under all conditions in C with a support s and a confidence c.

Example. the rule $(U_2 \to U_1)_{a_1 a_2}(0.2, 1)$ is a $BCAAR$ with a support 20% and a confidence 100%.

Definition 3.1.3. An *Attributional Condition Association Rules* according to *Biedermann* formalism (**BACAR**) is a triadic association rule the following notation $(A \to D)_C$ (s, c), where A and D are subsets of A, and C are subsets of U and means that A implies D under all conditions in C with a support s and a confidence c .

Example. the rule $(a_2 \to a_4)_{U_2 U_1}(0.4, 1)$ is a $BACAR$ with a support 40% and a confidence 100%.

These two types of triadic association rules (i.e., $BCAAR$ and $BACAR$) have the same notation but the sets of their premises, conclusions and conditions differ.

3.2   Proposed Approach

In what follows, we present our approach based on formal definitions and illustrative examples. Several approaches for researching and analysing triadic concepts have emerged in the literature [19], [15] and [5] for $n = 3$. [14], propose an effective approach based on the triadic context analysis for the extraction of triadic association rules. It consists of taking as input a formal triadic context which is flattened to produce a dyadic one. Then dyadic concepts and dyadic generators are extracted. After that, triadic concepts are then generated from dyadic concepts and triadic generators from dyadic ones. Once these two sets gathered, it is then possible to extract the triadic association rules. These operations give good results but can be avoided by our alternative which does not calculate these two sets. Our approach is based on the same theoretical basis that the one proposed by [14]. Nevertheless, our extraction process is different in terms of input data and our algorithms are applied rather on a set of dyadic association rules of type **RAA-C**. To extract these **RAA-C** we apply the algorithms of [16] on the dyadic context $\mathbb{K}^{(1)} := (R, U \times A, Y^{(1)})$ obtained from the projection of the set of properties U on the set of conditions A of the formal triadic context $\mathbb{K} := (R, U, A, Y)$. Then, from these latter and the definitions recalled in section 3.1.1, we apply the algorithms which we have proposed to search for triadic association rules in their various forms: *Biedermann* Conditional Attribute Association Rules $(BCAAR)$ and Biedermann Attributional Condition Association Rules $(BACAR)$.

3.3   Proposed Algorithms

The transition from dyadic association rules to the set of all triadic one, in their various forms, is performed using a main procedure called $TRIAR$. It permit to produce the triadic association rule through the two sub procedures $BCAAR$ and

$BACAR$. This choice of decomposition is motivated by the parallelization of these two procedures during implementation to have two types of personalization.

The main procedure $TRIAR$ (Algorithm 1) consists of three parts. The first one (lines 4-8) corresponds to a sorting procedure which identifies whether a dyadic association rule is eligible to become a triadic one or not. This step is justified by the mapping of the definition of triadic generator in [14] to a triadic rule. So as we collect the distinct values of attributes on the set A_L (line 6) of the rule premise LHS, the distinct values of conditions within the set M_L (line 7) of the rule premise LHS; and we check if their product corresponds to the size of LHS (line 8). The other two parts (lines 9 and 10) correspond to the procedures $BCAAR$ and $BACAR$ (Algorithms 2 and 3) which allow us to produce the set of triadic association rules.

Algorithm 1. Computation of Triadic Association Rule

1: **Procedure** TRIAR(D)
2: **In:** $D = \{(LHS, RHS, s, c)\}$
3: **Out:** $\Sigma = \{(L, R, C, t, s, c)\}$
4: $\Sigma \leftarrow \emptyset$;
5: **for** $RL = (LHS, RHS, s, c)$ in D **do**
6:    $A_L \leftarrow$ DISTINCT A(LHS)
7:    $M_L \leftarrow$ DISTINCT M(LHS)
8:    **if** $Size(A_L) \times Size(M_L) = Size(LHS)$ **then**
9:      $\Sigma \leftarrow \Sigma \cup \{(\text{BCAARs}(A_L, M_L, RHS), 1, s, c)\}$
10:      $\Sigma \leftarrow \Sigma \cup \{(\text{BACARs}(A_L, M_L, RHS), 2, s, c)\}$
11: **out** Σ

We have as input $TRIAR$ a set (D) of dyadic association rules (**RAA-C**) where each rule has the following form (LHS, RHS, s, c) representing respectively (the premise, the conclusion, the support and the confidence of the rule).

Example. the rule $U_3 - a_4, U_4 - a_4 \rightarrow U_2 - a_3, U_2 - a_2, U_2 - a_4, U_3 - a_1, U_3 - a_5, U_3 - a_2, U_1 - a_1, U_1 - a_5, U_1 - a_4, U_4 - a_5$(sup $= 0.20$; conf $= 1.00$) is written as follows ($\{U_3 - a_4, U_4 - a_4\}, \{U_2 - a_3, U_2 - a_2, U_2 - a_4, U_3 - a_1, U_3 - a_5, U_3 - a_2, U_1 - a_1, U_1 - a_5, U_1 - a_4, U_4 - a_5\}, 0.20, 1$).

The output of the procedure $TRIAR$, we have a set of triadic association rule (Σ), where each rule is presented in the following form (L, R, C, t, s, c), representing the premise, the conclusion, the condition, the type respectively (1: $BCAAR$; 2: $BACAR$), the support and the confidence.

Example. the $BCAAR$ ($U_3U_4 \xrightarrow{a_4} U_2U_1$ (sup $= 0.20$; conf $= 1.0$)) is written as follows ($U_3U_4, U_2U_1, a_4, 1, 0.20, 1.0$).

To expand $TRIAR$ algorithm, we take as an example the dyadic rule ($\{U_3 - a_4, U_4 - a_4\}, \{U_2 - a_3, U_2 - a_2, U_2 - a_4, U_3 - a_1, U_3 - a_5, U_3 - a_2, U_1 - a_1, U_1 - a_5, U_1 - a_4, U_4 - a_5\}, 0.20, 1$). Lines 5-7 of Algorithm 1, we create two sets A_L and M_L which respectively contain the distinct attributes and distinct conditions of the premise of the rule LHS $\{U_3 - a_4, U_4 - a_4\}$. Accordingly, $A_L = \{U_3, U_4\}$, $M_L = \{a_4\}$. This entails that the product $Size(A_L) \times Size(M_L) = 2$ (line 8) is equal to $Size(LHS)$, as the portion M_L will become a condition for the constructed rules. All the elements of A_L must verify this condition thus this rule is eligible to become a triadic association rule. Lines 9 and 10 of Algorithm 1 involve both procedures $BCAAR$ and $BACAR$ to produce both types of triadic rules.

The procedure $BCAARs$ (algorithm 2) takes as input three sets A_L, M_L and RHS. The set M_L represents the conditions which apply to all attributes in the set A_L and we want to find in RHS other attributes which are affected by the same conditions, from where the search of the conditions 5-7 lines. These attributes are isolated within line 9 (*group by* on attributes), to see whether their conditions meet the conditions of M_L (line 11), if they are identical to those of M_L we can build a rule.

Algorithm 2. Computation of $BCAAR$ (type =1)

1: **Procedure** BCAARs(A_L, M_L, RHS)
2: **In:** A_L, M_L, RHS
3: **Out:** BCAAR = (A_L, A_R, M_L)
4: $A_R \leftarrow \emptyset$; $Temp \leftarrow \emptyset$
5: **for** $e \in RHS$ **do**
6:     **if** MODUS(e) $\in M_L$ **then**
7:         $Temp \leftarrow Temp \cup \{e\}$
8: **if** $Temp \neq \emptyset$ **then**
9:     Creates containers $B = b_1, ..., b_n$ by grouping elements of $Temp$ having the same part of attributes in common
10: **for** $elem \in B$ **do**
11:     **if** $Size(elem) = Size(M_L)$ **then**
12:         $A_R \leftarrow A_R \cup \{Attr(elem)\}$
13: **if** $A_R \neq \emptyset$ **then**
14:     **out** (A_L, A_R, M_L)

The sequence of the algorithm is performed as follows: after the initialization of the parameters (lines 2-4), we take the conclusion of the rule RHS (line 5) which corresponds to $\{U_2 - a_3, U_2 - a_2, U_2 - a_4, U_3 - a_1, U_3 - a_5, U_3 - a_2, U_1 - a_1, U_1 - a_5, U_1 - a_4, U_4 - a_5\}$ in the example, and calculate the $Modus$ of each element which corresponds the condition. For the first component, $Modus(U_2 - a_3)=\{a_3\}$ The test shows that it is not included in the set $M_L = \{a_4\}$ the condition is not satisfied, the loop move to the next item. For the fourth element, $Modus(U_2-a_4)=\{a_4\}$ it is included in M_L the condition is satisfied. The variable $Temp$ gets this item $(U_2 - a_4)$, then in line 9, we group in a container denoted B the elements which have the same part attribute, in our example $(U_2 - a_4), (U_1 - a_4)$ will be contained in (B). The algorithm 2 checks in line 10-12, for each element contained in (B), if the size of the element is equal to the size of M_L. In our example, these two entities are equal for the two elements because they have a size equal to 1. The rule formed of triplet (A_L, A_R, M_L) is then constituted $(\{U_3, U_4\}, \{U_2, U_1\}, a_4)$ to which is added type, support and confidence. The result is: $BCAAR$ $(U_3U_4 \rightarrow U_2U_1)_{a_4}$, type = 1, Sup = 0.20 and Conf= 1.00. This is the exit point of the algorithm 2 and the rule is added to the set of $BCAAR$.

In the procedure $BACARs$ (algorithm 3), we input three sets A_L, M_L and RHS. The set A_L represents the attributes which apply to all conditions in the set M_L and we want to find in RHS other attributes which are affected by the same conditions, from where the search of the conditions 5-7 lines. These attributes will be isolated from line 9 (*group by* on conditions), to permit viewing if their attributes meet the attributes of M_L (line 11), if the attributes are identical to those of M_L we can build a rule. The others steps of the algorithm 3 are unrolled in the same way of algorithm 2.

Algorithm 3. Computation of $BACAR$ (type = 2)

1: **Procedure** BACARs(A_L, M_L, RHS)
2: **In:** A_L, M_L, RHS
3: **Out:** $BACAR = (M_L, M_R, A_L)$
4: $M_R \leftarrow \emptyset$; $Temp \leftarrow \emptyset$
5: **for** $e \in RHS$ **do**
6:     **if** $\text{ATTRIB}(e) \in A_L$ **then**
7:         $Temp \leftarrow Temp \cup \{e\}$
8: **if** $Temp \neq \emptyset$ **then**
9:     Creates containers $B = b_1, ..., b_n$ by grouping elements of $Temp$ having the same part of attributes in common
10: **for** $elem \in B$ **do**
11:     **if** $Size(elem) = Size(A_L)$ **then**
12:         $M_R \leftarrow M_R \cup \{Cond(elem)\}$
13: **if** $M_R \neq \emptyset$ **then**
14:     **out** (M_L, M_R, A_L)

3.4 Complexity Study

In what follows, we present the study of the complexity of our main algorithm $TRIAR$. It uses the procedures $BCAARs$ and $BACARs$. It takes as input D a set of dyadic association rules. Latter is obtained from a dyadic formal context $\mathbb{K}^{(1)} := (R, U \times A, Y)$. The maximum size of dyadic association rule is given by $|U| * |A|$. The overall complexity of the algorithm is linear in $|D|$ and is performed in $O(|D| * 2(|U| + |A|))$. This complexity is obtained by studying the loop "for" (line 5), which iterates through one time all the rules in D, it is given by: line 5 is performed in $O(|U|)$ because at worst, we have rules in all context properties, line 8 is performed in $O(|A|)$ because in the worst case we can have a rule in all properties of the context. The test in line 8 is performed $O(|D|)$ because this is the set of rules which is driven to test their eligibility to become triadic rules; Instructions 9 and 10, respectively, call the procedure $BCAARs$ and $BACARs$. Such appeals are made in the worst case $O(|D| * |U| + |A|)$, where all dyadic association rules are eligible to become triadic ones.

4   Architecture of *P-TRIAR*

P-TRIAR involves five steps (see Figure 1). In Section 2, we described the first three stages, namely: Modeling a triadic context data from query log of $OLAP$ analysis

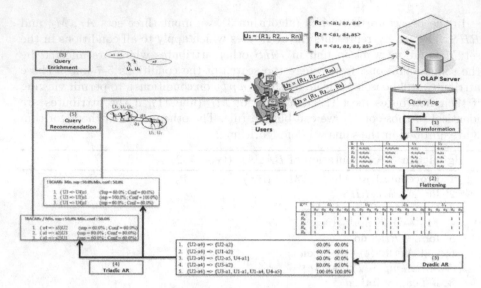

Fig. 1. Architecture of P-TRIAR

server; the transition of this triadic context to a dyadic one and finally the production of conventional dyadic association rules type *premise → conclusion*. Then, in section 3, we detailed the approach we propose to generate a set of triadic association rules type $(premise → conclusion)_{(condition)}$ by *factorization* of dyadic rules. In what follows, we describe the fifth stage of *P-TRIAR* regard to the exploitation of triadic association rules (*BCAAR* and *BACAR*) obtained by our algorithms.

4.1   Query Recommendation by *BCAAR*

The *BCAAR* determine the associations which exist between users that have as a condition attributes. In other words, this type of rules allows us to discover the relationship between users through the attributes involved in their queries. For example, the *BCAAR* $(U_4 → U_3U_1)_{a_1}$ (0.60,1) states that whenever a query is submitted by the user U_4 and contains the attribute a_1, the users U_3 and U_1 submit a query which contains the same attribute, with a support 60% and a confidence 100%. This rule highlights the similarity between user U_4 and users U_3 and U_1 but on condition to query the attribute a_1. Through this rule, we find the collaborative aspect because it allows forming a community link between three users. This community connection is conditioned by the involvement of the attribute a_1 in their queries and the degree of this link has a specific support and confidence.

The first personalization scenario, the user connects to *P-TRIAR* defines the initial parameters (the date from which he wants to explore the log, the threshold of support and confidence) and wants to know the links that he entertains with other users. *P-TRIAR* shows him the rules which satisfy these parameters. Assuming that U_4 choose the rule of our example, *P-TRIAR* will recommend a number of decisional queries that U_4 desires. These queries will be filtered and sorted: by frequencies, by users (U_3 and U_1) and by attributes (a_1). So as the user may choose the queries which are suitable for its analysis needs. If the user wants to directly access to queries, *P-TRIAR* recommend him a set of queries without having to choose among *BCAAR*, *P-TRIAR* detects which user is logged on and it offers a number of queries filtered by number of users and number of attributes, i.e., based on rules which have the largest number of users in the conclusion part rule and the largest number of attributes in the condition part.

4.2   Query Enrichment by *BACAR*

The *BACAR* determine the associations which exist between attributes which have users as a condition. This type of rules allows us to discover the relationships between attributes (descriptors and measures) involved in a query through users making it. For example the, *BACAR* $(a_2 \rightarrow a_4)_{U_2 U_1}(0.4, 1)$ is true when each time a request is submitted and which involves the attribute a_2, the attribute a_4 is also involved in the query on condition that users U_2 and U_1 formulate it.

The second scenario of personalization is based on *BACAR*. In this scenario, the user sets the same parameters of the first scenario and wishes to make a request for analysis taking inspiration the links which exist between the attributes of the warehouse. Assuming the user U_2 is connected and chooses the *BACAR* $(a_2 \rightarrow a_4)_{U_2 U_1}$ which means that each time a query is submitted and which contains the attribute a_2, the attribute a_4 is also involved in the query as long as users who formulates it are U_2 and/or U_1, with a support 40% and a confidence 100%. This rule highlights the similarity existing between the attributes a_2 and a_4 but under the condition that the users U_1 or U_2 formulate the query. *P-TRIAR* relies upon such rule to enrich the user U_2 query by recommendation of the attribute a_4 as element of its query.

5   Related Works

The personalization of queries has been the subject of several studies [3], [1], [12], [17]. It aims to help the user generally based on its behavior and its previous queries or those of other users. In the areas of databases and data warehouses the different personalization techniques have been classified into three categories [1], [17]: collaborative techniques [6] and [9] which exploits the similarity between users profiles and one for which the recommendation is determined; based on the content techniques [10] intended to recommend to a user attributes that frequently seeks; and finally hybrid techniques [18] which combine the two previous techniques. In literature, the recommender systems have as sources user data profiles, log files which are structured historization of queries for each user, or external sources such as ontologies, web pages, etc..

Several studies have exploited the idea of pattern extraction [11] and association rules [20], from log files, for the recommendation. However, their work was limited to a bi-dimensional framework. They represent the data log files across matching matrices (*users*× *query*) or (*attributes*× *query*) for association rules or patterns extraction. This modeling does not take into account the three-dimensionality of these data. In data warehouses, the association rules and the patterns they get are numerous and of dyadic type. This very large number of patterns and association rules makes more complicated the recommendation task and does not take into account at the same time the three sets of *users, attributes* and *queries*.

In addition, FCA [22], [8] and Galois lattices constitute a theoretical basis for solving many problems in the fields of artificial intelligence, software engineering and databases. The TCA was originally introduced by [23] and [13]. Their work focuses on the analysis of triadic contexts, concepts and lattices concepts called trilattices. They define the way, the theoretical basis for ATC. In this way, they defined the theoretical basis for the TCA. [4] provides a writing formalism of triadic implications and [21] defines polyadic concepts analysis and generalizes the work of [23] to polyadic formal contexts to produce polyadic formal concepts and n-lattice.

More recent work related to the TCA exist, [7] consider different types of triadic implications which he calls strong relying formalism stated by [4]. [15] propose an approach for mining rules applied to dynamic relational graphs which can be encoded in n-ary relationships ($n \geq 3$). The work of [14] offer not only an approach to triadic association rules production but also procedures to extract triadic concepts and generators from dyadic ones.

In [19], the authors deal with the calculation of generators and triadic association rules. However, the author provided a new definition of the latter which is different from that of [14] which, in turn, is based on the definition of *Biedermann*. In [15] and [5], the authors propose the generalization of the concept of association rules in a multidimensional context by working either on boolean matrices but on boolean tensors of arbitrary arity. They also provide measures of frequency and confidence to define the semantics of such rules.

Based on the literature review we conducted, our work is the first to model the log data through a triadic context. The proposed approach provides a personalization from triadic association rules. We show through our process, how to get triadic association rules from these triadic contexts, using only the dyadic association rules without calculating the triadic concepts and generators as proposed by the authors mentioned above.

6   Experiments

The tests we performed on the warehouse *PUBS* [3](Figure 2) focused on five users and 100 decisional queries, by user, composed of 34 distinct attributes that contains *PUBS*. It concerns the analysis of the turnover (CA) and quantity (Qty) of books sold. These measures are observed over the following dimensions: *Titles, Publishers, Stores, Times* and *Authors*.

Five users (U_1, U_2, U_3, U_4, U_5) logged on *PUBS* and submitted their different sequences of decisional queries denoted ($R_1,..., R_n$). Each query is formed in the SELECT clause of attributes (descriptors and measures) noted ($a_1, ...,a_n$).

Example. User U_4 launches a set of queries on the warehouse *PUBS*:

---

[3] Data warehouse constructed from the database *PUBS* provided by *Microsoft*: http://technet.microsoft.com/fr-fr/library/ms143221(v=sql.105).aspx

- R_1 = What is the turnover of the store *store_400* for the year *2013*. *PUBS* attributes involved in the SELECT clause of R_1 are (*CA, Stores.stor_id, Times.year*).
- R_2= Turnover realized on sales of books type *Computer Science* sold at stores in *Paris* during *2013*. (*CA, Titles.type, Stores.stor_id, Times.year*).
- R_3= the number of books written by *Parisian authors* and published by *Springer* in *2013*. (*Qty, Authors.city, Publishers.pub_name, Times.year*).

In this way, other users formulate other sequences of analytical queries which involve attributes already expressed in U_4 queries. Let us take for example, the user U_5 query:

- R_1 = What is the turnover (CA) of the store store_500 in Washington by month. R_1 (*CA, Stores.stor_id, Stores.city, Times.month*).

An example of triadic association rules extracted from of users U_4 and U_5 are:

- BCAAR: $(U_4 \longrightarrow U_5)_{(CA,Stores.stor\_id)}$ supp= 60%, conf= 80%.
- BACAR: $(CA \longrightarrow Stores.stor\_id)_{(U_4,U_5)}$ supp= 75%, conf= 100%.

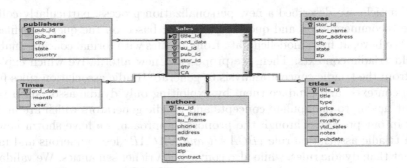

Fig. 2. PUBS data warehouse

In this paper, we propose a personalization based on these two types of rules. The user will interact with the interface *P-TRIAR* in two scenarios. According to the first scenario described in 4.1, the user U_4 wants to make new analysis on the data warehouse. He asks *P-TRIAR* and interrogates the log file from a specific date and requests all triadic association rules between him and other users with a condition on attributes, before setting a minimum threshold for the support and confidence. *P-TRIAR* will return him all the rules which satisfy these parameters. Then U_4 will choose according to the attributes he wants query the rules which suit him. Assuming he chooses the rule $(U_4 \longrightarrow U_5)_{(CA,Stores.stor\_id)}$, *P-TRIAR* will recommend him analysis queries the most frequent made by the user U_5 and having among their attributes CA and $Stores.stor\_id$. Unlike the dyadic rule $(U_4 \longrightarrow U_5)$ which would recommend all queries made by U_5, we add a condition on query attributes, so the rule is enriched and the number of queries to recommend is reduced considerably. So U_4 could choose from these queries which suits his analysis or by modifying it in part.

According to the second scenario described in 4.2, U_4 wants to make a new query on the attributes of the warehouse by exploiting $BACAR$ (rules between query attributes which have as a condition users). U_4 sets the initial parameters such as date, minimum support and confidence. Then U_4 will choose attributes he wants to involve in its query and $P\text{-}TRIAR$ will propose him triadic association rules associated with them. Assuming he chooses CA attribute $P\text{-}TRIAR$ would recommend him the attribute $Stores.stor\_id$ based on the rule $(CA \longrightarrow Stores.stor\_id)_{(U_4, U_5)}$. Contrary to the rule $(CA \longrightarrow Stores.stor\_id)$ will be proposed to all users, this rule will only be recommended to U_4 and U_5.

We obtained with a threshold of support and minimum confidence 50%, a total of 123 $BCAAR$ and 95 $BACAR$ from 42,638 AR dyadic. This result shows the triadic association rules compactness compared to dyadic ones. Then for personalisation, we take the example of user U_3, we obtain 14 $BCAAR$ and 12 $BACAR$ which would recommend queries and enrich its own ones according to his choices of analysis.

7   Conclusion

In this article, we described a new personalization process, particularly collaborative recommendation and query enrichment, based on the query log files of users. We have, at first, modelled data from *log* files with formal concept analysis to build triadic contexts. Then, we proposed a new alternative which exploits ideas from the triadic concept analysis to generate triadic association rules from triadic contexts, and produce them by exploiting only dyadic association rules without having to manipulate concepts and triadic generators which are unnecessary in our process. Through the proposed approach, we have shown how to obtain triadic association rule *(BCAAR and BACAR)* less numerous and more compact than dyadic rules, while also conveying a richer semantics. We validated our personalization approach by developing $P\text{-}TRIAR$ to extract these two types of rules from log files and personalize user queries according to each type.

This work opens up many opportunities for research. We plan in the short term to provide a system which collects user preferences through their choice of different personalization rules and queries recommended. Thus, they would be taken into account in their future choices. In the medium term, we plan to generalize the algorithms offered to polyadic association rules to deal with n-ary relationships to propose new methods for community detection in heterogeneous social networks.

Acknowledgement. We thank Rokia Missaoui for her collaboration in the work concerning the extraction of triadic associaton rules.

References

1. Adomavicius, G., Tuzhilin, A.: Toward the next generation of recommender systems: A survey of the state-of-the-art and possible extensions. IEEE Transactions on Knowledge and Data Engineering 17(6), 734–749 (2005)

2. Agrawal, R., Srikant, R., et al.: Fast algorithms for mining association rules. In: Proc. 20th Int. Conf. Very Large Data Bases, VLDB, vol. 1215, pp. 487–499 (1994)
3. Bellatreche, L., Giacometti, A., Marcel, P., Mouloudi, H., Laurent, D.: A personalization framework for olap queries. In: DOLAP, pp. 9–18 (2005)
4. Biedermann, K.: How triadic diagrams represent conceptual structures. In: ICCS, pp. 304–317 (1997)
5. Cerf, L., Besson, J., Nguyen, T.K.N., Boulicaut, J.-F.: Closed and Noise-Tolerant Patterrns in N-ary Relations. Data Mining and Knowledge Discovery 26(3), 574–619 (2013)
6. Chatzopoulou, G., Eirinaki, M., Polyzotis, N.: Query recommendations for interactive database exploration. In: Winslett, M. (ed.) SSDBM 2009. LNCS, vol. 5566, pp. 3–18. Springer, Heidelberg (2009)
7. Ganter, B., Obiedkov, S.A.: Implications in triadic formal contexts. In: ICCS, pp. 186–195 (2004)
8. Ganter, B., Wille, R.: Formal Concept Analysis: Mathematical Foundations. Springer-Verlag New York, Inc. (1999), Franzke, C. (trans.)
9. Golfarelli, M., Rizzi, S., Biondi, P.: myolap: An approach to express and evaluate olap preferences. IEEE Trans. Knowl. Data Eng. 23(7), 1050–1064 (2011)
10. Khemiri, R., Bentayeb, F.: Interactive query recommendation assistant. In: 2012 23rd International Workshop on Database and Expert Systems Applications (DEXA), pp. 93–97. IEEE (2012)
11. Khemiri, R., Bentayeb, F.: Fimioqr: Frequent itemsets mining for interactive olap query recommendation. In: DBKDA 2013, pp. 9–14 (2013)
12. Koutrika, G., Ioannidis, Y.: Personalized queries under a generalized preference model. In: Proceedings of 21st International Conference on Data Engineering, ICDE 2005, pp. 841–852. IEEE (2005)
13. Lehmann, F., Wille, R.: A triadic approach to formal concept analysis. In: ICCS, pp. 32–43 (1995)
14. Missaoui, R., Kwuida, L.: Mining triadic association rules from ternary relations. In: Valtchev, P., Jäschke, R. (eds.) ICFCA 2011. LNCS, vol. 6628, pp. 204–218. Springer, Heidelberg (2011)
15. Nguyen, T.K.N.: Generalizing Association Rules in N-ary Relations: Application to Dynamic Graph Analysis. Phd thesis, INSA de Lyon (October 2012)
16. Pasquier, N.: Data Mining: algorithmes d'extraction et de réduction des règles d'association dans les bases de données. PhD thesis (January 2000)
17. Patrick, M., Elsa, N., et al.: A survey of query recommendation techniques for datawarehouse exploration. In: EDA 2011 (2011)
18. Stefanidis, K., Drosou, M., Pitoura, E.: You may also like results in relational databases. In: PersDB 2009, pp. 37–42 (2009)
19. Trabelsi, C., Jelassi, N., Yahia, S.B.: Bgrt: une nouvelle base générique de règles d'association triadiques. application à l'autocomplétion de requêtes dans les folksonomies. Document Numérique 15(1), 101–124 (2012)
20. Veloso, A., de Almeida, H.M., Gonçalves, M.A., Meira Jr., W.: Learning to rank at query-time using association rules. In: SIGIR, pp. 267–274 (2008)
21. Voutsadakis, G.: Polyadic concept analysis. Order 19(3), 295–304 (2002)
22. Wille, R.: Restructuring lattice theory: An approach based on hierarchies of concepts. In: Rival, I. (ed.) Ordered Sets, pp. 445–470. Reidel, Dordrecht-Boston (1982)
23. Wille, R.: The basic theorem of triadic concept analysis. Order 12(2), 149–158 (1995)

An Event-Based Framework
for the Semantic Annotation of Locations

Anh Le, Michael Gertz, and Christian Sengstock

Database Systems Research Group, Heidelberg University, Germany
{anh.le.van.quoc,gertz,sengstock}@informatik.uni-heidelberg.de

Abstract. There is an increasing number of Linked Open Data sources that provide information about geographic locations, e.g., GeoNames or LinkedGeoData. There are also numerous data sources managing information about events, such as concerts or festivals. Suitably combining such sources would allow to answer queries such as '*When and where do live-concerts most likely occur in Munich?*' or '*Are two locations similar in terms of their events?*'. Deriving correlations between geographic locations and event data, at different levels of abstraction, provides a semantically rich basis for location search, topic-based location clustering or recommendation services. However, little work has been done yet to extract such correlations from event datasets to annotate locations.

In this paper, we present an approach to the discovery of semantic annotations for locations from event data. We demonstrate the utility of extracted annotations in hierarchical clustering for locations, where the similarity between two locations is defined on the basis of their common event topics. To deal with periodic updates of event datasets, we furthermore give a scalable and efficient approach to incrementally update location annotations. To demonstrate the performance of our approach, we use real event datasets crawled from the Website *eventful.com*.

1 Introduction

The main difference between a '*place*' and a position is that a place is represented as a human-readable description of a geographic location rather than just a geographic coordinate. Such descriptive information about locations is essential for location-based services (LBS), for instance, location recommendation or social event recommendation [5,6]. Typically, a data source managing information about locations provides various attributes of a location for an LBS application, including the name, address, description, and metadata such as tags. From a semantic perspective, such description or tags associated with a location are useful in semantic location search.

Unfortunately, such descriptive attributes detailing location information tend to be poor in many data sources. For example, based on our analysis, there are about one million locations in a dataset of events crawled for the years 2011 and 2012 from the Website *eventful.com*, but only 10% of them contain descriptions or tags. Moreover, querying based on simple text matching of descriptions and

Y. Manolopoulos et al. (Eds.): ADBIS 2014, LNCS 8716, pp. 248–262, 2014.

tags cannot take into account concept hierarchies that might exist for locations, time, or event topics. For example, using suitable concept hierarchies *'live jazz on Saturdays'* may be considered a match for *'live music on weekends'*. Therefore, enriching information about events at different levels of granularity and abstraction is necessary and useful.

Several methods have been proposed to extract semantic annotations for locations. However, some of them highly depend on the location data provided by external sources such as Wikipedia or the Google Maps API [2]. Other approaches exploit either user-tags (e.g., in the context of Flickr data) [11,12] or the user behavior (e.g., check-in data or user-interest-profiles) extracted from online social networks [6,15]. Such types of user generated content are often sparse, noisy and sometimes even inaccurate.

On the other hand, numerous data sources managing information about events are available on the Internet. This includes popular Websites such as *last.fm*, *eventim.de*, or *eventful.com*. Although in these sources the event data are less noisy (and more accurate) than in other georeferenced social media, there are still challenges in fully exploiting such information, as concept hierarchies, either explicitly or implicitly, exist for event topics, locations, and time.

Intuitively, some events occur more likely at some place/time than at other places/times. For example, events related to the topics *'live music'* or *'dance'* likely occur at a bar or club at weekends, whereas events related to *'conference'* or *'talk'* likely occur at a university on working days. Following this, we aim at extracting semantic annotations for locations from event data on the basis of exploiting correlations among geographic locations, time and event topics.

In this paper, we propose a framework to extract location annotations from event data. Our framework is based on the concept of a Location-Time-pair Class (LTC) to describe a group of location-time pairs that have the same location and time concepts, e.g., [*'Stadium'*,*'Weekend'*]. We define a measure to identify significant event topics with respect to an LTC, based on *Pointwise Mutual Information* [14]. A set of significant topics with respect to an LTC is called a Location-Time-pair Profile (LTP). LTPs are utilized to derive semantic annotations for locations, where an annotation is a pair of an event topic and a time concept, e.g., [*'Live-music'*,*'Weekend'*]. Figure 1 shows the components of our framework. The *LTProfile-Miner* component extracts LTPs from event datasets. To efficiently deal with (periodic) updates to event data, the component *LTProfile-Updater* updates the current set of location annotations. With the latter component, we provide a scalable and efficient approach to deal with large datasets that do not fit in main memory.

Based on external sources such as Wikipedia, extracted annotations of famous locations, e.g., stadiums or theatres, can be manually validated. Since there is no pre-existing ground-truth to validate all results obtained from a given dataset, we indirectly measure how good the extracted annotations are with location clustering. In summary, the contributions of this paper are as follows:

- We model semantic annotations for locations based on the concepts of events and event topics.

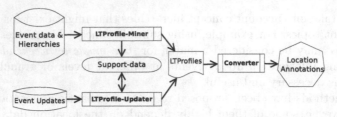

Fig. 1. Conceptual framework for annotating event locations

- We propose a measure based on *Pointwise Mutual Information* to identify significant event topics from a dataset of events.
- We develop two approaches: *LTProfile-Miner* to derive location annotations from an event dataset, and *LTProfile-Updater* to deal with periodic updates of that dataset.
- We demonstrate the utility of extracted annotations in semantic location search and clustering by using real event data.

In the following section, we discuss related work. In Section 3, we introduce the basic concepts and notations. We describe our method to extract semantic annotations for locations in Section 4. After presenting some experimental results in Section 5, we summarize the paper in Section 6.

2   Related Work

Basically, the term '*annotation*' means to attach information (metadata) to existing data. An example is that Flickr users add tags to photos to describe the photos. Since such human effort-based annotation systems are often noisy and incomplete, there have been many approaches to *automatically* annotate objects in different formats such as textual documents or photos, e.g., [4,7]. However, extracting annotations from spatio-temporal data like event data raises many challenges, e.g., annotations might differ among regions as well as over time, as discussed in [5]. Thus, such approaches cannot be directly applied to extract location annotations from event datasets. Nevertheless, the idea of *word-context matrices* and a statistical measure successfully used in annotating textual documents, called *Pointwise Mutual Information* [10,14], can be utilized to estimate correlations among locations, time, and event topics. This will be described in more detail in Section 3.2.

Several approaches are similar to our work in extracting annotations from spatio-temporal data. One direction of research relies on location information from external sources, such as the Google Maps API to annotate locations [2,3]. These approaches first extract points of interest (e.g., *stops* from trajectory data), and then annotate them with place categories (e.g., '*hotel*' or '*museum*') using external sources. Different from these approaches, we aim at extracting not only place categories but also relevant event topics w.r.t. a given location, important information that cannot be obtained from the above data sources.

Another direction of research aims at exploiting georeferenced social media to describe and annotate *geographic space*. Rattenbury and colleagues proposed several spatial clustering methods to identify Flickr tags corresponding to places and/or events [11,12]. Such tags can then be used to annotate geographic space on the basis of discovered clusters. Similarly, the approach in [13] aims at extracting latent geographic place semantics from Flickr data. Geographic space can then be annotated using spatial distributions and coefficients of extracted features. Although the above approaches focus on extracting annotations from spatio-temporal data, they are only able to annotate geographic space in general and not specific locations. Furthermore, these approaches do not explicitly model locations, in particular, they do not consider location hierarchies.

To the best of our knowledge, little work has been done yet to annotate locations with semantic tags. The most related work is [15], where a technique is proposed to annotate places with categorical tags such as '*restaurant*' or '*cinema*' by utilizing user check-in data. Similarly, the approach in [6] exploits check-in data to enrich places with semantic tags that are extracted from user interest-profiles on social networks. These approaches rely on characteristics of check-in data consisting of hidden user behaviors. Thus, they cannot be applied for event datasets for the following reasons. First, these approaches require a significant number of check-in records at a particular location and time to derive user behaviors. However, only few events occur at a particular location and time in an event dataset. Moreover, a check-in record as described in these approaches is a triple $\langle user\_id, time, location \rangle$ that does not contain semantic tags like an event description. Thus, in their approaches, a candidate set of tags for locations needs to be either predefined or obtained from an external source (user interest-profiles). Such a predefined set of tags is often small and only contains general, categorical tags. Rather than focusing on categorical tags like the above approaches, we aim at extracting more informative tags that can be used to discriminate one location from another. We also take concept hierarchies for time, locations and event topics into account, an important and useful piece of information not considered by the above approaches.

3   Basic Concepts and Notations

In the two following sections, we describe the concepts of events and event components to model semantic annotations for locations.

3.1   Events and Event Components

In this paper, an event is specified as a tuple $\langle e_{id}, C, T, L \rangle$, where the first component (e_{id}) is the event identifier and the last three are the event topic (context), time, and location, respectively, of that event. For example, a tuple $e = \langle$'#10202','*Borussia Dortmund vs Bayern Munich*', '*2013-06-27*', '*Signal Iduna Park*'$\rangle$ describes a football match. The topic, time, and location components of an event can be generalized to higher levels of abstraction and granularity, based on hierarchies, as detailed below.

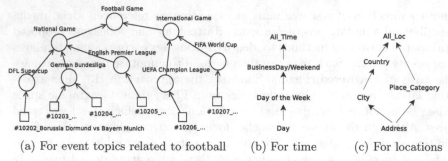

(a) For event topics related to football   (b) For time   (c) For locations

Fig. 2. Example of hierarchies for concepts (event topics), time, and locations

In data sources managing information about public activities (e.g., festivals or sports), an event topic (ET) is typically provided as a textual description, e.g., '*Borussia Dortmund vs Bayern Munich*' for a football game. An ET can be generalized to higher levels of abstraction, based on a concept hierarchy. For example, Figure 2(a) shows a simple hierarchy related to football, where the ET '*Borussia Dortmund vs Bayern Munich*' can be generalized to '*DFL Supercup*', '*National Game*', and then '*Football Game*'. Such a hierarchy might be explicitly provided, or it can be built using a learning approach.

We employ the operator ⇑ to compute the set of all generalizations for an ET in a given hierarchy. For example, '*DFL Supercup*'$^⇑$ is the set {'*National Game*', '*Football Game*'} based on the hierarchy shown in Figure 2(a). Event topics and their generalizations are key ingredients of semantic annotations for locations.

The time component of an event is typically specified as a time point. Since we focus on events such as festivals or sports, we assume that the time of an event is of granularity *Day*. Based on a predefined time hierarchy, an event time can be generalized to time concepts, e.g., a time point '*2013-06-27*' can be generalized to '*Friday*'→'*BusinessDay*'→'*All_Time*', based on the time hierarchy in Figure 2(b). We also use the operator ⇑ to compute the set of all generalizations of a time point, e.g., '*2013-06-27*'$^⇑$={'*Friday*', '*BusinessDay*', '*All_Time*'}.

Finally, the location component of an event is specified at a location granularity. Since an event like a concert or a football game takes place at a particular location, we assume that locations of events are of granularity *Address*. They can be generalized to a coarser granularity like *City* or to a location concept (place category) like '*Stadium*', again based on a predefined hierarchy. Also here we use the ⇑ operator to specify the generalizations of a location. For example, '*Signal Iduna Park*'$^⇑$ is the set {'*Dortmund*', '*Germany*', '*Stadium*', '*All_Loc*'}, based on the hierarchy in Figure 2(c).

Given an event topic f^1, a time concept T, and a location L, one might find some events whose respective components are related to f, T, and L from a given event dataset. The more such events are found, the more significant the association of f, T, and L is. In reality, some associations are more significant than others. For example, it is more likely to find events related to ice skating in

---

[1] In this paper, we often use 'e' to denote an event and 'f' to denote an event topic.

Winter than in Summer, or it is more likely to find a rock festival in some cities than in other cities. To model such associations, we introduce the concepts of *Location-Time-pair Instance* and *Location-Time-pair Class*.

3.2 Location-Time-Pair Instances and Classes

Let $\mathcal{D}$ be a dataset of events as $\langle e_{id}, C, T, L\rangle$ tuples, where the time and location components of each event are of granularity *Day* and *Address*, respectively. To formulate the probability to find an event topic (ET) at a given location and time later on, we define a *Location-Time-pair Instance* (LTI) as a pair $[l, t]$, where l and t are the location and time of some event in $\mathcal{D}$. We use $\mathcal{D}[l, t]$ to denote a subset of $\mathcal{D}$ consisting of events whose location and time components are l and t, respectively. The set of LTIs with respect to a given dataset $\mathcal{D}$ of events is defined as

$$\mathcal{I}(\mathcal{D}) := \{[l, t] \mid \exists e \in \mathcal{D}, e.L = l \wedge e.T = t\}. \tag{1}$$

As mentioned before, a location l can be generalized to a concept L based on a given location hierarchy. Similarly, a time point t can be generalized to a time concept T, based on a time hierarchy. The pair $[L, T]$ is called a *Location-Time-pair Class* (LTC) and the LTI $[l, t]$ is called an *instance* of that LTC. For example, one can infer that [*'Signal Iduna Park'*, *'2013-07-28'*] is an instance of [*'Stadium'*, *'Weekend'*]. The LTC set of a given event dataset $\mathcal{D}$ is defined as

$$\mathcal{C}(\mathcal{D}) := \{[L, T] \mid \exists [l, t] \in \mathcal{I}(\mathcal{D}), L \in l^{\Uparrow} \wedge T \in t^{\Uparrow}\}. \tag{2}$$

Given an LTC, it is straightforward to retrieve the set of its instances (LTIs). For example, {[*'Signal Iduna Park'*, *'2012-05-01'*], [*'Signal Iduna Park'*, *'2013-07-28'*], [*'Allianz Arena'*, *'2013-07-28'*],...} is the set of instances for [*'Stadium'*, *'Weekend'*]. For each LTI, event topics can then be derived from events in that LTI. Therefore, it is reasonable to determine the correlation between a given LTC and an ET based on the occurrences of that ET in the LTC. Clearly, ETs that are strongly related to an LTC are important to represent the characteristics of that LTC. For example, topics such as *'football'* or *'sport'* are expected for [*'Stadium'*, *'Weekend'*], whereas *'drink'* or *'live music'* are expected for [*'Bar/Club'*, *'Weekend'*].

To formulate correlations as the ones mentioned above, we employ the *Point-wise Mutual Information (pmi)* [14], commonly used in Computational Linguistics. The *pmi* value for an ET f with respect to an LTC Ω is computed based on the two following probabilities: (1) the probability to find f at any instance (LTI) of Ω, i.e., the conditional probability $P(f|\Omega)$, and (2) the probability to find f at any LTI in the dataset, i.e., $P(f)$. More precisely, the measure is computed as: $pmi(f; \Omega) = \log\left(\frac{P(f|\Omega)}{P(f)}\right) = \log\left(\frac{P(f,\Omega)}{P(f)P(\Omega)}\right).$

The *pmi* value of an ET f with respect to an LTC Ω represents the logarithmic difference between the two probabilities $P(f|\Omega)$ and $P(f)$. Thus, the *pmi* can be zero, positive or negative. If it is zero, i.e., $P(f|\Omega) = P(f)$, f and Ω are independent. If the value is positive, i.e., $P(f|\Omega) > P(f)$, the events related to f occur more likely at Ω than at other LTCs. If the value is negative, i.e., $P(f|\Omega)$

$< P(f)$, the events related to f more rarely occur at Ω than at other LTCs. The *pmi* measure can be normalized to a value between [-1,+1], where -1 means negatively correlated, 0 for independence, and +1 for perfectly correlated [1].

Definition 1. *(Normalized Pmi) Given an event topic f and an LTC Ω, the* **normalized pointwise mutual information** *(npmi) of f and Ω is defined as:*

$$npmi(f; \Omega) := \frac{pmi(f; \Omega)}{- \log\left(P(f, \Omega)\right)} = \frac{\log\left(\frac{P(f,\Omega)}{P(f)P(\Omega)}\right)}{- \log\left(P(f, \Omega)\right)} \quad \in [-1, 1]. \quad (3)$$

Since the *npmi* represents the difference between the probabilities $P(f|\Omega)$ and $P(f)$, it typically gives an ET a high score with respect to a given LTC if the ET frequently occurs at that LTC but rarely at other LTCs. For example, with sport events crawled from the Website *eventful.com*, the topics '*borussia*', '*dormund*', or '*bundesliga*' get higher *npmi* scores than the topics '*football*' or '*soccer*' with respect to an LTC ['*Signal Iduna Park*', '*Weekend*'][2]. One can see that the first three topics are better to identify that LTC, and thus, they have priority over the last two topics to annotate the location '*Signal Iduna Park*'.

Another advantage of the *npmi* measure is as follows. Since a frequency-based measure like *tf-idf* always gives a non-negative value, it is not trivial for the user to pick a good threshold in order to filter out irrelevant ETs. On the other hand, a non-positive *npmi* value indicates an insignificant correlation between an ET and an LTC. Thus, one can use any positive threshold δ to filter out irrelevant ETs (whose *npmi* values are zero or negative). Based on a positive threshold δ, one can select only ETs that have significant correlations to a given LTC. A set of such event topics is called a *Location-Time-Profile*.

Definition 2. *(Location-Time-Profile) Let $\mathcal{D}$ be a dataset of events and Ω be an LTC in $\mathcal{C}(\mathcal{D})$. The* **profile** *of Ω with respect to a given threshold $\delta > 0$ is a set of ETs, defined as $Profile(\Omega) := \{f \in e.C^{\Uparrow} | e \in \mathcal{D} \wedge npmi(f; \Omega) \geq \delta\}$.*

For a particular purpose such as location clustering where feature selection can be viewed as a form of weighting, both *npmi* and *tf-idf* can be used. However, as shown in our experiments later on, the *npmi* measure performs better than *tf-idf* when considering semantic similarity between locations.

We now present our method to compute the *npmi* for a given ET f with respect to an LTC Ω, based on a given event dataset $\mathcal{D}$. For this, we count the LTIs that support f, where an LTI $[l, t]$ *supports* f iff there exists an event $e \in \mathcal{D}[l, t]$ such that e is an instance of f. Based on that, we estimate the probabilities $P(f, \Omega)$, $P(f)$, and $P(\Omega)$ as follows.

Let N be the size of the LTI set (i.e., $\mathcal{I}(\mathcal{D})$), N_f the number of LTIs in $\mathcal{I}(\mathcal{D})$ that support f, N_Ω the number of LTIs in $\mathcal{I}(\mathcal{D})$ that are instances of Ω,

---

[2] Signal Iduna Park is the home stadium of the Borussia Dortmund football team playing in the German Bundesliga.

and $N_{f,\Omega}$ the number of instances of Ω that support f. The above probabilities are estimated as: $P(f,\Omega) = \frac{N_{f,\Omega}}{N}$, $P(f) = \frac{N_f}{N}$, and $P(\Omega) = \frac{N_\Omega}{N}$. Thus,

$$npmi(f;\Omega) = \frac{\log\left(\frac{P(f,\Omega)}{P(f)P(\Omega)}\right)}{-\log(P(f,\Omega))} = \frac{\log\left(\frac{N_{f,\Omega}N}{N_f N_\Omega}\right)}{-\log\left(\frac{N_{f,\Omega}}{N}\right)} = \frac{\log\left(\frac{N_{f,\Omega}N}{N_f N_\Omega}\right)}{\log\left(\frac{N}{N_{f,\Omega}}\right)}. \quad (4)$$

4   LT-Profiles and Applications

In this section, we first introduce a novel algorithm to generate LT-Profiles from an event dataset, and a scalable and efficient method to deal with periodic updates of the input data. We then show how to convert such profiles into location annotations. Finally, we describe how to exploit such information in semantic location search and clustering.

4.1   Generating Location-Time-Profiles

Given a dataset $\mathcal{D}$ of events, a set $\mathcal{H}$ of hierarchies for generating ETs from events and for generating location and time concepts, and a *npmi* threshold δ, this section describes a procedure called *LTProfile-Miner* to determine all profiles as defined in Definition 2.

Based on the Formulas (1) and (2), generating the set of LTIs ($\mathcal{I}(\mathcal{D})$) and the set of LTCs ($\mathcal{C}(\mathcal{D})$) is straightforward. For each LTC $\Omega \in \mathcal{C}(\mathcal{D})$, the set of ETs belonging to Ω is generated from all events that belong to any instance (LTI) of Ω. For each ET f in this set, the value of $npmi(f;\Omega)$ needs to be computed and compared with respect to the threshold δ. This can easily be done by counting LTIs in the set $\mathcal{I}(\mathcal{D})$ and then applying Formula (4). However, such a method is inefficient since the set $\mathcal{I}(\mathcal{D})$ will be scanned multiple times for all LTC-ET pairs. Thus, we propose a more efficient method as follows.

We utilize two data structures, called *Support_ET* and *Support_LTC*, where each one is a hash table mapping keys to LTI sets. Given an ET f, the set of LTIs that support f is retrieved by using the hash table *Support_ET*. This set is denoted *Support_ET[f]*. Let n_f, n_i, and n_{ei} be the number of ETs, the number of LTIs ($|\mathcal{I}(\mathcal{D})|$), and the average number of events of an LTI, respectively. The runtime complexity to build *Support_ET* is $O(n_f n_i n_{ei})$, since each element (with respect to an ET) is computed by scanning through all the LTIs and considering all events inside each LTI. Similarly, the hash table *Support_LTC* is used to retrieve the set of LTIs that are instances of a given LTC Ω, denoted *Support_LTC[Ω]*. The complexity to build *Support_LTC* is $O(n_c n_i)$, where n_c is the number of LTCs ($|\mathcal{C}(\mathcal{D})|$) and n_i is the number of LTIs ($|\mathcal{I}(\mathcal{D})|$).

Utilizing hash tables allows the values N_f, N_Ω, and $N_{f,\Omega}$ in Equation (4) to be computed with several set operations: $N_f = |Support\_ET[f]|$, $N_\Omega = |Support\_LTC[\Omega]|$, and $N_{f,\Omega} = |Support\_ET[f] \cap Support\_LTC[\Omega]|$. Thus, the value of $npmi(f;\Omega)$ for each pair of an LTC Ω and ET f can easily be computed. Finally, the profile of each LTC Ω in $\mathcal{C}(\mathcal{D})$ is obtained based on Definition 2.

4.2   Updating Location-Time-Profiles

In the previous section, we presented an approach to extract all *LTProfiles* from a given dataset of events. Such a dataset consists of events in a certain time-interval (e.g., [2011,2012]). Thus, the extracted profiles are only valid in this interval. In reality, datasets are incrementally updated. For example, events in 2013 are added to a dataset of events in [2011,2012]. Running again that procedure for the merged dataset is a possible solution, which, however, is neither efficient nor scalable. To adapt to periodic updates of event data, we propose another procedure, called *LTProfile-Updater*.

Assume that after executing *LTProfile-Miner*, the following intermediate values are stored on a secondary storage: N, N_f (as an element of a list), N_Ω (as an element of a list), $N_{f,\Omega}$ (as an element of a matrix). Such data, called *support-data*, contain sufficient information to extract profiles without considering the original (previous) dataset $\mathcal{D}$.

Let $\mathcal{D}^*$ be the dataset of new events to update. It is reasonable to assume that each event in $\mathcal{D}^*$ occurred after all events in $\mathcal{D}$, i.e., events in $\mathcal{D}^*$ are newer than events in $\mathcal{D}$. Therefore, there is no overlap between the LTI sets of the two datasets. Thus, the values of N, N_f, N_Ω and $N_{f,\Omega}$ can be updated as: $N = N + |\mathcal{I}(\mathcal{D}^*)|$, $N_f = N_f + |Support\_ET^*[f]|$, $N_\Omega = N_\Omega + |Support\_LTC^*[\Omega]|$, and $N_{f,\Omega} = N_{f,\Omega} + |Support\_ET^*[f] \cap Support\_LTC^*[\Omega]|$. Note that $Support\_ET^*$ and $Support\_LTC^*$ are two hash tables computed from the update ($\mathcal{D}^*$) with the method described in the previous section.

Summing up, *LTProfile-Updater* first loads the *support-data* and then combines it with the update ($\mathcal{D}^*$) to update the current location profiles. Since this procedure utilizes the *support-data*, only the update $\mathcal{D}^*$ is scanned.

Based on *LTProfile-Updater*, an *anytime* approach to deal with very large datasets works as follows. First, the events in a (large) dataset $\mathcal{D}$ are sorted by the time attribute and distributed in increasing order into sub-datasets $\mathcal{D}_0$, $\mathcal{D}_1,\ldots$ such that each $\mathcal{D}_i$ fits into main memory. *LTProfile-Miner* is then called to compute the *support-data* from $\mathcal{D}_0$. Finally, *LTProfile-Updater* is iteratively called for each $\mathcal{D}_i$ ($i \geq 1$). If the mining process is interrupted after processing $\mathcal{D}_i$, the results are valid until the latest time in $\mathcal{D}_i$.

4.3   Location Annotations

Location-Time-Profiles, each consisting of significant ETs at an LTC, can be exploited to annotate locations. Here, we define a location annotation as a set, where each element is a pair of an event topic and a time concept. For example, annotation elements for a specific bar/club might be ['*jazz*', '*Tuesday*'] or ['*dancing*', '*Weekend*']. The formal definition is given as follows.

Definition 3. *(Location Annotation) Let $\mathcal{D}$ be an event dataset. The **annotation** of a location (or location concept) L is a set defined as:*
   $Annotation(L) := \{[f,T] \mid \Omega = [L,T] \in \mathcal{C}(\mathcal{D}), \ f \in Profile(\Omega)\}.$

4.4 Similarity Measure for Location Search and Clustering

To determine how similar two locations are, we define a similarity measure for locations based on events. Basically, the more common event topics two locations have, the more similar they are. Given two locations L_1 and L_2, and their annotations AL_1 and AL_2, respectively, the similarity between the two locations is computed based on the Jaccard Index as: $sim(L_1, L_2) = \frac{|AL_1 \cap AL_2|}{|AL_1 \cup AL_2|} \in [0, 1]$.

This measure can be used to find locations that are similar to a given location or just to rank the results, for example, in the query '*Find all cities in the US like Munich (in Germany) in terms of beer festivals*'.

To apply clustering, the dissimilarity distance between two locations L_1, L_2 is computed as $dist(L_1, L_2){=}1{-}sim(L_1, L_2)$. Based on that, locations can be clustered with one of the various clustering algorithms, e.g., hierarchical clustering.

5 Experimental Evaluation

We demonstrate the utility and efficiency of our approach using datasets crawled from the Website *eventful.com* for different topics from 2011 to 2012. Our framework is implemented in Java and runs with 24GB heap size. All experiments were run on an Intel Xeon 2.27GHz with 48GB RAM, running Ubuntu 64bit. Before presenting the results, we first describe the experimental setup.

5.1 Datasets and Experimental Setup

We crawled from the Website *eventful.com* for events in Germany and only festivals in Europe to easily validate the results later on. As raw data, each event consists of an event identifier, title, time, location, and a list of tags. Based on tags, one can select events for a particular topic, e.g., '*sports*' or '*music*'.

As mentioned in Section 4.1, the runtime complexity of our algorithm depends on not only the number of events but also the numbers of locations (more precisely, LTIs). Hence, for evaluation purpose, we select different datasets in various topics and sizes in terms of the number of events and locations. Table 1 shows five datasets used in our experiments, where the first two datasets (DE-Festival and DE-Sports) are smaller than the last three. All events took place in Germany ('DE-') or Europe ('EU-') in the years 2011 and 2012.

Table 1. Properties of datasets used in experiments

Dataset	Topic	Area	Number of Events			Number of Locations
			2011	2012	Total	
DE-Sports	sports	Germany	1,335	1,673	3,008	960
DE-Festival	festival	Germany	1,278	1,654	2,932	1,515
EU-Festival	festival	Europe	13,592	20,561	34,143	18,018
DE-Music	music	Germany	24,756	32,398	57,154	12,591
DE-All	all topics	Germany	72,672	85,995	158,667	20,141

First, the raw data of events are transformed into the form $\langle e_{id}, C, T, L \rangle$, where e_{id} is the event identifier, and the last three components are the following attributes: the event identifier, start-time, and venue identifier, respectively. Note that here the event identifiers are utilized for two purposes: to distinguish an event from others and to link event contexts to tags. We built a hierarchy for tags based on the method described in [9]. The event location is of granularity *Address* and can be generalized to *City* or *Place_Category*. The time component of an event in *Day* is generalized to *Day of the Week* (Mon, Tue, etc.), then *Businessday/Weekend* (BD/WE), and finally *All_Time*(AT).

With the above settings, we conducted a series of experiments to evaluate our framework. In the following section, we present the results obtained from extracting location annotations for the five datasets. We then demonstrate the utility of these annotations in location clustering in Section 5.3. Finally, we show the efficiency of *LTProfile-Miner* and *LTProfile-Updater* in Section 5.4.

5.2 Annotation Extraction

We run *LTProfile-Miner* to obtain LT-Profiles for the five datasets for the two years (2011-2012). Then, annotations for locations are obtained using the method described in Section 4.3. For each dataset, we tried different *npmi* thresholds (δ). Basically, the larger the threshold δ, the less locations are annotated, but the more confident the annotations are. For example, when $\delta = 0.1$, about 70-90% of the locations were annotated, whereas less than 30% of the locations were annotated when $\delta > 0.5$. Table 2 shows typical annotations we obtained. Note that the words describing topics are stemmed, and an item of a location annotation is followed by its *npmi* value, e.g., socc_WE:0.39.

Based on these annotations, one can easily find locations related to some given event topics. For example, *NürnbergMesse* (Germany) will be found when we search for places related to '*technology*' and '*exhibition*', as shown in Table 2. This can be explained by annual events related to computer software/hardware or electronic systems that are located there, such as '*embedded world*'.

From the extracted annotations, one can see that some annotations are obvious, for instance, the annotation of a cinema (e.g., *Kino Babylon Mitte*) contains event topics related to film and movie festivals; or an exhibition center (e.g., *Messe Essen*) contains event topics related to '*expo*', '*industry*', or '*tradeshow*'. We also found some interesting relationships, such as a relationship between the exhibition center *Messe Essen* and the topic '*fashion*'. This relationship is explained by a series of Modatex Fashion Fair events frequently occurring at that location. From the dataset EU-Festival, we also discovered some cities in Europe that are famous for their annual festivals . For instance, Torre del Lago, Peraso (Italy), and Montpelier (France) are famous for opera festivals.

5.3 Location Clustering

We exploit the extracted annotations to cluster locations. Such clusters will be utilized further to assign higher level semantic tags to locations or to build taxonomies of locations, as described in Section 4.4. For this purpose, we employ hierarchical clustering. The performance of location clustering is evaluated based on the F-score measure, commonly used in document clustering [8]. First, we describe how to obtain datasets with ground-truth for clustering evaluation.

Table 2. Example annotations extracted from the experimental datasets. Items in each annotation are sorted by their *npmi* values.

Location/Granularity	Annotation
DE-Sports	
Signal Iduna Park - Dortmund (Address)	{borussia_Sat:0.66, borussia_WE:0.61, borussia_AT:0.56, bundesliga_Sat:0.46, bundesliga_WE:0.42, football_WE:0.32, socc_WE:0.29,...}
Oschersleben Sachsen-Anhalt (City)	{circuitracing_AT:0.81, circuitracing_WE:0.77, motorsport_AT:0.71, autosport_AT:0.70, racing_AT:0.69, motorsport_WE:0.68,...}
DE-Festival	
Kino Babylon Mitte - Berlin (Address)	{filmfestival_BD:0.63, filmfestival_Thu:0.63, movi_BD:0.59, movi_Thu:0.59, film_BD:0.55, film_Thu:0.55, filmfestival_AT:0.54, movi_AT:0.50,...}
Messe Essen GmbH (Address)	{expo_Thu:0.66, fashion_Sat:0.66, convention_Thu:0.66, fashion_WE:0.64, homeexhibition_Fri:0.58, industry_AT:0.49, expo_BD:0.48,...}
DE-Music	
Bar/Night Club (Place category)	{elektronic_WE:0.55, hardstyl_WE:0.55, nightlif_WE:0.52, tranc_WE:0.45, rhythmnblu_BD:0.43, elektronic_AT:0.42, hardstyl_AT:0.42,...}
Concert Hall (Place category)	{philharmonieess_AT:0.67, doommetal_Tue:0.49, epic_Tue:0.49, jamsession_Mon:0.48, monstrosity_Wed:0.46, greatesthit_Mon:0.46,...}
DE-ALL	
Nürnberg Messe (Address)	{softwar_AT:0.62,expopromot_AT:0.53,school&alumni_AT:0.52, tool_AT:0.42, tradeshow_AT:0.41,scienc_AT:0.41, business_AT:0.41,...}
Philharmonie Berlin (Address)	{klassischekonzert_AT:0.64, cultur_AT:0.54, klassisch_AT:0.54, classical_AT:0.53, cultur_WE:0.49, symphony_WE:0.44, violin_Mon:0.42,...}
EU-Festival	
Torre del Lago - Tuscany - Italy (City)	{art&theatr_AT:0.87, art&theatr_BD:0.84, art&theatr_WE:0.73, opera_AT:0.27, opera_BD:0.26, opera_Fri:0.24, opera_WE:0.22,...}
LilianBaylisTheatre - London (Address)	{ballet_AT:0.80, ballet_BD:0.77, ballet_WE:0.69, clubbing_WE:0.47, nightlif_WE:0.47, danc_AT:0.40, theatr_Wed:0.32, art_Wed:0.28,...}

Since a location in our dataset can be generalized to a place category (e.g., *'Hotel'*, *'Restaurant'*), we used such categories as ground-truth labels to evaluate location clustering, that is, locations of the same label are expected to be in the same cluster. We also removed locations with blank labels or non-categorical labels (e.g., *'postal code'* or *'named place'*) because they produce meaningless results in that clustering evaluation method. Since the number of locations of different categories varies a lot (e.g., more than 100 for *'Concert Hall'*, *'Bar/Club'*, but less than 10 for *'Hospital'*, *'Library'*), we finally selected only the top 7 categories (each category contains more than 10 locations) and generated datasets for clustering evaluation with a method as described below.

Let $L^k_{\{C_1,C_2,...,C_k\}}$ be a dataset consisting of locations of k categories C_1, C_2,..., C_k. Such a dataset is generated by choosing k categories from the top categories, e.g., $L^2_{\{Stadium,Theater\}}$. Let G_k be a group of generated datasets containing the same number of categories (i.e., k categories). For example, G_2 is a group of $\binom{7}{2} = 21$ datasets created by choosing 2 from the 7 categories, e.g., $L^2_{\{Stadium,Museum\}}$, $L^2_{\{Hotel,Museum\}}$. Instead of presenting the F-score for each individual dataset, we will show the mean F-score for each group G_k.

As mentioned in Section 3.2, an alternative to *npmi* is *tf-idf* that can be employed to weight event topics for location clustering. Here, we compare the performance of the *npmi* measure to the following versions of *tf-idf* that are widely used. Given an event topic f and an LTC Ω, two versions of *tf-idf*, called *tf-idf$_1$* and *tf-idf$_2$*, are defined as $\text{tf-idf}_1(f, \Omega) = N_{f,\Omega}*\log(\frac{N}{N_f})$ and $\text{tf-idf}_2(f, \Omega) = (1+\log(N_{f,\Omega}))*\log(\frac{N}{N_f})$. The values N, N_f, and $N_{f,\Omega}$ are the number of LTCs, the number of LTCs that contain ET f, and the number of instances of the LTC

Ω that support f, respectively. Similar to the *npmi* measure, profiles of LTCs can be computed with Definition 2, where *npmi* is replaced by either *tf-idf$_1$* or *tf-idf$_2$*. For a particular dataset and a particular measure, the threshold δ is selected so that the *F-score* is the largest. We use locations of the datasets DE-All and EU-Festival to assess the performance of location clustering since they covers all locations of the other datasets.

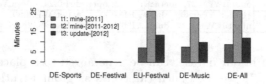

(a) Dataset DE-All   (b) Dataset EU-Festival

Fig. 3. Comparison of the measures *tf-idf$_1$*, *tf-idf$_2$*, and *npmi* in hierarchical clustering. The best result is achieved with the *npmi* measure.

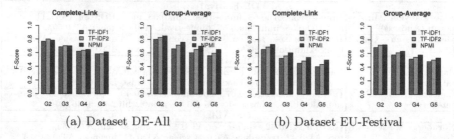

Fig. 4. Runtime of LTP-Miner and LTP-Updater

Figure 3 shows the comparison of the *npmi* measure with *tf-idf$_1$* and *tf-idf$_2$* in location clustering. Although different merging methods of hierarchical clustering were tried, due to space constraints, we present only results of two among the best methods: complete-link and group-average. In general, using the *npmi* measure gives the best result. A closer look at the generated profiles shows that a profile generated by the *npmi* measure contains more event topics presenting the characteristics of the corresponding location, as discussed in Section 3.2. In addition, one can see from Figure 3 that clustering on locations of the dataset DE-All gives better results than clustering on locations of the dataset EU-Festival. This is reasonable since it is more difficult to categorize locations of the latter dataset consisting of narrow topics, i.e., event topics related to '*festival*'.

We found many interesting clusters from the datasets. For example, we found clusters of bars/clubs regarding their music genres (e.g., jazz or r&b/soul); or a cluster of cities in Europe famous for opera festivals like Montpellier (France), Torre del Lago (Italy), or Pesaro (Italy).

5.4 Runtime and LTP-Updater Efficiency

We show the runtime of *LTProfile-Miner* for each dataset and also demonstrate the utility and efficiency of *LTProfile-Updater*. To do so, we split each dataset in Table 1 into two parts, each corresponding to one year. For example, from the dataset DE-Sports, we create two subdatasets DE-Sports$_{[2011]}$ and DE-Sports$_{[2012]}$, where the first one consists of events in 2011 and the latter one consists of events in 2012 of the dataset DE-Sports. From Table 1, one can see that the number of events in 2012 is larger than in 2011 (about 20 to 50%).

For each triple of datasets ($\mathcal{D}$, $\mathcal{D}_{[2011]}$ and $\mathcal{D}_{[2012]}$), we measure the runtime t_1 for *LTProfile-Miner* on $\mathcal{D}_{[2011]}$, the runtime t_2 for *LTProfile-Miner* on $\mathcal{D}$, and the runtime t_3 for *LTProfile-Updater* on $\mathcal{D}_{[2012]}$ (with support-data extracted from $\mathcal{D}_{[2011]}$). One can see the results in Figure 4. The first two datasets take only a few seconds to process. The cases of EU-Festival and DE-Music illustrate that the number of LTIs also affects the runtime. Although the number of events of the dataset EU-Festival is smaller than the dataset DE-Music, the number of locations of the dataset EU-Festival is larger, as shown in Table 1. In all cases, the runtime t_3 is larger than t_1, because the number of events in 2012 is larger than in 2011. However, in comparison to t_2, the runtime t_3 is much smaller for both datasets. This shows that using the *LTProfile-Updater* is an efficient and scalable approach to update the current location profiles with new data.

6   Conclusions and Ongoing Work

Event-based annotations of locations describe the event topics that are most related to a location, together with the time when the events of such topics most likely occur. We presented a comprehensive framework to extract such annotations from event datasets. Our approach is based on Location-Time-pair to associate a set of the most related event topics with a pair of a location and time. We also showed a scalable and efficient method to deal with periodic updates of event data. Our experimental results give a very good indication that the extracted annotations can be utilized well for semantic location search as well as clustering.

Using hierarchical clustering, taxonomies of locations can be built from annotated locations. We are currently developing a method to encode such taxonomies in RDF, and also to automatically link them to Linked Open Data sources. Another direction of current research focuses on exploiting negative *npmi* values in outlier detection. Such a method is important to detect errors or inaccurate information in event datasets.

References

1. Bouma, G.: Normalized (Pointwise) Mutual Information in Collocation Extraction. In: Proceedings of the Biennial GSCL Conference (2009)
2. Cao, X., Cong, G., Jensen, C.S.: Mining Significant Semantic Locations From GPS Data. Proceedings of the VLDB Endowment 3, 1009–1020 (2010)
3. Chakraborty, D., Spaccapietra, S., Parent, C.: SeMiTri: A Framework for Semantic Annotation of Heterogeneous Trajectories. In: EDBT, pp. 259–270 (2011)
4. Wang, C., Blei, D., Fei-Fei, L.: Simultaneous image classification and annotation. In: CVPR, pp. 1903–1910. IEEE (2009)

5. Derczynski, L.R.A., Yang, B., Jensen, C.S.: Towards context-aware search and analysis on social media data. In: EDBT, pp. 137–142. ACM Press (2013)
6. Hegde, V., Parreira, J.X., Hauswirth, M.: Semantic Tagging of Places Based on User Interest Profiles from Online Social Networks. In: Serdyukov, P., Braslavski, P., Kuznetsov, S.O., Kamps, J., Rüger, S., Agichtein, E., Segalovich, I., Yilmaz, E. (eds.) ECIR 2013. LNCS, vol. 7814, pp. 218–229. Springer, Heidelberg (2013)
7. Kulkarni, S., Singh, A., Ramakrishnan, G., Chakrabarti, S.: Collective annotation of Wikipedia entities in web text. In: KDD. ACM Press (2009)
8. Larsen, B., Aone, C.: Fast and effective text mining using linear-time document clustering. In: KDD, pp. 16–22. ACM Press (1999)
9. Le, A., Gertz, M.: Mining Spatio-temporal Patterns in the Presence of Concept Hierarchies. In: ICDM Workshops, pp. 765–772 (2012)
10. Pantel, P., Lin, D., Canada, A.T.H.: Discovering Word Senses from Text. In: KDD, pp. 613–619. ACM Press (2002)
11. Rattenbury, T., Good, N., Naaman, M.: Towards automatic extraction of event and place semantics from Flickr tags. In: SIGIR, pp. 103–110. ACM Press (2007)
12. Rattenbury, T., Naaman, M.: Methods for extracting place semantics from Flickr tags. ACM Transactions on the Web 3, 1–30 (2009)
13. Sengstock, C., Gertz, M.: Latent Geographic Feature Extraction from Social Media. In: SIGSPATIAL, pp. 149–158. ACM Press (2012)
14. Turney, P.D., Pantel, P.: From Frequency to Meaning: Vector Space Models of Semantics. Journal of Artificial Intelligence Research 37, 141–188 (2010)
15. Ye, M., Shou, D., Lee, W.-C., Yin, P., Janowicz, K.: On the semantic annotation of places in location-based social networks. In: KDD. ACM Press (2011)

Observing a Naïve Bayes Classifier's Performance on Multiple Datasets

Boštjan Brumen, Ivan Rozman, and Aleš Černezel

University of Maribor, Faculty of Electrical Engineering and Computer science,
Smetanova 17, Si-2000 Maribor, Slovenia
(bostjan.brumen,i.rozman,ales.cernezel)@uni-mb.si

Abstract. General theories describing the performance of artificial learners are of little help when a user is confronted with a selection of datasets and a given artificial classifier. The objective of this paper is to find out the best description of the learning curves produced by a Naïve Bayes classification. The performance of Naïve Bayes was measured on 121 datasets using k-fold cross-validation. Power, linear, logarithmic and exponential functions were fit to the data. The exponential function was a better descriptor of the error rate in 44 of 60 useful cases. Average mean squared error is significantly different at P=0,000 from power and linear and at P=0,001 from logarithmic function. The exponential function's rank is significantly different from the ranks of other models (P=0,000). The results can be used to forecast the future performance of the learner, or to check where on the learning curve the current measurement lies.

Keywords: Machine Learning, Power Law, Naïve Bayes, Error rate, Learning curve.

1   Introduction

Human cognitive performance was given quite a lot of attention in the research: the power function is generally accepted as an appropriate description in psychophysics, in skill acquisition, and in retention. Power curves have been observed so frequently, and in such varied contexts, that the term "power law" is now commonplace [1, 2].

The power law best describes the data in quite many real situations [3]. Nevertheless, recently some arguments arose against the power law [4]: the main argument is that it holds only on the aggregate level; on a specific learner's level the exponential law is much better. Thus, in human performance, the power law is a common description when describing a population of learners, and the exponential law when it comes to a single person.

In the area of artificial intelligence generally, and in the area of learning outcomes specifically, not much research is available [5]. Firstly, since the artificial learners are finite automata, a description of a learning problem could be a functional dependency between the data, the learning algorithm's internal specifics and its performance (e.g. error). The output (error rate) based on the input (data, selected learner) and learner's

Y. Manolopoulos et al. (Eds.): ADBIS 2014, LNCS 8716, pp. 263–275, 2014.

internal properties could have been determined analytically. Unfortunately, there is no such method yet developed, and this holds true for every single artificial learner so far devised. Secondly, standard numerical (and other statistical) methods become unstable when using large data sets [6]. Finally, estimates for the size of the confidence interval on the training error under various settings of the problem of learning from examples have been devised, i.e. the Vapnik-Chervonenkis theory [7] is the most appropriate for describing the artificial learners. Due to its limitations (the most obvious is that the oracle is never wrong) it cannot be used efficiently in the real life situations [8]. The literature displays some findings for a specific learner and for specific data can (e.g. [9]). All in all, there is no general analytical solution, nor is there a general estimative method available.

On the other hand, a learner's performance can be measured on as many tasks (data sets) as possible and the conclusion can be synthetized. However, not much research was conducted on the description of performance of Naïve Bayes on a large scale comparison using several different data sets. The classification trees were shown to be predicted by a power law [10-12], but these studies were not using a large number of datasets, nor were the results statistically significant. Due to the weaknesses of the mentioned studies Singh found the evidence against the power law [13]. Which law describes the performance of a selected artificial learner, in general, is thus generally unknown, although such knowledge would help in optimization within knowledge acquisition tasks [14].

In the paper we address the following research question: which of the following functions, the power, exponential, linear, or logarithmic, is best describing the performance of a selected artificial learner (in our case, the one based on the Naïve Bayes)? Our null hypotheses are as follows:

- The mean difference between the function's f_i () and function's f_j () average mean squared error equals 0.
- The median of differences between the function's f_i () and function's f_j () average rank equals 0.

Alternative hypotheses are that the mean squared error / median of differences are different.

The main contribution of this paper is the answer to the question: "Which mathematical function fits best a Naïve Bayes artificial classifier?"

2   Method

We have chosen a Naïve Bayes classifier [15] implemented in the Waikato Environment for Knowledge Analysis (WEKA) project toolkit [16, 17] version 3.6.8, with standard built-in settings and initial values.

For statistical analyses we used IBM SPSS version 21.

In the following we describe the methods for data collection, data processing, and measurements of the target values.

2.1 Data Collection

We used publicly available datasets from University of California at Irvine (UCI) Machine Learning Repository [18]. We selected the datasets where the problem task is classification; the number of records in a dataset was larger than 200 and the number of instances exceeded the number of attributes (i.e. the task was classification, not feature selection).

The UCI repository contains datasets in ".data" and ".names" format while Weka's native format is ARFF. Therefore we used files available from various sources, such as TunedIT [19], Håkan Kjellerstrand' weka page [20, 21] and Kevin Chai's page [22]. We gathered 121 datasets, listed in Table 1.

We used only the original or larger datasets where several ones were available and ignored any separate training or test set, or any associated cost model.

2.2 Data Pre-processing

We followed the following steps for obtaining the error rate curve (i.e. learning curve) [23]:

1. Data items in a data set are randomly shuffled
2. First, $n_{i=1}=50$ items are chosen
3. Build decision trees using k-fold cross-validation on sample size of n_i [24, 25]; k was set to 10 [12, 24-26];
4. Measure the error rate for each tree in 10-fold run and average the result over 10 runs
5. Store the pair (n_i=sample size, e_i=error)
6. The number of items in a data set is increased by 10; $n_{i+1}:=n_i+10$
7. Repeat steps 3-6 until all data items in a dataset are used.

2.3 Fitting a Curve Model to the Measured Data

The next step in our research was to fit a model to the error rate curves. We used four different functions, as in Equations (1)-(4):

$$\text{power (POW): } f(x) = p_1 + p_2 x^{p_3} \tag{1}$$

$$\text{linear (LIN): } f(x) = p_1 + p_2 x \tag{2}$$

$$\text{logarithm (LOG): } f(x) = p_1 + p_2 \log x \tag{3}$$

$$\text{exponential (EXP): } f(x) = p_1 + p_2 e^{p_3 x} \tag{4}$$

The functions do not have the same number of parameters (p_i). They all include the constant p_1 and coefficient p_2, in addition to potent p_3 for the power and the exponential function. Based on the specifics of the problem and the speed of convergence we limited the parameters to the following intervals:

- p_1 to interval [0, 1] (error rate cannot be less than 0 and more than 1)
- p_2 to interval [0, 100] for power function and to [-100, 0] for the others, and
- p_3 to interval [-100, 0] (error rate is decreasing hence p_3 needs to be negative)

We used the open-source GNU Octave software [27] and the built-in Levenberg-Marquardt's algorithm [28, 29], also known as the damped least-squares (DLS) method, for fitting the function parameters to the data.

The inputs to the algorithm were vector x (sample sizes n), vector y (error rates e), initial values of parameters p_i ([0,01; 1; -0,1] for POW, [0,1; -0,001] for LIN, [0,1; -0,01] for LOG and [0,01; 0,1; -0,01] for EXP), function to be fit to vectors x, y (power, linear, logarithm, or exponential), partial derivatives of functions with respect to parameters p_i, and limits of parameters p_i (as described above).

The algorithm's output were vector of functional values of fitted function for put x, vector of parameters p_i, where minimum mean squared error was obtained, and a flag whether the convergence was reached or not.

3   Results

For each dataset we tested the claim that the underlying data can be modeled by the probability density functions POW, LIN, LOG and EXP, respectively. We used the Pearson's chi-squared test (χ^2), also known as the chi-squared goodness-of-fit test or chi-squared test for independence, where the null hypothesis was H_0: $r_\mu = 0$ or there is no correlation between the population and the model [30], at $\alpha=0.05$. Table 1 lists the results: the values in bold are P values indicating that the null hypothesis is rejected, the number of degrees of freedom (df), and the coefficient of determination R^2 (indication how well a regression line fits a set of data). N/A indicates that the model was not calculated because the Levenberg-Marquardt's algorithm suggested a constant model and hence χ^2 cannot be computed.

Table 1. Datasets and respective P-values and r^2 values for goodness of fit of a model to data

Dataset	df	POW P	POW R^2	LIN P	LIN R^2	LOG P	LOG R^2	EXP P	EXP R^2
ada_agnostic	449	**0,000**	0,74	**0,000**	0,31	**0,000**	0,63	**0,000**	0,80
ada_prior	449	1,000	0,00	1,000	0,00	1,000	0,00	**0,000**	0,22
analcatdata_authorship	77	**0,000**	0,85	**0,000**	0,47	**0,000**	0,74	**0,000**	0,92
analcatdata_braziltourism	34	**0,000**	0,85	**0,000**	0,81	**0,000**	0,87	**0,000**	0,89
analcat data_broadwaymult	21	**0,008**	0,27	**0,017**	0,22	**0,008**	0,27	**0,006**	0,29
analcatdata_dmft	72	1,000	0,00	1,000	0,00	1,000	0,00	1,000	0,00
analcatdata_halloffame	126	1,000	0,00	1,000	0,00	1,000	0,00	1,000	0,00
analcatdata_marketing	29	**0,000**	0,35	**0,000**	0,34	**0,000**	0,35	**0,000**	0,35
analcatdata_reviewer	30	**0,000**	0,33	**0,000**	0,42	**0,000**	0,34	**0,000**	0,42
anneal	82	1,000	0,00	0,721	0,00	1,000	0,00	0,152	0,02
anneal.ORIG	82	**0,000**	0,32	**0,022**	0,06	**0,000**	0,21	**0,000**	0,39
audiology	15	**0,000**	0,81	**0,000**	0,88	**0,000**	0,84	**0,000**	0,87
australian	61	**0,000**	0,38	**0,013**	0,09	**0,002**	0,15	**0,000**	0,46
autos	13	**0,003**	0,45	**0,000**	0,61	**0,002**	0,47	**0,000**	0,58

Table 1. (*Continued*)

Dataset	df	POW P	POW R2	LIN P	LIN R2	LOG P	LOG R2	EXP P	EXP R2
badges_plain	22	**0,000**	0,93	**0,000**	0,73	**0,000**	0,87	**0,000**	0,92
balance-scale	55	**0,000**	0,80	**0,000**	0,52	**0,000**	0,73	**0,000**	0,84
baseball-hitter	25	**0,000**	0,47	**0,000**	0,63	**0,000**	0,34	**0,000**	0,62
baseball-pitcher	13	0,119	0,15	0,641	0,01	0,428	0,04	**0,034**	0,27
BC	21	0,827	0,00	0,277	0,05	0,822	0,00	0,289	0,05
Billionaires92	16	0,832	0,00	0,446	0,03	0,830	0,00	0,450	0,03
biomed	13	1,000	0,00	1,000	0,00	1,000	0,00	1,000	0,00
breast-cancer	21	**0,000**	0,78	**0,000**	0,60	**0,000**	0,73	**0,000**	0,79
breast-w	62	1,000	0,00	1,000	0,00	1,000	0,00	1,000	0,00
car	165	**0,000**	0,88	**0,000**	0,73	**0,000**	0,90	**0,000**	0,93
cars_with_names	33	**0,000**	0,60	**0,000**	0,68	**0,000**	0,62	**0,000**	0,68
CH	312	**0,000**	0,86	**0,000**	0,26	**0,000**	0,60	**0,000**	0,92
cmc	140	**0,000**	0,59	**0,000**	0,17	**0,000**	0,31	**0,000**	0,66
colic	29	**0,000**	0,65	**0,000**	0,46	**0,000**	0,60	**0,000**	0,68
colic.ORIG	29	1,000	0,00	1,000	0,00	1,000	0,00	1,000	0,00
cps_85_wages	46	1,000	0,00	1,000	0,00	1,000	0,00	1,000	0,00
credit-a	61	1,000	0,00	1,000	0,00	1,000	0,00	1,000	0,00
credit-g	92	**0,000**	0,48	**0,000**	0,38	**0,000**	0,44	**0,000**	0,40
credit	41	**0,000**	0,86	**0,000**	0,79	**0,000**	0,87	**0,000**	0,88
csb_ch12	153	1,000	0,00	1,000	0,00	1,000	0,00	1,000	0,00
csb_ch9	316	**0,000**	0,56	**0,000**	0,26	**0,000**	0,40	**0,000**	0,57
cylinder-bands	46	**0,000**	0,78	**0,000**	0,55	**0,000**	0,69	**0,000**	0,73
db3-bf	39	**0,000**	0,75	**0,000**	0,73	**0,000**	0,76	**0,000**	0,78
dermatology	29	**0,000**	0,81	**0,000**	0,59	**0,000**	0,73	**0,000**	0,80
diabetes	69	**0,000**	0,42	**0,000**	0,30	**0,000**	0,42	**0,000**	0,49
ecoli	26	**0,000**	0,66	**0,000**	0,47	**0,000**	0,59	**0,000**	0,68
eucalyptus	66	**0,000**	0,54	**0,000**	0,48	**0,000**	0,54	**0,000**	0,53
eye_movements	1086	**0,000**	0,22	**0,000**	0,11	**0,000**	0,16	**0,000**	0,14
genresTrain	1242	1,000	0,00	**0,001**	0,01	**0,000**	0,02	**0,000**	0,55
gina_agnostic	339	**0,000**	0,69	**0,000**	0,47	**0,000**	0,66	**0,000**	0,67
gina_prior	339	0,834	0,00	0,073	0,01	0,831	0,00	1,000	0,00
gina_prior2	339	**0,000**	0,70	**0,000**	0,07	**0,000**	0,20	**0,000**	0,74
GL	14	N/A	0,00	1,000	0,00	N/A	0,00	1,000	0,00
glass	14	1,000	0,00	N/A	0,00	N/A	0,00	1,000	0,00
haberman	23	N/A	0,00	1,000	0,00	1,000	0,00	1,000	0,00
HD	23	0,580	0,01	0,583	0,01	0,580	0,01	0,582	0,01
heart-c	23	1,000	0,00	1,000	0,00	1,000	0,00	1,000	0,00
heart-h	22	0,242	0,06	0,078	0,12	0,229	0,06	0,082	0,12
heart-statlog	19	**0,000**	0,69	**0,010**	0,27	**0,001**	0,44	**0,000**	0,74
HO	29	1,000	0,00	N/A	0,00	1,000	0,00	1,000	0,00
HY	309	**0,000**	-0,12	**0,000**	0,10	**0,000**	0,25	**0,000**	0,81
hypothyroid	370	0,992	0,00	0,996	0,00	1,000	0,00	1,000	0,50
ionosphere	28	1,000	0,00	1,000	0,00	1,000	0,00	1,000	0,00
irish	42	**0,000**	0,77	**0,000**	0,86	**0,000**	0,85	**0,000**	0,89
jEdit_4.0_4.2	20	1,000	0,00	N/A	0,00	1,000	0,00	N/A	0,00
jEdit_4.2_4.3	29	1,000	0,00	1,000	0,00	1,000	0,00	1,000	0,00
jm1	1081	**0,000**	0,48	**0,000**	0,12	**0,000**	0,35	**0,000**	0,66
kc1	203	**0,045**	0,02	0,088	0,01	**0,043**	0,02	**0,001**	0,06
kc2	45	**0,001**	0,00	1,000	0,00	1,000	0,00	1,000	-0,01
kc3	38	0,098	0,07	0,137	0,05	0,098	0,07	0,072	0,08

Table 1. (*Continued*)

Dataset	df	POW P	POW R2	LIN P	LIN R2	LOG P	LOG R2	EXP P	EXP R2
kdd_ipums_la_97-small	694	**0,000**	0,47	1,000	0,00	**0,001**	0,02	1,000	0,00
kdd_ipums_la_98-small	741	**0,000**	0,26	1,000	0,00	1,000	0,00	**0,000**	0,32
kdd_ipums_la_99-small	877	1,000	0,00	1,000	0,00	1,000	0,00	0,216	0,00
kdd_synthetic_control	52	**0,000**	0,70	**0,000**	0,53	**0,000**	0,67	**0,000**	0,69
kr-vs-kp	312	**0,000**	0,79	**0,000**	0,19	**0,000**	0,44	**0,000**	0,78
kropt	2798	**0,000**	0,90	**0,000**	0,41	**0,000**	0,83	**0,000**	0,95
landsat	636	1,000	0,00	1,000	0,00	1,000	0,00	**0,004**	0,01
letter	1992	**0,000**	0,94	**0,000**	0,09	**0,000**	0,39	**0,000**	0,82
liver-disorders	27	1,000	0,00	N/A	0,00	1,000	0,00	1,000	0,00
mc1	939	1,000	0,00	0,999	0,00	0,998	0,00	1,000	0,00
mfeat-factors	192	**0,000**	0,93	**0,000**	0,28	**0,000**	0,53	**0,000**	0,91
mfeat-fourier	192	**0,000**	0,70	**0,000**	0,18	**0,000**	0,31	**0,000**	0,74
mfeat-karhunen	192	0,995	-0,81	**0,000**	0,18	**0,000**	0,45	**0,000**	0,97
mfeat-morphological	192	0,187	0,01	0,235	0,01	0,190	0,01	**0,000**	0,08
mfeat-pixel	192	**0,000**	0,96	**0,000**	0,28	**0,000**	0,60	**0,000**	0,98
mfeat-zernike	192	**0,000**	0,78	**0,000**	0,12	**0,000**	0,32	**0,000**	0,84
monks-problems-1_test	36	N/A	0,00	1,000	0,00	1,000	0,00	1,000	0,00
monks-problems-2_test	36	1,000	0,00	1,000	0,00	1,000	0,00	1,000	0,00
monks-problems-3_test	36	1,000	0,00	1,000	0,00	0,429	0,00	1,000	0,00
mozilla4	1547	1,000	0,00	1,000	0,00	1,000	0,00	1,000	0,00
MU	805	**0,000**	0,94	**0,000**	0,74	**0,000**	0,95	**0,000**	0,92
mushroom	805	**0,000**	0,94	**0,000**	0,64	**0,000**	0,92	**0,000**	0,90
mw1	33	1,000	0,00	1,000	0,00	1,000	0,00	1,000	0,00
nursery	1288	**0,000**	0,51	**0,000**	0,21	**0,000**	0,44	**0,000**	0,54
optdigits	554	**0,000**	0,98	**0,000**	0,21	**0,000**	0,52	**0,000**	0,93
page-blocks	540	N/A	-0,77	1,000	0,00	**0,005**	0,01	0,985	0,00
pc1	103	**0,000**	0,94	**0,000**	0,44	**0,000**	0,74	**0,000**	0,92
pc3	149	1,000	0,00	**0,000**	0,00	0,994	0,00	0,934	0,00
pc4	138	**0,000**	0,09	**0,000**	0,25	**0,000**	0,10	**0,000**	0,24
pendigits	1092	**0,000**	0,88	**0,000**	0,22	**0,000**	0,57	**0,000**	0,79
primary-tumor	26	**0,000**	0,40	**0,006**	0,24	**0,003**	0,28	**0,000**	0,65
prnn_fglass	14	1,000	0,00	N/A	0,00	N/A	0,00	1,000	0,00
prnn_synth	17	1,000	0,00	1,000	0,00	1,000	0,00	1,000	0,00
rmftsa_propores	21	**0,001**	0,38	**0,000**	0,49	**0,001**	0,39	**0,000**	0,48
schizo	26	0,907	0,00	1,000	0,00	1,000	0,00	1,000	0,00
scopes-bf	55	0,916	0,00	0,944	0,00	0,962	0,00	**0,042**	0,07
SE	309	1,000	0,00	1,000	0,00	1,000	0,00	1,000	0,00
segment	223	1,000	0,00	1,000	0,00	0,947	0,00	1,000	0,00
sick	370	1,000	0,00	1,000	0,00	1,000	0,00	1,000	0,00
sonar	13	**0,045**	0,24	0,364	0,06	0,164	0,12	**0,028**	0,28
soybean	61	**0,000**	0,95	**0,000**	0,71	**0,000**	0,90	**0,000**	0,92
spambase	453	**0,000**	0,51	**0,000**	0,47	**0,000**	0,52	**0,000**	0,52
splice	311	**0,000**	0,96	**0,000**	0,47	**0,000**	0,81	**0,000**	0,90
sylva_agnostic	1432	0,996	0,00	0,999	0,00	1,000	0,00	**0,000**	0,46
sylva_prior	1432	1,000	0,00	1,000	0,00	1,000	0,00	**0,000**	0,20
tic-tac-toe	88	0,130	0,03	**0,002**	0,10	0,123	0,03	**0,002**	0,10
ticdata_categ	575	1,000	0,00	1,000	0,00	1,000	0,00	0,998	0,00
titanic	213	**0,000**	0,76	**0,000**	0,16	**0,000**	0,46	**0,000**	0,87
train	492	**0,000**	0,78	**0,000**	0,40	**0,000**	0,71	**0,000**	0,82
usp05	13	**0,000**	0,84	**0,000**	0,82	**0,000**	0,84	**0,000**	0,85

Table 1. (*Continued*)

Dataset	df	POW P	POW R2	LIN P	LIN R2	LOG P	LOG R2	EXP P	EXP R2
V1	36	1,000	0,00	1,000	0,00	N/A	0,00	0,206	0,04
vehicle	77	1,000	0,00	1,000	0,00	1,000	0,00	1,000	0,00
visualizing_fly	75	1,000	0,00	1,000	0,00	1,000	0,00	1,000	0,00
VO	36	1,000	0,00	0,775	0,00	1,000	0,00	N/A	0,00
vote	36	1,000	0,00	1,000	0,00	1,000	0,00	1,000	0,00
vowel	91	**0,000**	0,63	**0,000**	0,56	**0,000**	0,63	**0,000**	0,59
waveform-5000	492	**0,000**	0,35	**0,000**	0,20	**0,000**	0,28	**0,000**	0,31

It can be observed that out of 121 datasets, (only) 60 are such that all the models can be used to describe the data. The remaining datasets are such that the model does not describe the classifier's performance adequately, so we eliminated those from our further study because they cannot be compared. These datasets are such that the Naïve Bayes algorithm cannot be used, i.e. the algorithm is inappropriate for the problem domain. In these cases other algorithms and/or approaches need to be used [31].

From the vector of fitted function's values (f) and from the vector y we calculated the mean squared error (MSE) of jth dataset (DS), using Equation 5:

$$MSE_{DS_j} = \frac{\sum_{i=1}^{n}(y_i-f_i)^2}{n} \tag{5}$$

where n is the number of input points, i.e. the size of a vector, for each individual data set DS_j. MSE describes how well the observed points fit to the modeled function. The average MSEs for each dataset are listed in Table 2, together with the rank of function's model. The model with lowest average MSE gets assigned rank 1. It can be seen that EXP is the best fit for the data in 44 of 60 cases, POW is best in 9 out of 60 times, LOG in 7 out of 60 cases, and LIN in none out of 60 cases. Average MSEs across all datasets were 0,000437 for EXP, 0,000501 for POW, 0,000692 for LOG, and 0,000806 for LIN.

Table 2. Datasets and the average MSE across function models, and the model's rank (bold values indicate rank #1)

Dataset	Average MSE (power)	POW rank	Average MSE (linear)	LIN rank	Average MSE (logarithm)	LOG rank	Average MSE (exponent)	EXP rank
ada_agnostic	0,000216	2	0,000315	4	0,000230	3	0,000193	1
analcatda-ta_authorship	0,000178	3	0,000229	4	0,000151	2	0,000113	1
analcatda-ta_braziltourism	0,000514	4	0,000409	2	0,000466	3	0,000375	1
analcatda-ta_broadwaymult	0,002259	4	0,002217	2	0,002250	3	0,002186	1
analcatda-ta_marketing	0,000513	2	0,000507	1	0,000526	3	0,000539	4
analcatda-ta_reviewer	0,001129	4	0,001043	1	0,001123	3	0,001048	2
anneal.ORIG	0,000649	2	0,000690	4	0,000656	3	0,000619	1

Table 2. (*Continued*)

Dataset	Average MSE (power)	POW rank	Average MSE (linear)	LIN rank	Average MSE (logarithm)	LOG rank	Average MSE (exponent)	EXP rank
audiology	0,000615	2	0,000688	3	0,000725	4	0,000399	1
australian	0,000383	2	0,000604	4	0,000529	3	0,000331	1
autos	0,002046	4	0,001892	1	0,002026	3	0,001909	2
badges_plain	0,000066	2	0,000142	4	0,000089	3	0,000064	1
balance-scale	0,000325	3	0,000405	4	0,000324	2	0,000271	1
baseball-hitter	0,000801	3	0,000716	1	0,000838	4	0,000720	2
breast-cancer	0,001188	2	0,001410	4	0,001270	3	0,001080	1
car	0,000710	3	0,000740	4	0,000634	2	0,000537	1
cars_with_names	0,001191	4	0,001098	2	0,001177	3	0,001095	1
CH	0,000238	2	0,000655	4	0,000354	3	0,000162	1
cmc	0,002193	2	0,003457	4	0,003272	3	0,001845	1
colic	0,000318	3	0,000344	4	0,000312	2	0,000274	1
credit-g	0,000413	2	0,000589	4	0,000526	3	0,000388	1
credit	0,000258	4	0,000248	2	0,000250	3	0,000221	1
csb_ch9	0,000371	2	0,000600	4	0,000475	3	0,000353	1
cylinder-bands	0,001191	2	0,002058	4	0,001884	3	0,001026	1
db3-bf	0,000948	4	0,000845	2	0,000921	3	0,000818	1
dermatology	0,000128	2	0,000158	4	0,000129	3	0,000111	1
diabetes	0,000322	4	0,000308	2	0,000319	3	0,000295	1
ecoli	0,000326	2	0,000413	4	0,000354	3	0,000284	1
eucalyptus	0,000522	2	0,000575	4	0,000537	3	0,000512	1
eye_movements	0,000344	1	0,000384	4	0,000368	3	0,000359	2
gina_agnostic	0,000343	2	0,000409	4	0,000350	3	0,000327	1
gina_prior2	0,000228	2	0,000917	4	0,000678	3	0,000201	1
heart-statlog	0,000217	2	0,000367	4	0,000291	3	0,000186	1
HY	0,000125	4	0,000045	3	0,000033	2	0,000011	1
irish	0,000095	4	0,000062	1	0,000079	3	0,000064	2
jm1	0,000083	2	0,000104	4	0,000087	3	0,000069	1
kdd_synthetic_control	0,000308	4	0,000142	3	0,000125	2	0,000097	1
kr-vs-kp	0,000274	1	0,000649	4	0,000447	3	0,000283	2
kropt	0,000198	2	0,000512	4	0,000212	3	0,000144	1
letter	0,000080	1	0,001079	4	0,000675	3	0,000179	2
mfeat-factors	0,000536	2	0,002423	4	0,001696	3	0,000296	1
mfeat-fourier	0,000396	2	0,001417	4	0,001052	3	0,000337	1
mfeat-pixel	0,000081	2	0,000885	4	0,000399	3	0,000042	1
mfeat-zernike	0,000386	2	0,001299	4	0,001056	3	0,000209	1
MU	0,000010	1	0,000022	4	0,000010	2	0,000011	3
mushroom	0,000018	1	0,000043	4	0,000019	2	0,000019	3
nursery	0,000073	1	0,000082	3	0,000073	2	0,000125	4
optdigits	0,000292	2	0,001637	4	0,000891	3	0,000161	1
pc1	0,002179	1	0,006821	4	0,005871	3	0,002217	2
pc4	0,000561	4	0,000523	1	0,000542	2	0,000547	3
pendigits	0,000100	1	0,000404	4	0,000212	3	0,000125	2
primary-tumor	0,001002	2	0,001437	3	0,001546	4	0,000425	1
rmftsa_propores	0,000420	4	0,000400	1	0,000419	3	0,000401	2
soybean	0,000145	2	0,000467	4	0,000220	3	0,000143	1
spambase	0,000301	2	0,000325	4	0,000318	3	0,000282	1
splice	0,000095	1	0,000471	4	0,000193	3	0,000114	2

Table 2. (*Continued*)

Dataset	Average MSE (power)	POW rank	Average MSE (linear)	LIN rank	Average MSE (logarithm)	LOG rank	Average MSE (exponent)	EXP rank
titanic	0,000228	2	0,000503	4	0,000319	3	0,000157	1
train	0,000113	2	0,000182	4	0,000123	3	0,000105	1
usp05	0,000205	4	0,000203	3	0,000200	2	0,000195	1
vowel	0,000516	2	0,000653	4	0,000559	3	0,000497	1
waveform-5000	0,000096	2	0,000118	4	0,000108	3	0,000094	1
AVERAGE MSE	*0,000501*		*0,000806*		*0,000692*		*0,000437*	
RANK SUM		*144*		*201*		*172*		*83*

As can be observed, the EXP had rank-sum of 83 and an average MSE of 0,000437. Please note that the rank is an ordinal value and hence calculating its mean value is inappropriate [30, p.472].

Finally, the main research question was tested: which model was best? To rephrase, was EXP with the rank-sum of 83 and average MSE of 0,000437 significantly better than second-best POW with rank-sum of 144 and average MSE of 0,000501?

To test the significance of difference in MSE we used paired samples t-test for all combinations of models. The null hypotheses, the mean of differences between f_i (*MSE*) and f_j (*MSE*) equals 0, were as follows: $H1_0: \mu_{MSE/power} = \mu_{MSE/linear}$; $H2_0: \mu_{MSE/power} = \mu_{MSE/logarithmic}$; $H3_0: \mu_{MSE/power} = \mu_{MSE/exponential}$; $H4_0: \mu_{MSE/linear} = \mu_{MSE/logarithm}$; $H5_0: \mu_{MSE/linear} = \mu_{MSE/exponential}$; and $H6_0: \mu_{MSE/logaritmic} = \mu_{MSE/exponential}$.

Because we conducted six interrelated comparisons, we used the Bonferroni correction to counteract the problem of multiple comparisons [32]. The correction is based on the idea that if an experimenter is testing n dependent or independent hypotheses on a set of data, then one way of maintaining the family-wise error rate is to test each individual hypothesis at a statistical significance level of $1/n$ times what it would be if only one hypothesis were tested. We would normally reject the null hypothesis if $P < 0.05$. However, Bonferroni correction requires a modified rejection threshold for P, $\alpha = (0,05/6) = 0,008$. Table 3 lists the results of statistical analysis for all six comparisons, with values in bold indicating significance at modified α level.

The results show that exponential function's average mean squared error is significantly different at any reasonable threshold from average MSE power ($P = 0,000$) linear ($P = 0,000$) and logarithmic function ($P = 0,001$), regardless if using the Bonferroni correction or not. Thus, all hypotheses $H1_0$ - $H6_0$ need to be rejected.

Additionally, we tested whether the ranks of functions are statistically significantly different from each other. We used related samples Wilcoxon's signed rank test. The null hypotheses, the median of differences between f_i (*rank*) and f_j (*rank*) equals 0, were as follows: $H7_0: \mu_{1/2 RANK / power} = \mu_{1/2 RANK / linear}$; $H8_0: \mu_{1/2 RANK / power} = \mu_{1/2 RANK / logarithmic}$; $H9_0: \mu_{1/2 RANK / power} = \mu_{1/2 RANK / exponential}$; $H10_0: \mu_{1/2 RANK / linear} = \mu_{1/2 RANK / logarithm}$; $H11_0: \mu_{1/2 RANK / linear} = \mu_{1/2 RANK / exponential}$ and $H12_0: \mu_{1/2 RANK / logarithmic} = \mu_{1/2 RANK / exponential}$.

Table 3. Paired samples t-test for MSE

		Paired Samples Test							
		Paired Differences					t	df	Sig. (2-tailed)
		Mean	Std. Deviation	Std. Error Mean	95% Confidence Interval of the Difference				
					Lower	Upper			
Pair 1	POW (avg. MSE): LIN (avg. MSE)	-,0003047	,0007021	,0000906	-,0004861	-,0001233	-3,36	59	**,001**
Pair 2	POW (avg. MSE): LOG (avg. MSE)	-,0001910	,0005340	,0000689	-,0003289	-,0000530	-2,77	59	**,007**
Pair 3	POW (avg. MSE): EXP (avg. MSE)	,0000645	,0001028	,0000133	,0000379	,0000910	4,85	59	**,000**
Pair 4	LIN (avg. MSE): LOG (avg. MSE)	,0001137	,0002073	,0000268	,0000602	,0001673	4,24	59	**,000**
Pair 5	LIN (avg. MSE): EXP (avg. MSE)	,0003691	,0007168	,0000925	,0001840	,0005543	3,98	59	**,000**
Pair 6	LOG (avg. MSE): EXP (avg. MSE)	,0002554	,0005558	,0000718	,0001118	,0003990	3,56	59	**,001**

Table 4 lists the results of Wilcoxon's signed rank test analysis for all six comparisons, with values in bold indicating significance at the Bonferroni-modified $\alpha=0,008$ level.

Table 4. Wilcoxon signed rank test for different function models

Pair #	Pair	Sig. (2-tailed)
Pair 1	POW (rank) – LIN (rank)	**0,001**
Pair 2	POW (rank) – LOG (rank)	**0,005**
Pair 3	POW (rank) – EXP (rank)	**0,000**
Pair 4	LIN (rank) – LOG (rank)	**0,004**
Pair 5	LIN (rank) – EXP (rank)	**0,000**
Pair 6	LOG (rank) – EXP (rank)	**0,000**

The results show that exponential function's average rank is significantly different at any reasonable threshold from average rank of any other model (for all such cases $P=0,000$). Thus, all the above mentioned hypotheses $H7_0$ to $H12_0$ need to be rejected.

4   Conclusion

In this paper we conducted an analysis of an error rate curve produced by a selected Naïve Bayes classifier. The results show that, in average, the best mathematical description of a Naïve Bayes learner is the exponential function. The results were consistent when using the mean squared error measure ($P=0,000$ to $0,001$ for t-test) and the rank assignment ($P=0,000$ for Wilcoxon's test). Logarithmic and power functions can, however, be superior in a limited number of specific cases whereas linear model cannot be considered as appropriate at all. We observed the learner on 60 different tasks and the exponential function was superior in 44 cases. Of the remaining 16 cases, its average mean squared error was within 95 % of the winner's in 9 cases and within 90 % in additional 3 cases. In only 4 cases out of 60 (6,6 %) the other functions performed much better. These findings are in line with outcomes observed by Heathcote et al. [4] in the measurements of human cognitive performance.

The contribution of the presented work is important in many respects: firstly, the exponential model can be used to forecast the future performance of a Naïve Bayes learner based on a small training sample. Sometimes it is prohibitive expensive to conduct a full scale analysis due to limited resources [33], e.g. in measuring the losses in synchronous motor [34-38]. Secondly, the findings are important in an on-line learning scenario where algorithms should act in dynamic environments with continuous data flow; the actual decision model must first make a prediction and then update the current model with new data. When to update depends on optimizing the cost of updating of the model [39, 40]. The drift detection in the learning process [41] can be additionally verified by checking a deviation from the appropriate model. Thirdly, the results can help the operator to check where on the learning curve the current measurement lies (steeply sloping portion early in the curve, a more gently sloping middle portion, and a plateau late in the curve), thus enabling her to get an early insight in the possible future data needs [42, 43].

Fourthly, early in the learning phase one can fit the model's parameters and estimate the final error rate. In case the estimated final performance is lower than required, one can modify the learner's parameters early in the process. Lastly, the results of our experiment show that some datasets exist where modelling of the artificial learner's performance is not successful due to the inability of a learner to properly capture the data interrelations. This too could be detected early in the learning process to avoid unnecessary algorithm runs, involving sometimes expensive additional data collection.

Acknowledgements. This work was partially supported by the Slovenian Research Agency under grant number 1000-11-310138.

References

1. Anderson, J.R., Schooler, L.J.: Reflections of the Environment in Memory. Psychological Science 2(6), 396–408 (1991)
2. Anderson, R.B.: The power law as an emergent property. Memory & Cognition 29(7), 1061–1068 (2001)
3. Clauset, A., Shalizi, C.R., Newman, M.E.J.: Power-Law Distributions in Empirical Data. SIAM Review 51(4), 661–703 (2009), doi:10.1137/070710111
4. Heathcote, A., Brown, S., Mewhort, D.J.K.: The power law repealed: The case for an exponential law of practice. Psychonomic Bulletin & Review 7(2), 185–207 (2000), doi:10.3758/bf03212979
5. Kotsiantis, S.B.: Supervised Machine Learning: A Review of Classification Techniques. Informatica (Ljubljana) 31(3), 249–268 (2007)
6. Dzemyda, G., Sakalauskas, L.: Large-Scale Data Analysis Using Heuristic Methods. Informatica (Lithuan.) 22(1), 1–10 (2011)
7. Vapnik, V.N.: Estimation of Dependences Based on Empirical Data. Springer, NY (1982)
8. Brumen, B., Jurič, M.B., Welzer, T., Rozman, I., Jaakkola, H., Papadopoulos, A.: Assessment of classification models with small amounts of data. Informatica (Lithuan.) 18(3), 343–362 (2007)

9. Dučinskas, K., Stabingiene, L.: Expected Bayes Error Rate in Supervised Classification of Spatial Gaussian Data. Informatica (Lithuan.) 22(3), 371–381 (2011)

10. Frey, L.J., Fisher, D.H.: Modeling decision tree performance with the power law. In: Seventh International Workshop on Artificial Intelligence and Statistics. Morgan Kaufmann, Ft. Lauderdale (1999)

11. Last, M.: Predicting and Optimizing Classifier Utility with the Power Law. In: 7th IEEE International Conference on Data Mining, ICDM Workshops 2007. IEEE, Omaha (2007), doi:10.1109/icdmw.2007.31

12. Provost, F., Jensen, D., Oates, T.: Efficient progressive sampling. In: Fifth International Conference on Knowledge Discovery and Data Mining. ACM, San Diego (1999)

13. Singh, S.: Modeling Performance of Different Classification Methods: Deviation from the Power Law. Project Report. Vanderbilt University, Nashville, Tennessee, USA, Department of Computer Science (2005)

14. Dzemyda, G., Sakalauskas, L.: Optimization and Knowledge-Based Technologies. Informatica (Lithuan.) 20(2), 165–172 (2009)

15. John, G.H., Langley, P.: Estimating Continuous Distributions in Bayesian Classifiers. In: Eleventh Conference on Uncertainty in Artificial Intelligence, August 18-20. Morgan Kaufmann, San Francisco (1995)

16. Witten, I.H., Frank, E.: Data Mining: Practical machine learning tools and techniques, 2nd edn. Morgan Kaufmann, San Francisco (2005) ISBN: 0120884070

17. Hall, M., Frank, E., Holmes, G., Pfahringer, B., Reutemann, P., Witten, I.H.: The WEKA data mining software: an update. ACM SIGKDD Explorations Newsletter 11(1), 10–18 (2009)

18. Asuncion, A., Newman, D.: UCI Machine Learning Repository (2010), http://archive.ics.uci.edu/ml/datasets.html (Archived by WebCite® at http://www.webcitation.org/6C2hgsRrX)

19. TunedIT. TunedIT research repository (2012), http://tunedit.org/search?q=arff&qt=Repository (accessed: December 12, 2012) (Archived by WebCite® at http://www.webcitation.org/6CqplN6Xr)

20. Kjellerstrand H.: My Weka page (2012), http://www.hakank.org/weka/ (accessed: December 12, 2012) (Archived by WebCite® at http://www.webcitation.org/6Cqq5pQtZ)

21. Kjellerstrand, H.: My Weka page/DASL (2012), http://www.hakank.org/weka/DASL/ (accessed: December 12, 2012) (Archived by WebCite® at http://www.webcitation.org/6CqqCwPmy)

22. Chai, K.: Kevin Chai Datasets (2012), http://kevinchai.net/datasets (accessed: December 12, 2012) (Archived by WebCite® at http://www.webcitation.org/6CqqWlQEp)

23. Brumen, B., Hölbl, M., Harej Pulko, K., Welzer, T., Heričko, M., Jurič, M.B., Jaakkola, H.: Learning Process Termination Criteria. Informatica (Lithuan.) 23(4), 521–536 (2012)

24. Cohen, P.R.: Empirical methods for artificial intelligence. MIT Press, Cambridge (1995) ISBN: 9780262032254

25. Weiss, S.M., Kulikowski, C.A.: Computer systems that learn: classification and prediction methods from statistics, neural nets, machine learning, and expert systems. Morgan Kaufmann, San Mateo (1991) ISBN: 978-1558600652

26. McLachlan, G.J., Do, K.-A., Ambroise, C.: Analyzing microarray gene expression data. Wiley, Hoboken (2004) ISBN: 0471226165

27. Eaton, J.W.: GNU Octave (2012), http://www.gnu.org/software/octave/ (accessed: December 12, 2012) (Archived by WebCite® at http://www.webcitation.org/6CqyEvDKU)
28. Marquardt, D.W.: An Algorithm for Least-Squares Estimation of Nonlinear Parameters. Journal of the Society for Industrial and Applied Mathematics 11(2), 431–441 (1963), doi:10.2307/2098941
29. Levenberg, K.: A Method for the Solution of Certain Non-Linear Problems in Least Squares. Quarterly of Applied Mathematics 2, 164–168 (1944)
30. Argyrous, G.: Statistics for research: With a guide to SPSS, 3rd edn. SAGE Publications Ltd., Thousand Oaks (2011) ISBN: 1849205957
31. Medvedev, V., Dzemyda, G., Kurasova, O., Marcinkevicijus, V.: Efficient Data Projection for Visual Analysis of Large Data Sets Using Neural Networks. Informatica (Lithuan.) 22(4), 507–520 (2011)
32. Abdi, H.: The Bonferonni and Šidák Corrections for Multiple Comparisons. In: Salkind, N.J. (ed.) Encyclopedia of Measurement and Statistics. SAGE Publications, Inc., Thousand Oaks (2007) ISBN: 9781412916110
33. Pragarauskaite, J., Dzemyda, G.: Markov Models in the Analysis of Frequent Patterns in Financial Data. Informatica (Lithuan.) 24(1), 87–102 (2014)
34. Pišek, P., Štumberger, B., Marčič, T., Virtič, P.: Design analysis and experimental validation of a double rotor synchronous PM machine used for HEV. IEEE Transactions on Magnetics 49(1), 152–155 (2013), doi:10.1109/TMAG.2012.2220338
35. Virtič, P.: Determining losses and efficiency of axial flux permanent magnet synchronous motor. Przegląd Elektrotechniczny 89(2b), 13–16 (2013)
36. Virtič, P., Pišek, P., Hadžiselimović, M., Marčič, T., Štumberger, B.: Torque analysis of an axial flux permanent magnet synchronous machine by using analytical magnetic field calculation. IEEE Transactions on Magnetics 45(3), 1036–1039 (2009), doi:10.1109/TMAG.2009.2012566
37. Virtič, P., Pišek, P., Marčič, T., Hadžiselimović, M., Štumberger, B.: Analytical analysis of magnetic field and back electromotive force calculation of an axial-flux permanent magnet synchronous generator with coreless stator. IEEE Transactions on Magnetics 44(11), 4333–4336 (2008)
38. Hadžiselimović, M., Virtič, P., Štumberger, G., Marčič, T., Štumberger, B.: Determining force characteristics of an electromagnetic brake using co-energy. Journal of Magnetism and Magnetic Materials 320(20), e556-e561 (2008), doi: 10.1016/j.jmmm.2008.04.013
39. Castillo, G., Gama, J.: Adaptive Bayesian network classifiers. Intelligent Data Analysis 13(1), 39–59 (2009), doi:10.3233/IDA-2009-0355
40. Castillo, G., Gama, J.: An adaptive prequential learning framework for Bayesian network classifiers. In: Fürnkranz, J., Scheffer, T., Spiliopoulou, M. (eds.) PKDD 2006. LNCS (LNAI), vol. 4213, pp. 67–78. Springer, Heidelberg (2006)
41. Gama, J., Medas, P., Castillo, G., Rodrigues, P.: Learning with drift detection. In: Bazzan, A.L.C., Labidi, S. (eds.) SBIA 2004. LNCS (LNAI), vol. 3171, pp. 286–295. Springer, Heidelberg (2004)
42. Cipresso, P., Carelli, L., Solca, F., Meazzi, D., Meriggi, P., Poletti, B., Lulé, D., Ludolph, A.C., Silani, V., Riva, G.: The use of P300-based BCIs in amyotrophic lateral sclerosis: from augmentative and alternative communication to cognitive assessment. Brain and Behavior 2(4), 479–498 (2012), doi:10.1002/brb3.57
43. Cipresso, P., Paglia, F., Cascia, C., Riva, G., Albani, G., La Barbera, D.: Break in volition: a virtual reality study in patients with obsessive-compulsive disorder. Experimental Brain Research 229(3), 443–449 (2013), doi:10.1007/s00221-013-3471-y

A Parallel Algorithm for Building iCPI-trees*

Witold Andrzejewski and Pawel Boinski

Poznan University of Technology, Institute of Computing Science,
Piotrowo 2, 60-965 Poznan, Poland
{witold.andrzejewski,pawel.boinski}@cs.put.poznan.pl

Abstract. In spatial databases *collocation pattern discovery* is one of
the most interesting fields of data mining. It consists in searching for
types of spatial objects that are frequently located together in a spatial
neighborhood. With the advent of data gathering techniques, huge volumes of spatial data are being collected. To cope with processing of such
datasets a GPU accelerated version of the collocation pattern mining
algorithm has been proposed recently [3]. However, the method assumes
that a supporting structure that contains information about neighborhoods (called iCPI-tree) is given in advance. In this paper we present a
GPU-based version of iCPI-tree generation algorithm for the collocation
pattern discovery problem. In an experimental evaluation we compare
our GPU implementation with a parallel implementation of iCPI-tree
generation method for CPU. Collected results show that proposed solution is multiple times faster than the CPU version of the algorithm.

1 Introduction

Huge volumes of spatial data result from advances in sensing technologies and
mass storage devices. For example, weather and climate monitoring based on
satellite observations can produce terabytes of spatial data each day. As human abilities to interpret such data are limited, automatic methods known as
Knowledge discovery in databases (KDD) are required. KDD has been defined
as a non-trivial process of discovering valid, novel, and potentially useful, and
ultimately understandable patterns in large data volumes [9]. A crucial step in
this process is called *data mining*. It consists in application of specially designed
algorithms to find particular patterns in data.

In spatial data mining, i.e., mining data with spatial components, one of the
most popular types of patterns is called a *spatial collocation pattern* or in short
a *collocation*. Shekhar and Huang defined a collocation [16] as a set of spatial
features that are frequently located together in a spatial proximity. A spatial
feature is a class of a spatial object and can be interpreted as a characteristic of
space in a given location. Typical examples of spatial features include species,
business types or points of interest (e.g., hospitals, schools, airports etc.). As the
concept of the spatial feature is not limited to variables measured using sensors

---

* This work was partially supported from the Polish National Science Center (NCN),
grant No. 2011/01/B/ST6/05169.

Y. Manolopoulos et al. (Eds.): ADBIS 2014, LNCS 8716, pp. 276–289, 2014.

(i.e., raster images can be used) a collocation pattern discovery can be applied in a wide range of domains, e.g., marketing, ecology and public health, meteorology, mobile advertising, astronomy etc.

In recent years many algorithms have been developed to improve the efficiency of the collocation pattern mining. One of the popular and viable means to achieve better performance is to utilize parallel processing capabilities of modern processors. In [3] GPU-CM algorithm has been proposed. It is designed to exploit the power of graphics processing units (GPUs). GPU-CM is based on the state of the art solution in the field of collocation pattern mining [17] that takes advantage of the specially designed structure, called iCPI-tree, to store neighborhood information. The authors of GPU-CM assume that this supporting structure is given in advance, i.e., precomputed before the execution of the collocation pattern mining algorithm on the GPU.

In this paper we propose a method for building iCPI-tree structure directly on the GPU, which complements the work from [3]. The advantages of such approach are twofold: (1) massive parallel processing offered by GPUs can be utilized, (2) data transfers between CPU and GPU are reduced as the iCPI-tree structure, that can be multiple time larger than an input dataset, is built in the GPU memory from which it is accessed during the execution of the collocation pattern mining algorithm. To the best of authors' knowledge, no alternative solutions for building of iCPI-trees on GPUs have been developed.

The structure of this paper is as follows. In Section 2, we present the current solution used for generating iCPI-tree structure and introduce basics of general processing on GPUs. Section 3 presents our contribution - the GPU-based version of method for iCPI-tree construction. Section 4 contains the results of performed experiments. Summary and plans for future work are presented in Section 5.

2 Related Work

2.1 Collocation Pattern Mining

Let f be a spatial feature. An object x is an *instance* of the feature f, if x is a type of f and is described by a location and unique identifier. Let F be a set of spatial features and S be a set of their instances. Given a neighbor relation R, we say that the *collocation* C is a subset of spatial features $C \subseteq F$ whose instances $I \subseteq S$ form a clique w.r.t. the relation R.

The most popular overall schema for collocation pattern mining has been introduced in [16] and consists of three steps: (1) generate candidates for collocations, (2) search for instances of candidates, (3) remove candidates without required prevalence threshold. These steps are repeated as long as there are candidates for the next iteration. Candidates can be generated using Apriori [1] approach as well as other algorithms (e.g., [18]). The second step of this overall approach is the most computationally demanding as it requires to find all instances of collocations, i.e., sets of objects that are neighbors and simultaneously have appropriate spatial features. Consequently, many optimizations and techniques to execute this task were presented in the literature (e.g., [5,17,19]).

Particularly noteworthy are [19] and [17]. In [19] the authors introduced a structure to efficiently generate all possible instances of candidates. For each feature instance a list of its neighbors with a spatial feature greater than a spatial feature of this particular instance is stored. Such an entry is called a star neighborhood. During the execution of the algorithm, instances of candidates are generated from star neighborhoods, however some of these instances might violate the definition, i.e., some pairs of objects are not neighbors. In an additional step such incorrect instances are removed. In [17] the authors improved the aforementioned solution. A tree structure to hold star neighborhoods as well as a new method for generating instances has been proposed. This new structure, called *improved Candidate Pattern Instance tree (iCPI-tree)*, is an index to answer the question regarding neighbors with a required feature of a given object. For example, in Fig. 1 a sample dataset is presented. There are seven instances of features A, B and C. Assuming a user defined neighborhood threshold we can build an iCPI-tree presented in Fig. 2. The iCPI-tree structure can easily provide an answer to the following question: what are the neighbors of object A2 that are instances of feature C? The answer is C1 and C2. Such information is used to generate all clique instances of candidates. For details please refer to [17].

Although the structure is called a tree, in a practical implementation [5] it is sufficient to store it as a hashmap where keys consist of feature instance identifiers and features of neighbors whereas neighbors are stored as the corresponding values. We will refer to this structure as the *iCPI-hashmap*. For the mentioned example, for object A2 there are two entries in the map: key A2,B maps to B1, B2 and key A2,C maps to C1 and C2.

To build iCPI-hashmap it is necessary to detect all neighbors. An efficient method is based on a plane sweep technique [7]. The aim is to avoid testing all pairs of objects as there is no need to test objects that are far apart. Let us assume that r is the neighborhood distance threshold. We can imagine a vertical line sweeping over the plane starting from the left most position. During the sweep, we maintain a tracking list of all objects that are in the distance not greater than r from the sweeping line. Only objects that are on tracking list must be checked for neighborhood relationship.

2.2   General Processing on Graphics Processing Units

Modern graphics cards can work as co-processors performing highly parallel single instruction multiple data computations. APIs such as NVIDIA CUDA [13] and OpenCL [12] facilitate development of programs utilizing *graphics processing units* (GPUs) of graphics cards for acceleration of their operations. In our solutions we utilize NVIDIA CUDA, though some of the solutions presented in this paper are also applicable to OpenCL API. Below we give a short description of NVIDIA CUDA API and its capabilities.

Computation tasks for GPUs are implemented in a form of *kernels*. A kernel is a function which is performed concurrently in multiple threads. Threads are grouped into blocks - arrays of at most 1024 threads, where each thread can be uniquely identified by its position in this array. The set of blocks forms a *com-*

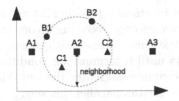

Fig. 1. Running example dataset 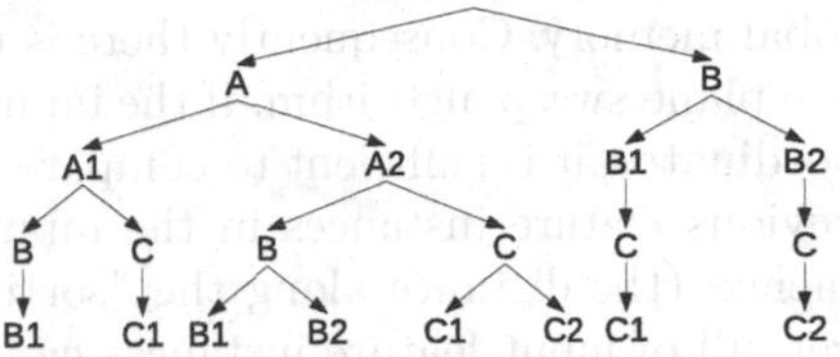   **Fig. 2.** Exemplary iCPI-tree

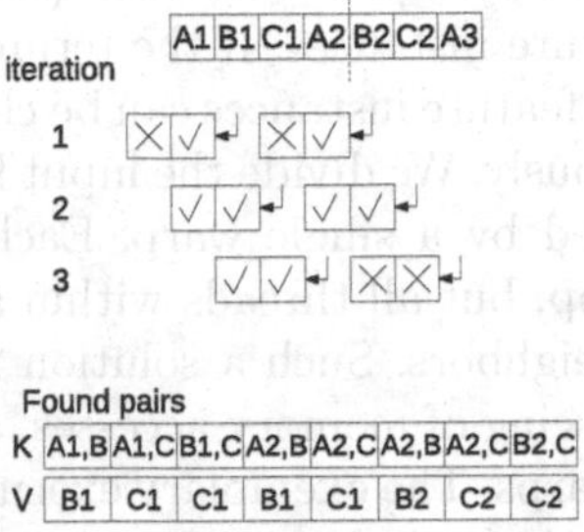

Fig. 3. Illustration of finding neighbor pairs 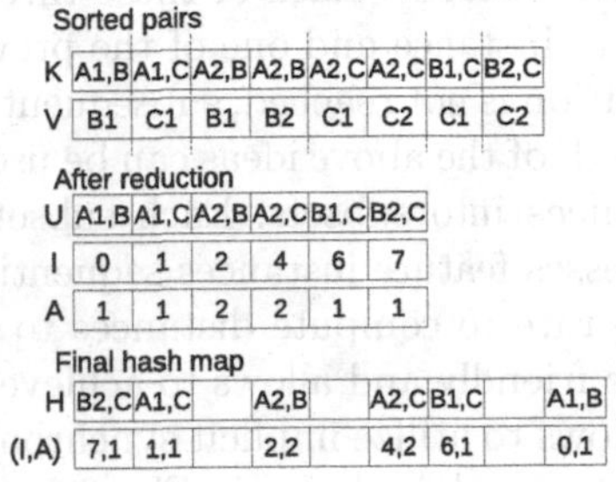   **Fig. 4.** Processing of obtained pairs

putation grid. Threads in a single block may communicate via a very fast *shared memory.* Threads running in different blocks may communicate via a slow *global memory* of the graphics card. Synchronization capabilities of the threads are limited. Threads in a single block may be synchronized via a *barrier synchronization primitive,* however global synchronization of threads is achievable only by means of a costly workaround. Threads in a block are executed in 32 thread SIMD groups called *warps.*

To simplify implementation of programs performing parallel computations, parallel primitives have been developed. Such primitives include: *sort* and *sort by key* for sorting data, *reduce* and *reduce by key* for aggregating data, *exclusive scan* for computing agregations of prefixes of an array and *compact* for removing of chosen entries from arrays. Our solution uses parallel primitives implemented in the Thrust library [4]. Moreover, we also had to implement special versions of *exclusive scan, reduce* and *compact* primitives which work on small shared memory arrays and perform their operations in each warp independently.

Recently, graphics cards started receiving recognition from the database scientific community. Graphics cards are utilized as co-processors for performing typical database [6,10,14] and data mining [8,11] operations.

3 GPU-Based Algorithm for iCPI-hashmap Construction

3.1 Main Algorithm

The algorithm is based on the following observations. Due to specificity of processing data on GPUs (no hard disk access), all input data must be stored in the global memory. Consequently there is no need to utilize a tracking list as in the base plane sweep algorithm. If the input arrays are sorted according to one of the coordinates, it is sufficient to compare current feature instance with consecutive previous feature instances in the input arrays until a termination condition is reached (the distance along the "sorting" coordinate is greater than r). Moreover, all of input feature instances can be processed in parallel this way.

Another observation, which allows to leverage the SIMD capabilities of GPUs, is as follows. Assume we utilize a group of N threads to find neighbors of a single feature instance. Each of these threads can find distance between the processed feature instance and one of the previous N feature instances. If the termination condition is not reached, subsequent previous N feature instances can be checked.

Both of the above ideas can be used simultanously. We divide the input feature instances into subsets. Each subset is processed by a single warp. Each warp processes feature instances sequentially in a loop, but all threads within a warp cooperate to compute distances to potential neighbors. Such a solution is very cache friendly and allows to achieve good coalescing of memory accesses, as well as allows to utilize implicit synchronization of warps. The exemplary execution of this approach is shown in Fig. 3 for the dataset shown in Fig. 1. In this example two warps process two subsets of feature instances. For the sake of clarity, warps in Fig. 3 consist of only 2 threads (normally 32 threads).

Algorithm 1 presents the main steps required to build an iCPI-hashmap. The input of the algorithm is composed of: (1) arrays X, Y storing x and y coordinates respectively and an array F with corresponding feature instance identifiers, (2) a neighborhood distance threshold r and (3) a grid configuration which is used to start appropriate number of warps in the neighbor finding step. As a result, the algorithm produces a hashmap H (via open addresing hashing scheme [2]) which maps pairs of feature instance identifiers and feature numbers (called *extended feature instance identifiers*) to lists of neighbors of the instances which have the specified feature numbers. The lists of neighbors are concatenated and stored in an array V. Each list is described in H by a starting index and a number of neighbors.

For the sake of clarity, we introduce the term: a *neighbor pair*. By a *neighbor pair* we understand a pair of feature instance identifiers (a, b) such that the distance between these feature instances is less than the neighborhood distance threshold. Moreover, the feature instance identifier a should store a feature number smaller than the feature number stored in the feature instance identifier b. Finally, the feature instance identifier a should be extended with the feature number of the feature instance b.

The first step of the algorithm (line 1) performs sorting of the input arrays by a chosen coordinate. Without the loss of generality, we assume x is the chosen sorting coordinate. This stage is performed via *sort by key* primitive. Sorted input arrays can be now used to find neighbor pairs. As mentioned earlier, to find neighbor pairs, we start an arbitrary number of warps and each warp is assigned a subset of input feature instances. Threads within each warp cooperate to find neighbor pairs, where at least one of the neighbors is in the assigned subset.

The finding of neighbor pairs is split into two phases. In the first phase each warp only counts the number of neighbor pairs it is able to find and stores the appropriate number in an array C. The first phase is performed by the kernel *CountNeighbors*. This kernel is detailed in Algorithm 2 and will be described in Section 3.2. First phase allows to determine the amount of memory required to store all of the neighbor pairs, as well as positions in the output arrays where these pairs should be stored by each of the warps. Total number of the neighbor pairs is equal to the sum of all values in the array C, whereas the starting positions can be computed by an *exclusive scan* of this array. The number of pairs (variable *total*) and starting positions are computed in lines 4-6.

Next, arrays K and V are allocated. These arrays will be used to store neighbor pairs found by the kernel *FindNeighbors* (line 9). This kernel is detailed in Algorithm 3 and described in Section 3.2. Exemplary K and V arrays are shown in Fig. 3. The obtained arrays K and V are sorted lexicographically via *sort* parallel primitive. Note that after sorting, the sets of consecutive entries in array V form neighbor lists of the corresponding extended feature instance identifiers in the array K. The array V is one of the output structures, whereas the array K is the basis for building the hashmap H. In order to build the hash map, we need to find a set of unique entries in the array K and, for each of these unique entries: (1) the first index in the array K at which they appear and (2) the number of times they appear in the array K. This is performed in line 11. As a result we obtain arrays: U, I and A, which store unique feature instance identifiers, positions and counts respectively. This step can be performed via *reduce by key* parallel primitive. Reduction of an auxiliary array filled with ones, by keys from the array K by using sum operator allows to obtain the arrays U and A. Reduction of an auxiliary array filled with consecutive numbers, by keys from the array K by using the *min* operator allows to obtain the arrays U and I. Thrust implementation of *reduce by key* parallel primitive used in our solution, allows to perform both of the aforementioned operations in one step via usage of zip iterators. Moreover, materialization of auxiliary arrays can be omitted via usage of constant and counting iterators. Results of sorting and reduction of exemplary arrays K and V are shown in Fig. 4.

Values from the array U form hash map keys, whereas pairs of corresponding values from arrays I and A form hash map values (list positions and list sizes respectively). Such key value pairs are inserted into a hash map in parallel via open addressing hashing scheme [2]. The obtained hash map H as well as the array V form the result of the algorithm. Exemplary result is shown in Fig. 4.

3.2 Neighbor Finding Kernels

In this section we provide a detailed description of the kernels *CountNeighbors* (see Algorithm 2) and *FindNeighbors* (see Algorithm 3) used in Algorithm 1. The kernel *CountNeighbors* computes the number of neighbors each warp can find in the assigned input data subset. The kernel *FindNeighbors* materializes these neighbors into output arrays. Both of these kernels are very similar. Consequently, we will first describe the *CountNeighbors* kernel in detail and then

---

Algorithm 1. Main algorithm

Require:
    Input arrays X, Y and F ▷ see Section 3.1
    Neighborhood distance threshold r
    Size of block bs and number of blocks nb
Ensure:
    An iCPI-hashmap H and an array with neighbor lists V ▷ see Section 3.1

1: Sort arrays X, Y and F according to values of X array
2: $C \leftarrow$ empty array of size $bs * nb/32$ ▷ # of neighbor pairs found by each warp
3: $CountNeighbors <<< nb, bs >>> (X, Y, F, r, C)$ ▷ see Algorithm 2
4: $total \leftarrow C[|C| - 1]$ ▷ Get the number of pairs found by the last warp
5: $C \leftarrow exclusiveScan(C)$ ▷ Obtain starting output positions for each warp
6: $total \leftarrow total + C[|C| - 1]$ ▷ Add the number of pairs found by other warps
7: $K \leftarrow$ new array of size $total$ ▷ Array storing future hash map keys
8: $V \leftarrow$ new array of size $total$ ▷ Array storing neighbor lists
9: $FindNeighbors <<< nb, bs >>> (X, Y, F, r, C, K, V)$ ▷ see Algorithm 3
10: Lexicographically sort arrays K and V
11: $(U, I, A) \leftarrow ReduceByKey(K)$ ▷ Find unique keys, positions and sizes of neighbor lists
12: $H \leftarrow$ new hash map able to store $|U|$ key-value pairs   ▷ Build a resulting hash map
13: Insert into H pairs $(U[j], (I[j], A[j]))$ for $j = 0, \ldots, |U| - 1$ (in parallel)
14: Hashmap H as well as array V form a result

Algorithm 2. Kernel used for computing how much neighbor pairs are found by each warp

---

1: **kernel** CountNeighbors(X,Y,F,r,C)
2:    $(gtid, gwid, wtid, btid, bwid, start, stop) \leftarrow ThreadNumbersAndSubsets()$
3:    $flags \leftarrow$ shared memory array of size $[blockDim.x/32]$ ▷ A flag per warp
4:    $found \leftarrow$ shared memory array of size $[blockDim.x]$ ▷ A counter per thread
5:    $found[btid] \leftarrow 0$ ▷ Clear array $found$ in parallel
6:    **for** $i \in start, \ldots, stop$ **do**   ▷ Sequentially find neighbors of every assigned feature instance
7:       **if** $wtid = 0$ **then** $flags[bwid] \leftarrow false$ ▷ Initialize $flags$ array
8:       $j \leftarrow i - 32$ ▷ j is a base index for all threads within a warp
9:       **while** $j >= -32$ **do**   ▷ Dont read beyond start of the input array
10:          **if** $wtid + j >= 0$ **then**   ▷ If the thread can access an existing array entry...
11:             **if** $X[i] - X[wtid + j] > r$ $flags[bwid] \leftarrow true$ ▷ Check for stop condition
12:             **if** $d(i, wtid + j) \leq r$ **and** $F[i] \neq F[wtid + j]$ **then**   ▷ Detect neighbor
13:                $found[btid] \leftarrow found[btid] + 1$ ▷ Increase counter if it was found
14:             **end if**
15:          **end if**
16:          **if** $flags[bwid]$ **break**   ▷ If any thread reached stop condition, abort the loop
17:          $j \leftarrow j - 32$ ▷ Otherwise move warp to next set of feature instances in array
18:       **end while**
19:    **end for**
20:    $reduceWithinWarp(found)$ ▷ Find total number of neighbors found by the warp
21:    **if** wtid=0 **then**   ▷ First thread in a warp should...
22:       $C[gwid] \leftarrow$ reduction result   ▷ ...store computed number into array C
23:    **end if**
24: **end kernel**

Algorithm 3. Kernel used for finding neighbor pairs

1: **kernel** FINDNEIGHBORS(X,Y,F,r,C,K,V)
2:     $(gtid, gwid, wtid, btid, bwid, start, stop) \leftarrow ThreadNumbersAndSubsets()$
3:     $flags \leftarrow$ shared memory array of size $[blockDim.x/32]$ ▷ A flag per warp
4:     $found \leftarrow$ shared memory array of size $[blockDim.x]$ ▷ A counter per thread
5:     $scanBuf \leftarrow$ shared memory array of size $[blockDim.x]$ ▷ A temporary array
6:     $found[btid] \leftarrow 0$ ▷ Clear array $found$ in parallel
7:     **for** $i \in start, \ldots, stop$ **do**   ▷ Sequentially find neighbors of every assigned feature instance
8:         **if** $wtid = 0$ **then** $flags[bwid] \leftarrow false$ ▷ Initialize $flags$ array
9:         $j \leftarrow i - 32$ ▷ j is a base index for all threads within a warp
10:        **while** $j >= -32$ **do**   ▷ Dont read beyond start of the input array
11:            $found[btid] \leftarrow 0$ ▷ Clear array $found$ in parallel
12:            **if** $wtid + j >= 0$ **then**   ▷ If the thread can access an existing array entry...
13:                **if** $X[i] - X[wtid + j] > r$ $flags[bwid] \leftarrow true$ ▷ Check for stop condition
14:                **if** $d(i, wtid + j) \leq r$ **and** $F[i] \neq F[wtid + j]$ **then**   ▷ Detect neighbor
15:                    $found[btid] \leftarrow 1$ ▷ Mark that it was found
16:                    **if** $F[i].f < F[wtid + j].f$ **then**   ▷ Store neighbor pair into a and b
17:                        $a \leftarrow ((F[i].f, F[i].id), F[wtid + j].f)$
18:                        $b \leftarrow (F[wtid + j].f, F[wtid + j].id)$
19:                    **else**
20:                        $a \leftarrow ((F[wtid + j].f, F[wtid + j].id), F[i].f)$
21:                        $b \leftarrow (F[i].f, F[i].id)$
22:                    **end if**
23:                **end if**
24:            **end if**
25:            $scanBuf[btid] \leftarrow found[btid]$ ▷ Copy array $found$ in parallel
26:            $intraWarpExclusiveScan(scanBuf)$ ▷ Find output position for each thread
27:            **if** $found[btid] = 1$ **then**   ▷ Store found neighbor pair in the output arrays
28:                $K[scanBuf[btid] + C[gwid]] \leftarrow a$ ▷ C stores base output indices for warps
29:                $V[scanBuf[btid] + C[gwid]] \leftarrow b$
30:            **end if**
31:            **if** $gtid = gwid * 32$ **then**   ▷ Update base indices in C
32:                $C[gwid] \leftarrow C[gwid] + scanBuf[bwid * 32 + 31] + found[bwid * 32 + 31]$
33:            **end if**
34:            **if** $flags[bwid]$ **break**   ▷ If any thread reached stop condition, abort the loop
35:            $j \leftarrow j - 32$ ▷ Otherwise move warp to next set of feature instances in array
36:        **end while**
37:    **end for**
38: **end kernel**

---

just introduce the differences between this kernel and the *FindNeighbors* kernel. Please also note, that we take advantage of the fact that all threads within a single warp are implicitly synchronized. Consequently, no *synchronization primitives* need to be used in these algorithms. The presented algorithms also use such identifiers as: *blockDim*, *blockIdx* and *threadIdx*, which represent size of blocks, block position within a grid and thread position within a block respectively.

Algorithm 4. A function for computing thread numbers and the work subset assigned to the thread's warp

1: **function** THREADNUMBERSANDSUBSETS
2:     $warpCount \leftarrow blockDim.x * blockSize.x$
3:     $gtid \leftarrow blockIdx.x * blockDim.x + threadIdx.x$ ▷ Global thread number
4:     $gwid \leftarrow \lfloor gtid/32 \rfloor$ ▷ Global warp number
5:     $wtid \leftarrow gtid \bmod 32$ ▷ Warp thread number
6:     $btid \leftarrow threadIdx.x$ ▷ Block thread number
7:     $bwid \leftarrow \lfloor btid/32 \rfloor$ ▷ Block warp number
8:     $start \leftarrow round((|X| - 1) * gwid/warpCount) + 1$ ▷ Start of subset
9:     $stop \leftarrow round((|X| - 1) * (gwid + 1)/warpCount)$ ▷ End of subset
10:    **return** $(gtid, gwid, wtid, btid, bwid, start, stop)$
11: **end function**

CountNeighbors kernel starts with the execution of the *ThreadNumbersAnd-Subsets* function. This function (shown in Algorithm 4) computes for each thread such values as: global thread number ($gtid$), global number of the warp the thread is in ($gwid$), number of the thread within the warp ($wtid$), number of the thread within a block ($btid$) and number of the warp within the block ($bwid$). Moreover, based on the global warp number, this function computes indices $start$ and $stop$ which point to the input data subset associated with the warp. Next, two shared memory arrays are allocated. The array $flags$ stores a single value for each warp within the current block, and is used to pass information among all of the threads within a warp that the stop condition was detected. The array $found$ stores a single value for each thread within the current block. Each such value is a counter of all of the neighbors found by the corresponding thread. These counters are initialized in line 5. Next, each thread iterates in a loop over all of input feature instances in the assigned input data subset. Each such iteration compares a feature instance number i with consecutive previous feature instances until a stop condition is reached. In line 7, first thread of each warp initializes the $flags$ array entry to mark that the warp has not reached the stop condition. Next, the j index is introduced. This index is equal to the position, the first thread within a warp should compare to the feature instance number i. Based on this index, each thread in the warp computes its corresponding index in the input arrays: $j + wtid$. The subsequent while loop in each iteration decrements the j index by 32 until either stop condition is reached, or start of the input array is reached (see lines 9 and 16). In the while loop each thread determines whether $j + wtid$ is a valid index (line 10). If it is valid each thread checks for the stop condition (line 11). If the stop condition is reached, this fact is saved in the $flags$ array. Next, the distance between the two compared feature instances is computed. If the computed distance is less than the neighborhood distance threshold r then a neighbor pair is found and appropriate counter is incremented (lines 12-14). After the loop is finished, each warp performs a *reduction* (sum) of all of the computed counters (line 20). Finally, the first thread of each warp stores the computed sum in the appropriate position in the output array C (line 22).

Basic structure of the *FindNeighbors* kernel is similar. The first difference is that an additional shared memory array called *scanBuf* is allocated (line 5). This is an auxiliary array which is used for computing positions of found neighbor pairs in the output arrays K and V. The next difference is the addition of the line 11, which zeroes counters of the found neighbor pairs. Consequently, this changes the meaning of the array *found* which now stores only the number of found neighbor pairs in each iteration of the while loop (either 0 or 1). The most important difference however is in the processing of the found neighbor pairs (lines 15-22). The fact that a neighbor pair is found is denoted in the array *found* as in *CountNeighbors* kernel, but assignment is used instead of increment. After this, the extended feature instance identifier and feature instance identifier, which will be eventually stored in the arrays K and V respectively, are computed and stored in local variables a and b. The obtained neighbor pairs now need to be stored into the result arrays K and V. To store found neighbor pairs without gaps, a compact step is needed. This is performed in lines 25-30. First, the array *found* is copied into *scanBuf*. Next, every warp performs an *exclusive scan* over its corresponding part of the array *scanBuf*. The resulting array stores, for every thread within a warp, the number of neighbor pairs found by the previous threads within a warp ($scanBuf[btid]$). This is used in lines 27-30 to store the found neighbor pairs in the appropriate locations in the output arrays. Let us now recall the array C that stores for every warp the location in the output arrays at which the warp should start storing the found neighbor pairs. As each warp can find more than one neighbor pair in each iteration, the output index here is computed as $scanBuf[btid] + C[gwid]$. After results are stored, the positions in array C are updated in lines 31-33 so that next iterations of the while loop do not overwrite the previously saved neighbor pairs.

4 Experiments

4.1 Implementation and Testing Environment

To evaluate the performance of our new method we have compared it with a CPU implementation. The CPU version uses a similar approach as the one described in Section 3 and utilizes Intel® Threading Building Blocks technology [15] to parallelize computations on a multi-core CPU. Due to divergent code paths encountered while finding neighbor pairs, SIMD extensions could not be used in the CPU version. Consequently, potential neighbors of processed instances are checked sequentially, not in batches. The implementation starts a user specified number of threads to detect neighbors and build a hashmap. In experiments we have used 1, 2, 4 and 8 threads.

In the GPU-based method a user has to specify the number of warps. Basically it is a hardware dependent parameter, which should be at least equal to the maximal number of active warps on a graphics card. However, to achieve better balancing of workload this value should be even larger. Our experiments have shown that for sufficiently large number of warps further increase of this parameter does not affect the performance as long as it is less than the number of feature instances in the input dataset.

All experiments were performed on an Intel Core i7 930@2.8Ghz CPU (4 core CPU with Hyperthreading) with 24GB of RAM and NVIDIA Geforce GTX TITAN graphics card (2688 CUDA Cores) with 6GB of RAM working under Microsoft Windows 7 operation system.

4.2 Data Sets

To evaluate our GPU-based method we have used synthetic and real world datasets. To prepare synthetic datasets we have used a synthetic generator for collocation pattern mining similar to the one described in [19]. The desired neighborhood distance threshold was set to 5 units. We have generated 60 datasets that contain up to one million of feature instances per dataset. The number of spatial features ranges from 30 to 200. Up to 80 percent of instances are instances of noise. Additionally we have prepared one dataset DS11 with 11 million of feature instances and 252 features to examine scalability of our solution.

We have also performed experiments on real world dataset based on the spatial data acquired from the OpenStreetMap project. We have processed the data and extracted potentially interesting locations based on the user-provided tags, which after categorization and filtering became spatial features. Then we have chosen a point on the map corresponding to the location of the Central Park in New York and selected 1.5 million of nearest neighbors representing 241 different spatial features.

In the next subsection we present results from performed experiments. We measured a *speedup*, i.e., the ratio of the execution time of CPU version of the algorithm to the execution time of GPU implementation. In the first series of experiments we have examined how the size of input dataset affects performance. We also present measured absolute processing times. In the second series we compared results for different neighbor distance thresholds as well as for varying number of spatial features. Finally, in the last series we have tested influence of the size of real world datasets on the algorithm performance.

Processing times of GPU version are denoted by TGPU while processing times of 1, 2, 4 and 8-threaded CPU versions are denoted by TCPU1, TCPU2, TCPU4 and TCPU8 respectively.

4.3 Results of Experiments

Figure 5 presents the results of the first experiment in which we tested how the performance of our new method is affected by the increasing size of the input dataset. The experiment was conducted on the DS11 dataset. Using sliding window, with variable size and constant position of the left border we selected from 1000 up to 10 million of feature instances. On such datasets we have compared GPU-based version of the algorithm with a CPU implementation using 1, 2, 4 and 8 threads. Please note that logarithmic scale on x axis is used. Although it is not clearly visible in the chart the speedup, in comparison with eight-threaded

CPU version, surpasses the value of 1 when the size of the input dataset reaches 5000 of feature instances. The highest speedup (~ 9) is for input datasets with sizes exceeding 1 million of feature instances, however even for 100K feature instances the speedup is approx. 8. It is important that our solution scales very well and the speedup is preserved even for 10 million of feature instances. In Fig. 6 we present the results of the same experiment, but rather than a speedup we visualize absolute processing times. It is clearly visible that in all cases processing times increase linearly with the increasing size of the input dataset. For the biggest dataset we used, the time required by single-threaded CPU version is equal to 397 seconds, eight-threaded CPU version needed 112 seconds while GPU implementation required only 12.5 seconds.

Figure 7 shows the results of the second experiment which tested the influence of the neighborhood distance threshold on the algorithm performance. This time we performed the experiment on the 60 synthetic datasets. Most of them are datasets with sizes not exceeding $200K$ of feature instances. Presented speedups are average values gathered for all tested datasets. We have examined speedups for values of neighborhood distance threshold r ranging from 1 to 14 units. Recall that all input datasets were generated using the value of r equal to 5 as an assumed distance for discovered collocation patterns. We can observe that for values of r higher than 5, the performance of the GPU version is very stable. For lower values of r the performance gap between CPU and GPU version is reduced. It is a result of significant reduction of the number of neighbors and from our sidetrack tests we know that the neighbor finding step is the most time consuming part of the algorithm. Therefore the speedup, especially versus single-threaded CPU version, for small values of r is quite low as there is a very low number of neighbors. This causes warps to perform unnecessary computations.

We have also tested how the speedups are affected by the varying number of spatial features. To perform this test we took a dataset that contained one million feature instances. Then we counted the number of instances per feature and we merged two features with the least number of instances into one feature. We repeated this step to obtain datasets with $50, 90, 130, 170, 210$ and 250 features. The speedups are almost perfectly consistent across the whole experiment (due to the limited space we omit the chart for this experiment). This can be explained as follows. The number of neighbor checks does not change with respect to the number of features, as both: the number of points and neighborhood distance threshold are constant. The only difference is in the step of sorting and building hashmap, however the participation of these steps in the overall processing time is very low in comparison to the neighbor detection step.

Finally, we have performed experiement on the real world dataset. From the acquired 1.5 million points from OpenStreetMap, we prepared 15 datasets such that the n-th dataset contains $n*100000$ feature instances nearest to the Central Park. The results, presented in Fig. 8, are very similar to the results of experiments on the synthetic datasets and the highest speedup (more than 7) is for input datasets with sizes exceeding approximately 1 million of feature instances.

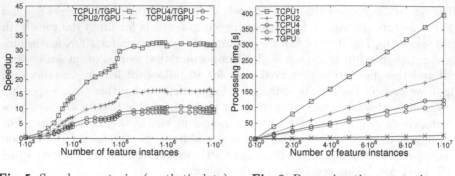

Fig. 5. Speedup w.r.t. size (synthetic data)   **Fig. 6.** Processing time w.r.t. size

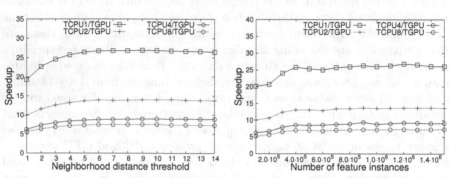

Fig. 7. Speedup w.r.t distance threshold **Fig. 8.** Speedup w.r.t. size (real world data)

5   Summary and Future Work

In this paper we present an efficient GPU based algorithm for construction of iCPI-hashmaps. We have performed numerous experiments and have achieved speedups of up to 9 times w.r.t. the multithreaded CPU implementation. The obtained speedups are stable and mostly independent on input data parameters. Our solution was able to process datasets composed of milions of feature instances in times measured in tens of seconds.

While the proposed solution already offers very high performance, there are still more interesting things to be done. We plan on extending the algorithm to process datasets that require more memory than available on graphics card. Moreover, we also plan on designing methods of distributing the workload to multiple graphics cards.

References

1. Agrawal, R., Srikant, R.: Fast Algorithms for Mining Association Rules in Large Databases. In: Proceedings of the 20th International Conference on Very Large Data Bases, pp. 487–499. Morgan Kaufmann Publishers Inc., San Francisco (1994)
2. Alcantara, D.A.F.: Efficient Hash Tables on the GPU. PhD thesis, University of California, Davis (2011)

3. Andrzejewski, W., Boinski, P.: GPU-accelerated collocation pattern discovery. In: Catania, B., Guerrini, G., Pokorný, J. (eds.) ADBIS 2013. LNCS, vol. 8133, pp. 302–315. Springer, Heidelberg (2013)

4. Bell, N., Hoberock, J.: GPU Computing Gems: Jade edition, chapter Thrust: A Productivity-Oriented Library for CUDA, pp. 359–371. Morgan-Kauffman (2011)

5. Boinski, P., Zakrzewicz, M.: Collocation Pattern Mining in a Limited Memory Environment Using Materialized iCPI-Tree. In: Cuzzocrea, A., Dayal, U. (eds.) DaWaK 2012. LNCS, vol. 7448, pp. 279–290. Springer, Heidelberg (2012)

6. Bress, S., Beier, F., Rauhe, H., Sattler, K.-U., Schallehn, E., Saake, G.: Efficient co-processor utilization in database query processing. Information Systems 38(8), 1084–1096 (2013)

7. de Berg, M., van Kreveld, M., Overmars, M., Schwarzkopf, O.: Computational Geometry: Algorithms and Applications. Springer-Verlag New York, Inc., Secaucus (1997)

8. Fang, W., Lu, M., Xiao, X., He, B., Luo, Q.: Frequent itemset mining on graphics processors. In: Proceedings of the Fifth International Workshop on Data Management on New Hardware, DaMoN 2009, pp. 34–42. ACM, New York (2009)

9. Fayyad, U., Piatetsky-Shapiro, G., Smyth, P.: From Data Mining to Knowledge Discovery in Databases. AI Magazine 17, 37–54 (1996)

10. He, B., Lu, M., Yang, K., Fang, R., Govindaraju, N.K., Luo, Q., Sander, P.V.: Relational query coprocessing on graphics processors. ACM Trans. Database Syst. 34(4), 21:1–21:39(2009)

11. Jian, L., Wang, C., Liu, Y., Liang, S., Yi, W., Shi, Y.: Parallel data mining techniques on graphics processing unit with compute unified device architecture (cuda). The Journal of Supercomputing 64(3), 942–967 (2013)

12. Khronos Group. The OpenCL Specification Version: 1.2 (2012), http://www.khronos.org/registry/cl/specs/opencl-1.2.pdf/

13. NVIDIA Corporation. Nvidia cuda programming guide (2014), http://docs.nvidia.com/cuda/cuda-c-programming-guide/

14. Przymus, P., Kaczmarski, K.: Dynamic compression strategy for time series database using GPU. In: Catania, B., Cerquitelli, T., Chiusano, S., Guerrini, G., Kämpf, M., Kemper, A., Novikov, B., Palpanas, T., Pokorny, J., Vakali, A. (eds.) New Trends in Databases and Information Systems. Advances in Intelligent Systems and Computing, vol. 241, pp. 235–244. Springer, Heidelberg (2014)

15. Reinders, J.: Intel Threading Building Blocks, 1st edn. O'Reilly & Associates, Inc., Sebastopol (2007)

16. Shekhar, S., Huang, Y.: Discovering spatial co-location patterns: A summary of results. In: Jensen, C.S., Schneider, M., Seeger, B., Tsotras, V.J. (eds.) SSTD 2001. LNCS, vol. 2121, pp. 236–256. Springer, Heidelberg (2001)

17. Wang, L., Bao, Y., Lu, J.: Efficient Discovery of Spatial Co-Location Patterns Using the iCPI-tree. The Open Information Systems Journal 3(2), 69–80 (2009)

18. Yoo, J.S., Bow, M.: Mining Maximal Co-located Event Sets. In: Huang, J.Z., Cao, L., Srivastava, J. (eds.) PAKDD 2011, Part I. LNCS, vol. 6634, pp. 351–362. Springer, Heidelberg (2011)

19. Yoo, J.S., Shekhar, S.: A joinless approach for mining spatial colocation patterns. IEEE Transactions on Knowledge and Data Engineering 18(10), 1323–1337 (2006)

SemIndex: Semantic-Aware Inverted Index

Richard Chbeir[1], Yi Luo[2], Joe Tekli[3,*], Kokou Yetongnon[2],
Carlos Raymundo Ibañez[4], Agma J. M. Traina[5], Caetano Traina Jr.[5],
and Marc Al Assad[3]

[1] University of Pau and Adour Countries, Anglet, France
[2] University of Bourgogne, Dijon, France
[3] Lebanese American University, Byblos, Lebanon
joe.tekli@lau.edu.lb, jtekli@gmail.com
[4] Universidad Peruana de Ciencias Aplicadas, Lima, Peru
[5] University of São Paulo, São Carlos-SP, Brazil

Abstract. This paper focuses on the important problem of semantic-aware search in textual (structured, semi-structured, NoSQL) databases. This problem has emerged as a required extension of the standard containment keyword based query to meet user needs in textual databases and IR applications. We provide here a new approach, called *SemIndex*, that extends the standard inverted index by constructing a tight coupling inverted index graph that combines two main resources: a general purpose semantic network, and a standard inverted index on a collection of textual data. We also provide an extended query model and related processing algorithms with the help of *SemIndex*. To investigate its effectiveness, we set up experiments to test the performance of *SemIndex*. Preliminary results have demonstrated the effectiveness, scalability and optimality of our approach.

Keywords: Semantic Queries, Inverted Index, NoSQL indexing, Semantic Network, Ontologies.

1 Introduction

Processing keyword-based queries is a fundamental problem in the domain of Information Retrieval (IR). Several studies have been done in the literature to provide effective IR techniques [10,9,6]. A standard containment keyword-based query, which retrieves textual identities that contain a set of keywords, is generally supported by a full-text index. Inverted index is considered as one of the most useful full-text indexing techniques for very large textual collections [10], supported by many relational DBMSs, and recently extended toward semi-structured [9] and unstructured data [6] to support keyword-based queries.

Besides the standard containment keyword-based query, semantic-aware or knowledge-aware (keyword) query has emerged as a natural extension encouraged by real user demand. In semantic-aware queries, some knowledge[1] needs

---

[*] Corresponding author.
[1] Also called domain, collaborative, collective knowledge or semantic network

Y. Manolopoulos et al. (Eds.): ADBIS 2014, LNCS 8716, pp. 290–307, 2014.
© Springer International Publishing Switzerland 2014

to be taken into consideration while processing. Let's assume having a movie database, as shown in Table 1. Each movie, identified with an *id*, is described with some (semi-structured) text, including movie `title`, `year` and `plot`. For queries "sound of music", "Maria nun" and "sound Maria", the query result is movie O_3. However, if the user wants to search for a movie but cannot recall the exact movie title, it is natural to assume that (s)he may modify the query terms to some semantically similar terms, for example, "voice of music". Also, it is common that the terms provided by users are not exactly the same, but are semantically relevant to terms that the plot providers use. Clearly, the standard inverted index which only supports exact matching cannot deal with these cases.

Table 1. A Sample Movie Data Collection

ID	Textual Contents
O_1	*When a Stranger Calls (2006)*: A young high school student babysits for a very rich family. She begins to receive strange phone calls threatening the children...
O_2	*Code R (1977)*: This CBS adventure series managed to combine elements of "Adam-12", "Emergency" and "Baywatch" at the same time...
O_3	*Sound of Music, The (1965)*: Maria had longed to be a nun since she was a young girl, yet when she became old enough discovered that it wasn't at all what she thought...
...	...

Various approaches combining different types of data and semantic knowledge have been propose to enhance query processing (cf. Related Works). In this paper, we present a new approach integrating knowledge into a semantic-aware inverted index called *SemIndex* to support semantic-aware querying. Major differences between our work and existing methods include:

- **Pre-processing the Index**: Existing works use semantic knowledge to pre-process queries, such as query rewriting/relaxation and query suggestion [2,5], or to post-process the query results, such as semantic result clustering [16,17,25]. Our work can be seen as another alternative to consider the semantic gap by enclosing semantic knowledge directly into an inverted index, so that main tasks can be done before query processing,
- **User Involvement**: Most existing works introduce some predefined parameters (heuristics) to rewrite queries such that users are only involved in the query refinement (expansion, filtering, etc.) process after providing the first round of results [3,4,14,20]. In our work, we aim at allowing end-users to write, using the same framework, classical queries but also semantically enriched queries according to their needs. They are involved in the whole process (during initial query writing and then query rewriting).
- **Providing More Results**: Most existing works focus on understanding the meaning of dat/queries through semantic disambiguation [2,13,16], which: i) is usually a complex process requiring substantial processing time [15], and ii) depends on the query/data context which is not always sufficiently available, and thus does not guarantee correct results [7,23]. The goal of our work is, with the help of semantic knowledge, to find *more* semantically relevant results than what a traditional inverted index could provide, while doing it more efficiently than existing disambiguation techniques.

In order to build *SemIndex*, we create connections between two data resources, *a textual data collection* and *a semantic knowledge base*, and map them into a single data structure. An extended query model with different levels of semantic awareness is defined, so that both semantic-aware queries and standard containment queries are processed within the same framework. Figure 1 depicts the overall framework of our approach and its main components. The *Indexer* manages *SemIndex*, while the *Query Processor* accepts semantic-aware queries from the user and processes the queries with *SemIndex*.

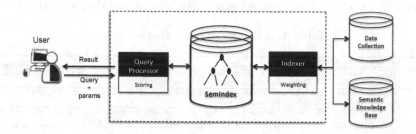

Fig. 1. SemIndex Framework

The rest of this paper is organized as follows. Designing and building *SemIndex* will be presented in Section 2. Section 3 will introduce our query model for semantic-aware queries. We will also present the algorithms for processing semantic-aware queries using *SemIndex*. We provide preliminary experimental results of executing different queries on a set of textual data collections in Section 4. Related works are discussed in Section 5 before concluding the paper and providing some future works.

2   Index Design

In the following, we analyze the (textual and semantic) input resources required to build our index structure, titled *SemIndex*. We also show how to create connections between the input resources and how to design the logical structure of *SemIndex*. The physical structure will be detailed in a dedicated paper.

2.1   Representation and Definitions

Textual Data Collection: In our study, a textual data collection could be a set of documents, XML fragments, or tuples in a relational or NoSQL database.

Definition 1. *A Textual Data Collection Δ is represented as a table defined over a set of attributes $\mathcal{A} = \{A_1, \ldots, A_p\}$ where each A_i is associated with a set of values (such as strings, numbers, BLOB, etc.) called the domain of A_i and denoted by $dom(A_i)$. A semantic knowledge base (KB) can be associated to one or several attribute domains $dom(A_i)$. Thus, given a table Δ defined over $\mathcal{A}$, objects t in Δ are denoted as $\langle a_1, \ldots, a_p \rangle$, where $a_i \in dom(A_i)$. Each a_i from t is denoted as $t.a_i$.*

Semantic Knowledge-Base: We adopt graph structures for modeling semantic knowledge. Thus, entities are represented as vertices, and semantic relationship between entities are modeled as directed edges [2]. In this work, we will illustrate the design process of *SemIndex* using WordNet version 3.0 [8] as the semantic knowledge resource. Part of the WordNet ontology is shown in Figure 2. Each synset represents a distinct concept, and is linked to other synsets with semantic relations (including *hypernymy*, *hyponymy*, *holonymy*, etc.). Note that multiple edges may exist between each ordered pair of vertices, and thus the knowledge graph is a multi-graph.

Definition 2. *A Semantic Knowledge Base KB, such as WordNet, is a graph $G_{kb}(V_{kb}, S_{kb}, E_{kb}, L_{kb})$ such that:*

- *V_{kb} is a set of vertices/nodes, denoting entities in the given knowledge base. For WordNet, V_{kb} includes synsets and words*
- *S_{kb} is a function defined on V_{kb}, representing the string value of each entity*
- *E_{kb} is a set of directed edges; each has a label in L_{kb} and is between a pair of vertices in V_{kb}*
- *L_{kb} is a set of edge labels. For WordNet, L_{kb} includes hyponymy, meronymy, hypernymy, holonymy, has-sense and has-sense-inverse, etc.*

In Figure 2, W_4, W_6, W_7, W_8 and W_9 represent words, and their string values (lemma of the words) are shown aside of the nodes. S_1, S_3 and S_4 are synsets, and their string values are their definitions. If one sense of a word belongs to a synset, it is represented with two edges between the synset node and word node with opposite directions, labeled *has-sense* and *has-word*, that we represent here with only one left-right arrow.

SemIndex **Graph:** To combine our resources, we define a *SemIndex* graph.

---

[2] We use the terms "edge" and "directed edge" interchangeably in this paper.

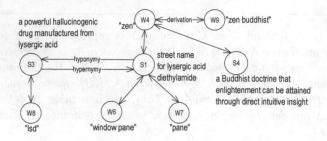

Fig. 2. Part of the Semantic Knowledge Graph of WordNet

Definition 3. *A SemIndex graph $\tilde{G}$ is a directed graph $(V_i, V_d, L, E, S_v, S_e, W)$:*

- *V_i is a set of index nodes (denoting entities in a knowledge-base, index or searchable terms in a textual collection) represented visually as circles $\bigcirc$*
- *V_d is the set of data nodes (belonging to the textual collection) represented visually as squares $\square$*
- *L is a set of labels*
- *E is a set of ordered pairs of vertices in $V_i \bigcup V_d$ called edges. Edges between index nodes are called index edges (represented visually as $\rightarrow$), while edges between index nodes and data nodes are called data edges (represented visually as $--\rightarrow$)*
- *S_v is a function defined on $V_i \bigcup V_d$, representing the value of each node*
- *S_e is a function defined on E, assigning a label $\in L$ to an edge*
- *W is a weighting function defined on nodes in $V_i \bigcup V_d$ and edges in E.*

2.2   Logical Design

In this part, we introduce the logical design techniques in building *SemIndex*.

Building *SemIndex*: *SemIndex* adapts *tight coupling* techniques to index a textual data collection and a semantic knowledge base in the same framework, and directly create a posting list for all searchable contents. In the following, we describe how to construct *SemIndex*.

1- Indexing Input Resources: Given a textual data collection Δ, a multi-attributed inverted index (associated to one or several attributes A) is a mapping $ii : V_A^s \rightarrow V^t$, where V_A^s is a set of values for attributes A, which we also call a *searchable terms*, and V^t is the set of textual data objects. A multi-attributed inverted index of a set of textual data objects Δ is represented as a *SemIndex* graph $\tilde{G}_A$ such that:

- V_i is a set of *index nodes*, representing all searchable terms which appear in the attribute set A of the collection
- V_d is the set of textual data objects in Δ
- L includes *contained* label indicating the containment relationship from a searchable term in V_i to a data object in V_d
- E is a set of ordered pairs of vertices in $V_i \bigcup V_d$
- S_v assigns a term to an index node, and its text contents to a data node
- S_e assigns the *contained* label to each edge
- W assigns a weight (according to the importance/frequency of the term within the text content) to an edge E.

An example of a *SemIndex* graph inverted index $\tilde{G}_A$ based on the textual collection provided in Table 1 is shown in Figure 3 (upper part).

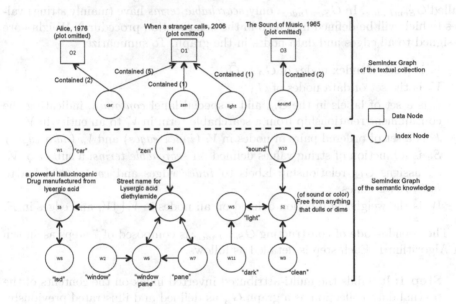

Fig. 3. Part of the *SemIndex* Graph of the textual collection

Similarly, indexing the semantic knowledge base G_{kb} is also represented as a *SemIndex* graph $\tilde{G}_{kb}$ that inherits the properties of G_{kb} where:

- V_d is an empty set
- V_i is a set of vertices/nodes, denoting entities in the given knowledge base and all other searchable terms. For WordNet, V_i includes synsets and words as well as other terms that appear in the string value in G_{kb}. Thus, V_i is a superset of vertices in the knowledge graph G_{kb}
- L is a set of edge labels, including those inherited from G_{kb} (e.g., *hyponymy*, *meronymy*, etc.), and a special label *meronymy** indicating the *containment* relation from a searchable term to an entity in V_i

- S_e assigns, in addition to previous edges of G_{kb}, the *meronymy** label from a searchable term to an entity in V_i
- W is the weighting function assigning a weight (default weight is 1) to all nodes and edges.

We assume that connections between searchable terms and entities in V_i can be found by the same Natural Language Processing techniques used when indexing the textual collection. For example, in Figure 3 (lower part), we find that word "window" (W_2) is contained in the word "window pane" (W_6). Thus, an extra edge labeled *meronymy** from W_2 to W_6 is inserted into the graph as shown in Figure 3. Note that the NLP algorithms run only on synsets and multi-term words, in order to prevent duplicated nodes to be produced.

2- Coupling Resources: When coupling both indexes, we get one *SemIndex* graph called $\tilde{G}_{SemIndex}$. In $\tilde{G}_{SemIndex}$, only *searchable terms* have (mainly string) values (which will be defined in Step 3 in the construction procedure). Weights are assigned to all edges and data nodes in the graph. To summarize:

- V_i is a set of index nodes of $\tilde{G}_A \cup \tilde{G}_{kb}$
- V_d is the set of data nodes of $\tilde{G}_A$
- L is a set of labels in the $\tilde{G}_{kb}$ and a special label *contained*, indicating the containment relationship from a searchable term in V_i to an entity in V_d
- E is a set of ordered pairs of nodes in V_i (*index edges*) and V_d (*data edges*)
- S_v is a function of string values defined on *searchable terms*: a subset of V_i
- S_e assigns $\tilde{G}_{kb}$ relationship labels to *index* edges and *contained* label to *index/data* edges
- W is the weighting function defined on all nodes in $V_i \bigcup V_d$ and edges in E.

The pseudo-code of constructing $\tilde{G}_{SemIndex}$ is composed of 7 steps as shown in Algorithm 1. Each step is detailed as follows.

- **Step 1:** It builds the multi-attributed inverted index on the contents of the textual data collection as a graph $\tilde{G}_A$ as defined and illustrated previously.
- **Step 2:** Given a semantic knowledge graph G_{kb} representing the semantic knowledge base KB given as input, it builds an inverted index for string values of each knowledge base entity, and construct the graph $\tilde{G}_{kb}$.
- **Step 3:** it combines the two *SemIndex* graphs. Data nodes in the result graph $\tilde{G}_{SemIndex}$ are the set V_d in $\tilde{G}_A$ (denoted as V_d^A), while all other nodes are index nodes. This step denotes the searchable terms of $\tilde{G}_{kb}$ as V_i^{kb} (which are vertices with one or more outgoing edges labeled *contained*) and then merges the two sets of searchable terms V_i^{kb} and V_i^A (representing the index nodes of $\tilde{G}_A$) as follows: if string values of two vertices are equal, remove one of them and merge all the connected edges. We use V_i^+ to denote the conjunctive set $V_i^{kb} \bigcup V_i^A$, which is the set of all *searchable terms* in *SemIndex*. Figure 4 shows the result of combining the two *SemIndex* graphs of the sample textual collection and the WordNet extract provided here.

Algorithm 1 $\tilde{G}_{SemIndex}$ Construction

Input:

KB: a semantic knowledge base;

Δ: a textual data collection;

ω: a weighting schema;

c_1: a constant (used in Step 4) to delimit the co-occurrence window

c_2: a constant (used in Step 4) to select top terms

Output:

$\tilde{G}_{SemIndex}$: a *SemIndex* graph instance

1: Build inverted index for Δ to construct $\tilde{G}_A$
2: Build inverted index for G_{kb} to construct $\tilde{G}_{kb}$
3: Merge $\tilde{G}_A$ and $\tilde{G}_{kb}$ into $\tilde{G}_{SemIndex}$ and find searchable terms
4: For each missing term, find the most relevant terms in $\tilde{G}_{kb}$
5: Assign weights to all edges in $\tilde{G}_{SemIndex}$ and all data nodes according to ω
6: Aggregate edges between each ordered pair of nodes
7: Remove from $\tilde{G}_{SemIndex}$ all edge labels, and string values of all nodes except searchable terms

---

- **Step 4:** For the missing term problem, we create links from each missing term to one or more closely related terms, with a new edge label *refers-to*, using a distributional thesaurus[3] based on the textual collection to mine relativeness between missing terms and used index. We cover the *missing term problem* in more detail in a dedicated paper.
- **Step 5:** It assigns weights to edges and textual objects, according to ω. The weight will be used to rank query results. Different weighting schemes can be adopted in our approach. We propose below the principles of a simple weighting schema for computing edge and node weights:
  - *Containment edges*: For a (data) edge from a term to a textual object, its weight is an IR score, such as term frequency. If the textual collection is formatted, this IR score could also be assigned to reflect the importance of a term, e.g., in large font size, in capitalized form, etc. When the textual collection is structured, higher weights are given to terms which appear in important places, e.g., title, author's name, etc.
  - *Structural edges*: The weight of a structural (index) edge is determined by edge label and by the number of edges with the same label from its starting node [21]. Please note that if the knowledge base is hierarchical (which is not the case for WordNet), the level of the edge in the hierarchy can also be taken into consideration [19].
  - *Nodes*: Assign "object rank" to all object nodes, based on metadata of objects, including text length, importance or reliability of data source, its publishing date, query logs, and so on. A PageRank-style weighting schema could also be adapted for Web documents.

---

[3] A distributional thesaurus is a thesaurus generated automatically from a corpus by finding words that occur in similar contexts to each other [11,24].

- **Step 6:** If an ordered pair of vertices is connected with two or more edges, it merges the edges and aggregates the weights. This means that $\tilde{G}_{SemIndex}$ becomes a graph rather than a multi-graph, which simplifies processing.
- **Step 7:** It removes edge labels and string values of all nodes except V_i^+ (searchable terms), since they are not required for processing semantic queries, which helps improve *SemIdex*'s scalability.

Figure 4 illustrates an instance of $\tilde{G}_{SemIndex}$ (without edge and node weights) which is based on the knowledge graph depicted in Figure 2.

3   Executing queries with *SemIndex*

In this section, we define our query model and present a processing algorithm to perform semantic-aware search with the help of *SemIndex*.

3.1   Queries

The *semantic-aware queries* considered in our approach are *conjunctive projection-selection* queries over Δ of the form $\pi_X \sigma_P \ell(\Delta)$ where X is a non empty subset of $\mathcal{A}$, $\ell \in \mathbb{N}$ is a query-type threshold, and P is a selection projection predicate defined as follows.

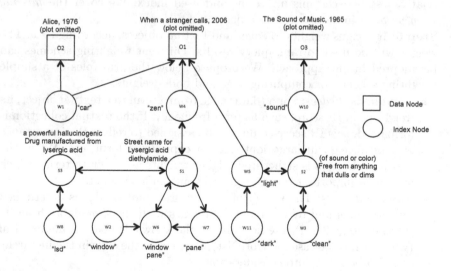

Fig. 4. *SemIndex* graph integrating the textual collection and semantic knowledge

Definition 4. *A selection projection predicate P is an expression, defined on a string-based attribute A in A, of the following forms: (A θ a), where a is a user-given value (e.g., keyword), and θ ∈ {=, like} whose evaluation against values in dom(A) is defined. A conjunctive selection projection query is made of a conjunction of selection projection predicates.*

Following the value of ℓ, we consider four semantic-aware query types:

- **Standard Query**: When $\ell = 1$, the query is a standard containment query and no semantic information is involved.
- **Lexical Query**: When $\ell = 2$, besides from the previous case, lexical connections, i.e., links between terms, may be involved in the result.
- **Synonym-based Query**: When $\ell = 3$, synsets are also involved. Note that there is no direct edge between textual object and synset node.
- **Extended Semantic Query**: When $\ell \geqslant 4$, the data graph of *SemIndex* can be explored in all possible ways. When ℓ grows larger, the data graph is explored further to reach even more results.

3.2   Query Answer

The answer to q in Δ, denoted as $q(\Delta)$, is defined as follows.

Definition 5. *Given a SemIndex graph $\tilde{G}$, the query answer $q(\Delta)$ is the set of distinct root nodes of all answer trees. We define an answer tree as a connected graph T satisfying the following conditions:*

- *(tree structure) T is a subgraph of $\tilde{G}$. For each node in T, there exists exactly one directed path from the node to the root object.*
- *(root object) The tree root is a data node, and it is the only data node in T, which corresponds to the textual object returned to the user.*
- *(conjunctive selection) For each query term in S, its corresponding index node is in the answer tree.*
- *(height boundary) Height of the tree, i.e., the maximal number of edges between root and each leaf, is no greater than the threshold ℓ.*
- *(minimal tree) No node can be removed from T without violating some of the above conditions.*

It can be proven that all leaves in the answer tree are query terms, and the number of leaves in T is smaller or equal to k, where k is the number of query terms. Also, the maximal in-degree of all nodes in T is at most k.

According to the value of ℓ which serves as an interval radius in the *SemIndex* graph, various answer trees can be generated for a number of query types:

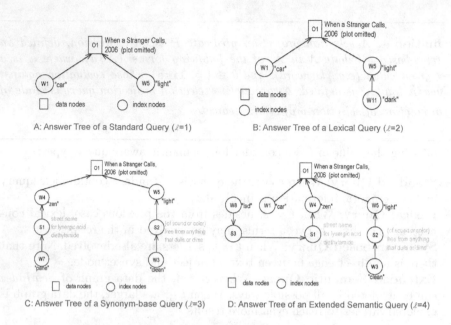

Fig. 5. Sample query answer trees

- **Standard Query**: When $\ell = 1$, the root of the answer tree is linked directly to all leaves, representing the fact that the result data object contains all query selection terms directly. An example answer tree is shown in Figure 5-A for the query q: $\pi_{\mathcal{A}}\sigma_{A\in(\text{"car"},\text{"light"})}\ell_{=1}(\Delta)$.
- **Lexical Query**: When $\ell = 2$, the answer tree contains also lexical connections between selection terms. Figure 5-B is an example answer tree of q: $\pi_{\mathcal{A}}\sigma_{A\in(\text{"car"},\text{"dark"})}\ell_{=2}(\Delta)$.
- **Synonym-based Query**: When $\ell = 3$, the answer tree contains, in addition to the two previous cases, the synsets. Note that due to the "minimal tree" restriction, a synset cannot be a leaf node of an answer tree. Thus, if an answer tree contains a synset, the height of the tree is no less than 3. A sample answer tree is shown in Figure 5-C for q: $\pi_{\mathcal{A}}\sigma_{A\in(\text{"pane"},\text{"clean"})}\ell_{=3}(\Delta)$. Synonyms of the two query terms, "zen" and "light", are also contained in the result object O_1.
- **Extended Semantic Query**: When $\ell \geqslant 4$, the answer tree contained additional nodes according to the provided value. An example answer tree is shown in Figure 5-D for q: $\pi_{\mathcal{A}}\sigma_{A\in(\text{"lsd"},\text{"clean"})}\ell_{=4}(\Delta)$.

3.3 Query Processing

Algorithm 2 is the procedure to process semantic-aware queries, given a set of query terms, *terms*, and a query-type threshold ℓ. Function expandNode(n,ℓ) performs the expansion of a node n. Basically, it explores the *SemIndex* graph with Dijkstra's algorithm from multiple starting points (multiple query terms). For each visited node n, we store its shortest distances from all starting points (query terms). The path score of a node n to a query term t is the sum of all weights on index edges along the path between t and n, thus the shortest distances of n are also the minimal path scores of n to all query terms.

Algorithm 2 SemSearch(terms[], ℓ)

Input:
terms: a set of selection terms
ℓ : a query-type threshold
Output:
out: a set of data nodes

1: **for** each $i \in terms$ **do**
2:     rs = Selecting nodeid from *SemIndex* with value = i); //selecting index nodes from the knowledge base as well as missing terms from the textual collection
3:     **for** each nodeid $\in$ rs **do**
4:         n = nodes.cget(nodeid); //retrieve or create a node with given id
5:         n.initPathScores(); //initialize the path scores of n
6:         todo.insert(n); //insert n into todo list
7:     **end for**
8: **end for**
9: **while** todo.isNotEmpty() **do**
10:     n = todo.pop(); //retrieve node with minimal structural score
11:     expandNode(n,ℓ);
12: **end while**
13: return out;

For example in Figure 5-C, the query terms are "pane" and "clean", and the algorithm starts to expand from two nodes W_7 and W_3. Path scores of W_7 are initialized to be a vector $< 0, \infty >$, since the shortest distance from W_7 to "pane" is 0, but the node is not reachable from "clean". Similarly, the path scores of W_3 are initially $< \infty, 0 >$. The minimal path scores must be updated when an edge is explored in the graph. For example, before finding the tree in Figure 5-B, the path scores of node O_1 is $< 2, \infty >$ (assume all edge weights are equal to 1), and the path scores of W_5 are $< \infty, 2 >$. After exploring the edge from W_5 to O_1, the path scores value of O_1 becomes $< 1, 2 >$, and O_1 is reachable from all query terms. The algorithm also keeps a *todo* list, which contains all nodes to be further expanded. The *todo* list is ordered on structural scores of the nodes. We define the structural score of a node n to be the maximal path score in the tree rooted on n, as shown in the following formula.

$$sscore_n = \max_{t \in rterms} pathscore_n(t) \qquad (1)$$

where *rterms* is the set of reachable query terms of n. It can be proven that if n is reachable from every term and all path scores of n are minimal, the structural score of n must be minimal among all trees rooted on n. For each result textual object, the algorithm always returns the answer tree with minimal structural score, thus it is not necessary to prune duplicated query results.

4   Experiments

We conducted a set of preliminary experiments to observe the behavior of *SemIndex* in weighting, scoring, and retrieving results. In this paper, we only present results related to processing time. We are currently working on experimentally comparing our approach with existing methods.

4.1   Experimental Setup

We ran our experiments on a PC with Intel 2GHz Dual CPU and 2GB memory. *SemIndex* was physically implemented in a MySQL 5.1 database with the query processor written in Java. We downloaded 90,091 movie plots from the IMDB database[4], and used WordNet 3.0 as our semantic knowledge base. We build *SemIndex* on plot contents and movie titles, which means that each textual object is a movie title combined with its plot (cf. Table 2).

Table 2. *SemIndex* Database Size

Database Name	Table Name	Table Size	Table Cardinality (#Row)
IMDB	Data IMDB	56 M	90K
WordNet	Data Adjective	3,2M	18K
	Data Verb	2,8M	13K
	Data Noun	15,3M	82K
	Data Adverb	0,5K	3K
	Index Verb	0,5M	11k
	Index Adverb	0,2M	3,6K
	Index Noun	4,8M	117K
	Index Adjective	0,8M	21k
	Index Sense	7,3M	207K
SemIndex	Lexicon	5.8M	146K
	Neighbors	116M	230K
	PostingList	340M	740K

4.2   Query Processing

In order to test the performance of *SemIndex*, we manually picked two groups of queries, shown in Table 3. In the first query group $Q1$, from $Q1-1$ to $Q1-8$, the height of the answer trees is bounded, the number of returned results is limited to 10, and each query contains from 2 to 5 selection terms. In the second group $Q2$ group, from $Q2-1$ to $Q2-4$, queries share the same query terms with different levels of tree height boundary and with an unlimited number of query

---

[4] http://imdb.com

Table 3. Sample Queries

Query Id	Height Boundary	Max. N# of Results	Selection Terms
Q1-1	4	10	car window
Q1-2	4	10	reason father
Q1-3	4	10	car window clean
Q1-4	4	10	apple pure creative
Q1-5	4	10	car window clean music
Q1-6	4	10	death piano radio war
Q1-7	4	10	car window clean music Tom
Q1-8	4	10	sound singer stop wait water
Q2-1	1	unlimited	car window clean
Q2-2	2	unlimited	car window clean
Q2-3	3	unlimited	car window clean
Q2-4	4	unlimited	car window clean

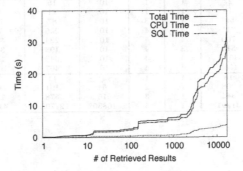

Fig. 6. Processing Time (Q2-4)

results. All queries were processed 5 times, retaining average processing time. Detailed statistics are shown in Table 4.

Figure 6 shows query processing time to retrieve the 10000 results of $Q2 - 4$. Initial response time is the processing time to output the first result. We break the latter into CPU and I/O time in order to better evaluate the processing costs of the algorithm. Minimal height is the height of the first returned answer tree while the maximal one is the height of the last answer tree allowing to reach the number of expected results. k, N_E and N_O are respectively the number of query terms, visited entity nodes, and object nodes during query processing.

From Table 4, we see that most queries are processed within 2 seconds, which is positively encouraging, except for Q2-4, since it is of an extended semantic type, which retrieves 16830 results in about a half minute. In fact, while the total number of textual objects in our plot dataset is around 90K, Q2-4 visits about 80K of object nodes which explains the significant increase in time. Also, we realize that the initial response times of Q1-5 and Q1-6 are quite different. Both of them contain 4 terms and all statistics of the two queries are similar except the minimal tree height. We also observe that for the second query group, the CPU time is dominated by the SQL time, while for the first query group, CPU cost is dominant. We analyzed the difference and found that, for the second query group, the algorithm cannot stop until all visited entity nodes are queried and the whole search space is examined. However for the first query group, since the number of query results is limited, the algorithm stops whenever it has found enough answer trees. Thus, for the first group of queries, although large N_E and N_O suggest large CPU time to create those nodes in memory, yet most of the nodes are not revisited for expansion before the algorithm stops, which significantly reduces overall query processing time.

Table 4. Processing Statistics

Queryid	Initial Response Time (ms)	Total Process Time (ms)	Min. Height	Max. Height	# of Results Returned	SQL Time (ms)	CPU Time (ms)	k	N_E	N_O
Q1-1	26	27	1	1	10	15	12	2	399	3346
Q1-2	39	40	1	1	10	21	19	2	252	9386
Q1-3	25	148	1	3	10	87	60	3	2528	7646
Q1-4	90	315	4	4	10	236	79	3	2812	11543
Q1-5	694	700	4	4	10	224	475	4	7849	32776
Q1-6	175	511	3	4	10	193	318	4	5071	38988
Q1-7	1223	1371	4	4	10	643	728	5	9505	50747
Q1-8	1469	1555	4	4	10	467	1076	5	12449	45715
Q2-1	18	18	1	1	2	11	7	3	3	3803
Q2-2	25	208	1	2	6	191	16	3	533	4394
Q2-3	21	842	1	3	515	678	164	3	1564	34564
Q2-4	200	31991	1	4	16830	27793	4198	3	19007	79997

5   Related Work

Including semantic processing in inverted indexes to enhance data search capabilities has been investigated in different approaches: i) including semantic knowledge into an inverted index, ii) including full-text information into the semantic knowledge base, and iii) building an integrated hybrid structure.

The first approach consists in adding additional entries in the index structure to designate semantic information. Here, the authors in [12] suggest extending the traditional (*term, docIDs*) inverted index toward a (*term, context, docIDs*) structure where contexts designate synsets extracted from WordNet, associated to each term in the index taking into account the statistical occurrences of concepts in Web document [1]. An approach in [26] extends the inverted index structure by adding additional pointers linking each entry of the index to semantically related terms, (*term, docIDs, relatedTerms*). Yet, the authors in [12,26] do not provide the details on how concepts are selected from WordNet and how they are associated to each term in the index.

Another approach to semantic indexing is to add words as entities in the ontology [1,18,22]. For instance, adding triples of the form *word occurs-in-context concept*, such that each word can be related to a certain ontological concept, when used in a certain context. Following such an approach: i) the number of triples would naturally explode, given that ii) query processing would require reaching over the entire left and the right hand sides of this *occurs-in-context* index, which would be more time consuming [1] than reading on indexed entry such as with the inverted index.

A third approach to semantic indexing consists in building an integrated hybrid structure: combining the powerful functionalities of inverted indexing with semantic processing capabilities. To our knowledge, one existing method in [1] has investigated this approach, introducing a joint index over ontologies and text. The authors consider two input lists: containing text postings (for words or occurrences), and lists containing data from ontological relations (for concept relations). The authors tailor their method toward incremental query construction with context-sensitive suggestions. They introduce the notion of *context lists* instead of usual inverted lists, where a prefix contains one index item per occurrence of a word starting with that prefix, adding an entry item for each occurrence of an ontological concept in the same context as one of these words, producing an integrated 4-tuples index structure (*prefix, terms*) ↔ (*term, context, concepts*). The method in [1] is arguably the most related to our study, with one major difference: the authors in [1] target semantic full-text search and indexing with special emphasis on IR-style incremental query construction, whereas we target semantic search in textual databases: building a hybrid semantic inverted index to process DB-style queries in a textual DB.

6   Conclusions and Future Work

In this paper, we introduce a new semantic indexing approach called *SemIndex*, creating a hybrid structure using a tight coupling between two resources: a general purpose semantic network, and a standard inverted index defined on a collection of textual data, represented as (multi)graphs. We also provide an extended query model and related processing algorithms, using *SemIndex* to allow semantic-aware querying. Preliminary experimental results are promising, demonstrating the scalability of the approach in querying a large textual data collection (IMBD) coined with a full-fledge semantic knowledge base (WordNet). We are currently completing an extensive experimental study to evaluate *SemIndex*'s properties in terms of: i) genericity: to support different types of textual (structured, semi-structured, NoSQL) databases, ranking schema, and knowledge-bases, ii) effectiveness: evaluating the interestingness of semantic-aware answers from the user's perspective, and iii) efficiency: to reduce index's building and updating costs as well as query processing cost. The system's physical structure (in addition to the logical designs provided in this paper) will also be detailed in an upcoming study.

Acknowledgements. This study is partly funded by: Bourgogne Region program, CNRS, and STIC AmSud project Geo-Climate XMine, and LAU grant SOERC-1314T012.

References

1. Bast, H., Buchhold, B.: An index for efficient semantic full-text search. In: 22nd ACM Int. Conf. on CIKM, pp. 369–378 (2013)

2. Burton-Jones, A., Storey, V.C., Sugumaran, V., Purao, S.: A heuristic-based methodology for semantic augmentation of user queries on the web. In: Song, I.-Y., Liddle, S.W., Ling, T.-W., Scheuermann, P. (eds.) ER 2003. LNCS, vol. 2813, pp. 476–489. Springer, Heidelberg (2003)
3. Carpineto, C., et al.: Improving retrieval feedback with multiple term-ranking function combination. ACM Trans. Inf. Syst. 20(3), 259–290 (2002)
4. Chandramouli, K., et al.: Query refinement and user relevance feedback for contextualized image retrieval. In: 5th International Conference on Visual Information Engineering, pp. 453–458 (2008)
5. Cimiano, P., et al.: Towards the self-annotating web. In: 13th Int. Conf. on WWW, pp. 462–471 (2004)
6. Das, S., et al.: Making unstructured data sparql using semantic indexing in oracle database. In: IEEE 29th ICDE, pp. 1405–1416 (2012)
7. de Limaand, E.F., Pedersen, J.O.: Phrase recognition and expansion for short, precision-biased queries based on a query log. In: 22nd Int. Conf. ACM SIGIR, pp. 145–152 (1999)
8. Fellbaum, C.: Wordnet an electronic lexical database. MIT Press (May 1998)
9. Florescu, D., et al.: Integrating keyword search into xml query processing. Comput. Netw. 33(1-6), 119–135 (2000)
10. Frakes, W.B., Baeza-Yates, R.A. (eds.): Information retrieval: Data structures and algorithms. Prentice-Hall (1992)
11. Grefenstette, G.: Explorations in automatic thesaurus discovery. Kluwer Pub. (1994)
12. Kumar, S., et al.: Ontology based semantic indexing approach for information retrieval system. Int. J. of Comp. App. 49(12), 14–18 (2012)
13. Li, Y., Yang, H., Jagadish, H.V.: Term disambiguation in natural language query for XML. In: Larsen, H.L., Pasi, G., Ortiz-Arroyo, D., Andreasen, T., Christiansen, H. (eds.) FQAS 2006. LNCS (LNAI), vol. 4027, pp. 133–146. Springer, Heidelberg (2006)
14. Mishra, C., Koudas, N.: Interactive query refinement. In: 12th Int. Conf. on EDBT, pp. 862–873 (2009)
15. Navigli, R.: Word sense disambiguation: A survey. ACM Comput. Surv. 41(2), 10:1–10:69 (2009)
16. Navigli, R., Crisafulli, G.: Inducing word senses to improve web search result clustering. In: Int. Conf. on Empirical Methods in Natural Language Processing, pp. 116–126 (2010)
17. Nguyen, S.H., Świeboda, W., Jaśkiewicz, G.: Semantic evaluation of search result clustering methods. In: Bembenik, R., Skonieczny, Ł., Rybiński, H., Kryszkiewicz, M., Niezgódka, M. (eds.) Intell. Tools for Building a Scientific Information. Studies in Computational Intelligence, vol. 467, pp. 393–414. Springer, Heidelberg (2013), http://dx.doi.org/10.1007/978-3-642-35647-6_24
18. Navigli Paola, R., et al.: Extending and enriching wordnet with ontolearn. In: Int. Conf. on GWC 2004, pp. 279–284 (2004)
19. Resnik, P.: Using information content to evaluate semantic similarity in a taxonomy. In: 14th Int. Conf. on Artificial intelligence, pp. 448–453 (1995)
20. Salton, G., Buckley, C.: Improving retrieval performance by relevance feedback. In: Readings in Information Retrieval, pp. 355–364 (1997)
21. Sussna, M.: Word sense disambiguation for free-text indexing using a massive semantic network. In: 2nd Int. ACM Conf. on CIKM, pp. 67–74 (1993)
22. Velardi, P., et al.: Ontolearn reloaded: A graph-based algorithm for taxonomy induction. Computational Linguistics 39, 665–707 (2013)

23. Voorhees, E.M.: Query expansion using lexical-semantic relations. In: 17th Int. ACM Conf. on SIGIR, pp. 61–69 (1994)
24. Weeds, J., et al.: Characterising measures of lexical distributional similarity. In: 20th Int. Conf. on Computational Linguistics (2004)
25. Wen, H., et al.: Clustering web search results using semantic information. In: 2009 Int. Conf. on Machine Learning and Cybernetics, vol. 3, pp. 1504–1509 (2009)
26. Zhong, S., et al.: A design of the inverted index based on web document comprehending. JCP 6(4), 664–670 (2011)

Entity Resolution with Weighted Constraints

Zeyu Shen and Qing Wang

Research School of Computer Science, The Australian National University, Australia
errand2901@163.com, qing.wang@anu.edu.au

Abstract. Constraints ubiquitously exist in many real-life applications
for entity resolution (ER). However, it is always challenging to effectively specify and efficiently use constraints when performing ER tasks.
In particular, not every constraint is equally effective or robust, and using
weights to express the "confidences" on constraints becomes a natural
choice. In this paper, we study entity resolution (ER) (i.e., the problem
of determining which records in a database refer to the same entities)
in the presence of weighted constraints. We propose a unified framework
that can interweave positive and negative constraints into the ER process, and investigate how effectively and efficiently weighted constraints
can be used for generating ER clustering results. Our experimental study
shows that using weighted constraints can lead to improved ER quality
and scalability.

1 Introduction

Constraints ubiquitously exist in many real-life applications for entity resolution
(ER), which can be obtained from a variety of sources: background knowledge
[20], external data sources [21], domain experts, etc. Some constraints may be
captured at the instance level [19,20], e.g., "PVLDB" refers to "VLDB Endowment" and vice versa, and some can be specified at the schema level [2,8,17],
e.g., two paper records refer to different papers if they do not have the same
page numbers. In general, constraints allow us to leverage rich domain semantics for improved ER quality. Nevertheless, not all constraints can be completely
satisfied due to the existence of dirty data, missing data, exceptions, etc. In such
cases, common approaches are to conduct manual reviews of conflicts, or relax
the satisfaction requirement of constraints by allowing some constraints to be violated in terms of a predefined cost model. This helps to produce a solution, but
often also requires additional (and often expensive) computational resources for
finding an optimal solution. Such burden has prevented constraints from being
widely applied to solve the ER problem in the past.

In this paper, we study two questions on ER constraints: (1) How to specify
constraints that can effectively improve the quality of ER solutions? (2) How
to use such constraints efficiently in the ER process? We attempt to establish a
uniform framework that can incorporate semantic capabilities (in form of constraints) into existing ER algorithms to improve the quality of ER, while still
being computationally efficient. A key ingredient in achieving this is to associate

Y. Manolopoulos et al. (Eds.): ADBIS 2014, LNCS 8716, pp. 308–322, 2014.
© Springer International Publishing Switzerland 2014

A database schema
PAPER := $\{pid, authors, title, journal, volume, pages, tech, booktitle, year\}$
AUTHOR := $\{aid, pid, name, order\}$
VENUE := $\{vid, pid, name\}$

Views	
TITLE := $\pi_{pid,title}$PAPER	HASVENUE := $\pi_{pid,vid}$VENUE
PAGES := $\pi_{pid,pages}$PAPER	VNAME := $\pi_{vid,name}$VENUE
PUBLISH := $\pi_{aid,pid,order}$AUTHOR	ANAME := $\pi_{aid,name}$AUTHOR

Constraints	Weights
$r_1:$ PAPER$^*(x,y) \leftarrow$ TITLE(x,t), TITLE(y,t'), $t \approx_{0.8} t'$	0.88
$r_2:$ PAPER$^*(x,y) \leftarrow$ TITLE(x,t), TITLE(y,t'), $t \approx_{0.6} t'$, SAMEAUTHORS(x,y)	0.85
$r_3:$ PAPER$^*(x,y) \leftarrow$ TITLE(x,t), TITLE(y,t'), $t \approx_{0.7} t'$, HASVENUE(x,z), HASVENUE(y,z'), VENUE$^*(z,z')$	0.95
$r_4:$ $\neg$PAPER$^*(x,y) \leftarrow$ PAGES(x,z), PAGES(y,z'), $\neg z \approx_{0.5} z'$	1.00
$r_5:$ VENUE$^*(x,y) \leftarrow$ HASVENUE(z,x), HASVENUE(z',y), PAPER$^*(z,z')$	0.75
$r_6:$ VENUE$^*(x,y) \leftarrow$ VNAME(x,n_1), VNAME(y,n_2), $n_1 \approx_{0.8} n_2$	0.70
$r_7:$ $\neg$AUTHOR$^*(x,y) \leftarrow$ PUBLISH(x,z,o), PUBLISH(y,z',o'), PAPER$^*(z,z')$, $o \neq o'$	0.90
$r_8:$ AUTHOR$^*(x,y) \leftarrow$ COAUTHORML(x,y), $\neg$CANNOT(x,y)	0.80

Fig. 1. An ER model with weighted constraints

each constraint with a weight that indicates the confidence on the robustness of semantic knowledge it represents.

Recently, a good number of works have studied constraints in ER [2,8,11,12]. However, little work has been carried out on how to efficiently and effectively deal with weighted constraints. Our work was motivated by Dedupalog [2], a declarative framework for resolving entities using constraints without weights. The authors of Dedupalog defined the ER clustering as the correlation clustering problem over clustering graphs [4], and discussed computational difficulties of adding weights to constraints. Nevertheless, their study has some limitations: (1) Their clustering graphs are complete, which is often not the case in real-life applications, and the inference rules based on complete graphs for solving clustering conflicts do not scale well to very large data sets. (2) Their constraints are unweighted, which makes it difficult to fine-tune the confidences on constraints in order to resolve knowledge uncertainty in reality. For instance, when two records u and v are both identified as a match by a constraint r_1 and as a non-match by a constraint r_2, how can we decide whether (u,v) is a match or non-match? If r_1 and r_2 have the weights 0.9 and 0.6 respectively, one may decide that (u,v) should be a match since r_1 has a higher weight than r_2. Therefore, weights can provide us with useful insights concerning ambiguity or conflicting information.

In Figure 1 we present: (a) an example database schema that consists of three relation schemas: PAPER, AUTHOR and VENUE, (b) some views that are defined using relational algebra [1], and (c) several weighted constraints $r_1 - r_8$ in an ER model built on this database schema. We adopt the notation in [2] to use PAPER*, VENUE* and AUTHOR* as the equivalence relations storing the records that are resolved for entities of the entity types PAPER, VENUE and AUTHOR, respectively. Among these constraints, r_4 and r_7 are negative; while the others are positive. Each constraint is associated with a weight in the range [0,1]. More specifically, $r_1 - r_3$ describe that two paper records likely refer to the same paper if they have similar titles (with varied similarity thresholds) plus some additional conditions required in r_2 and r_3, i.e., they must have the same authors in r_2 or the same venue in r_3; r_4 describes that two paper records do not likely refer to the same paper if their pages are not similar; r_5 is defined across two different entity types *author* and *paper*, describing that if two paper records refer to the same paper, then their venue records must also refer to the same venue; r_6 describes that two venue records likely refer to the same venue if they have similar names; r_7 describes that two author records do not likely refer to the same author if they appear in the same paper but are placed at different orders.

In this framework, constraints can be defined broadly to include existing ER-related algorithms. For example, we may define a relation COAUTHORML to store the result returned by a machine-learning based ER algorithm, which resolves authors based on the similarity of their attributes and the number of same authors with whom they have co-authorships. We can also use constraints to capture how such results from ER algorithms are combined with other knowledge in resolving entities. r_8 in Figure 1 describes that the effects of COAUTHORML is restricted by another relation CANNOT that contains a set of cannot-be-matched authors identified by the domain expert.

Outline. The remainder of the paper is structured as follows. We introduce ER constraints and models in Section 2. Then we discuss how to learn constraints in Section 3, and how to use constraints in Section 4. Our experimental results are presented in Section 5. We present the related works in Section 6 and conclude the paper in Section 7.

2   ER Constraints and Models

A *database schema* $\mathbf{R}$ consists of a finite, non-empty set of relation schemas. A *relation schema* contains a set of attributes. $\mathbf{R}_e \subseteq \mathbf{R}$ is a set of equivalence relation schemas, each $R^* \in \mathbf{R}_e$ corresponding to an entity type $R \in \mathbf{R} - \mathbf{R}_e$. A *database instance* is a finite, non-empty set of relations, each having a finite set of records. Each record is *uniquely identifiable* by an identifier. A *cluster* c is a set of records, denoted by their identifiers in the form of $\langle k_1, \ldots, k_n \rangle$. We may also have a set $\mathbf{R}_v$ of views defined by relational algebra, SQL or other database languages [1] over $\mathbf{R}$.

ER Constraints. We specify constraints as rules using a declarative Datalog-style language [1]. There are two types of atoms: a *relation atom* $R(x_1, \ldots, x_n)$ for a n-ary $R \in \mathbf{R} \cup \mathbf{R}_v$, and a *similarity atom* $x \approx_a y$ for a fixed *similarity threshold* $a \in [0, 1]$, and two variables x and y. The higher the similarity threshold a is, the greater the similarity between x and y is required. For instance, $x \approx_1 y$ means that $x \approx_1 y$ is true if x and y are exactly the same, and it can also be referred as an *equality atom* $x = y$. A *negated atom* is an expression of the form $\neg A$ where A is an atom. A *rule* r has the form of

$$\texttt{head(r)} \leftarrow \texttt{body(r)},$$

where the *head* of r, denoted as $\texttt{head(r)}$, is a relation atom or a negated relation atom over $\mathbf{R}_e$, and the *body* of r, denoted as $\texttt{body(r)}$, is a conjunction of atoms and negated similarity atoms. We do not allow any negated relation atoms occurring in $\texttt{body(r)}$. This restriction is important for guaranteeing the termination of computation and improving efficiency of using rules. Each variable in $\texttt{head(r)}$ must also occur in $\texttt{body(r)}$ to ensure the safety of evaluation as in Datalog programs [1]. A rule r is *positive* if $\texttt{head(r)}$ is an atom, and is *negative* if $\texttt{head(r)}$ is a negated atom.

Let I be a database instance and r be a rule over the same database schema, and ζ be a valuation over variables of $\texttt{body(r)}$ in I. Then a relation atom $R(x_1, \ldots, x_n)$ is true under ζ if $(\zeta(x_1), \ldots, \zeta(x_n))$ is a record in I, and a similarity atom $x \approx_a y$ is true if $f_{sim}(\zeta(x), \zeta(y)) > a$, where f_{sim} is a string-similarly function defined for measuring the similarity of x and y, otherwise false. A negated atom $\neg A$ is true under ζ if A is false under ζ. The body of r is true under ζ if the conjunction of atoms and negated atoms in $\texttt{body(r)}$ are true under ζ. The *interpretation* of r with $\texttt{head(r)} = R^*(x, y)$ is a set of *matches* of R^*:

$$\{(u, v)^= \,|\, u = \zeta(x), v = \zeta(y), \zeta \text{ is a valuation over variables}$$
$$\text{of } \texttt{body(r)} \text{ in } I, \text{and } \texttt{body(r)} \text{ is true under } \zeta\}.$$

The *interpretation* of r with $\texttt{head(r)} = \neg R^*(x, y)$ is a set of *non-matches* of R^*:

$$\{(u, v)^{\neq} \,|\, u = \zeta(x), v = \zeta(y), \zeta \text{ is a valuation over variables}$$
$$\text{of } \texttt{body(r)} \text{ in } I, \text{and } \texttt{body(r)} \text{ is true under } \zeta\}.$$

We use $r(I)$ to denote the matches or non-matches identified by r over I, and $r(I)$ is assumed to be symmetric, e.g., if $(u, v)^= \in r(I)$, then $(v, u)^= \in r(I)$.

Different similarity functions [9] may be used for evaluating different similarity atoms, depending on specific data characteristics of attributes, e.g., using edit distance for computing the similarity of names in AUTHOR, and Jaccard for computing the similarity of pages in PAPER. This kind of flexibility is important for developing a built-in library that can incorporate various ER-related algorithms in a declarative manner.

In practice, not all constraints are equally important. For example, a constraint r_a stating that two paper records refer to the same paper if they have same title and same authors is usually more "robust" than a constraint r_b stating that two paper records refer to the same paper if they have same title. Therefore, we associate each rule r with a real-valued weight $\omega(r) \in [0, 1]$ to express the confidence of its robustness. A rule with $\omega(r) = 1$ is a *hard rule* that must be satisfied and a rule with $\omega(r) < 1$ is a *soft rule* that can be violated.

ER Models. Given a database schema $\mathbf{R}$, an *ER model* $\mathcal{M}$ over $\mathbf{R}$ consists of a non-empty finite set of weighted rules. Let $\mathcal{M}(I) = \bigcup_{r \in \mathcal{M}} r(I)$, $\mathcal{M}_s^+$ and $\mathcal{M}_s^-$ be the sets of positive and negative soft rules in $\mathcal{M}$ which yield $(u, v)^=$ and $(u, v)^{\neq}$ of the same entity type R^*, respectively, $w_1 = \frac{1}{|\mathcal{M}_s^+|} \sum_{r \in \mathcal{M}_s^+} \omega(r)$ and $w_2 = \frac{1}{|\mathcal{M}_s^-|} \sum_{r \in \mathcal{M}_s^-} \omega(r)$. Then the *matching result* $\langle \pi, \ell, \tau \rangle$ of applying $\mathcal{M}$ over I is a set π of matches and non-matches, each $e \in \pi$ being associated with a weight $\ell(e)$ and an entity type $\tau(e) = R^*$. That is, for each $(u, v) \in \pi$, (1) if $(u, v) \in r(I)$ (resp. $(u, v)^{\neq} \in r(I)$) for some hard rule $r \in \mathcal{M}$ with $\mathtt{head(r)} = R^*(x, y)$, then $\ell(u, v) = \top$ (resp. $\ell(u, v) = \bot$); (2) otherwise $\ell(u, v)$ is defined as $w_1 - w_2$ if both $|\mathcal{M}_s^+| > 0$ and $|\mathcal{M}_s^-| > 0$, as w_1 if $|\mathcal{M}_s^+| > 0$ and $|\mathcal{M}_s^-| = 0$, as $-w_2$ if $|\mathcal{M}_s^+| = 0$ and $|\mathcal{M}_s^-| > 0$, or as 0 if both $|\mathcal{M}_s^+| = 0$ and $|\mathcal{M}_s^-| = 0$.

Given a matching result, a set of clusters can be generated by applying clustering algorithms such that each cluster corresponds to a distinct entity. With such clusters, each equivalence relation is updated, i.e., $(k_1, k_2) \in R^*$ iff two records k_1 and k_2 of R are in the same cluster. The updates on these equivalence relations iteratively lead to discovering new matches or non-matches, and as a result the clustering result is expanded. This process repeats until no more updates on the equivalent relations. To ensure the termination of computation, the rules in an ER model are interpreted in the similar way as a Datalog program under the inflationary semantics [1].

3   Learning Constraints

In this section we discuss how to learn an ER model by leveraging domain knowledge into the "best-fit" constraints for specific applications.

In practice, constraints are commonly used to capture domain knowledge, e.g., "if two paper records have similar titles, then they likely represent the same paper" can be expressed as:

$$g_1 : \quad \mathrm{PAPER}^*(x, y) \leftarrow \mathrm{TITLE}(x, t), \mathrm{TITLE}(y, t'), t \approx_\lambda t',$$

where λ is a threshold variable indicating that the similarity between t and t' is undefined yet. We call such a constraint, in which the threshold of each similarity atom (if any) is undefined, a *ground rule*. Intuitively, a ground rule provides the generic description for *a family of rules* that capture the same semantic relationship among elements of a database but may differ in the interpretation of similarity atoms, e.g., the interpretation of "to which extent the titles of two publication records are considered as being similar". Let $\boldsymbol{\lambda}$ denote a vector of threshold variables, and $g[\boldsymbol{\lambda} \mapsto \boldsymbol{a}]$ denote a rule obtained by substituting $\boldsymbol{\lambda}$ in a ground rule g with the vector $\boldsymbol{a}$ of real numbers in the range $[0.1]$. The following

Table 1. Metrics α and β

	POSITIVE RULES	NEGATIVE RULES
α	$\dfrac{tp}{tp+fp}$	$\dfrac{tn}{tn+fn}$
β	$\dfrac{tp}{tp+fn}$	$\dfrac{tn}{fp+tn}$

two rules $g_1[\lambda \mapsto 0.8]$ and $g_1[\lambda \mapsto 0.7]$ are in the same family because they both associate with g_1:

- $\text{PAPER}^*(x,y) \leftarrow \text{TITLE}(x,t), \text{TITLE}(y,t'), t \approx_{0.8} t'$;
- $\text{PAPER}^*(x,y) \leftarrow \text{TITLE}(x,t), \text{TITLE}(y,t'), t \approx_{0.7} t'$.

A *ground ER model* is constituted by a set of ground rules provided by domain experts. As different applications may have different data requirements, we need to refine a ground ER model into the one that is the most effective for solving the ER problem in a specific application. This requires a learning mechanism to be incorporated into the specification process of an ER model. There are many aspects to consider when developing an effective learning mechanism for ER models, including: (1) the availability of training data; (2) the suitability of learning measure; (3) the complexity of learning model. In our study, we assume that the user provides some (positive and negative) labeled examples as training data, and identifies a number of potential options for similarity functions and thresholds. Then an ER model is learned by finding the optimal combination of similarity functions and thresholds for each ground rule. In order to measure the goodness of a rule, we develop learning metrics for both positive and negative rules. Conceptually, positive rules should be measured by their "positive effects", while negative rules be measured by their "negative effects". We thus use two different criteria for the metrics α and β shown in Table 1, where tp, fp, tn and fn refer to true positives, false positives, true negatives and false negatives, respectively. These metrics are widely used in binary classification [3]. We define an objective function ξ for learning constraints, which takes two input parameters: α and β, and returns a real number. Then, given a ground rule g, a soft rule $g[\lambda \mapsto a]$ can be learned with the "best" similarity thresholds a in terms of

$$max_\lambda \xi(\alpha, \beta) \text{ subject to } \alpha \geq \alpha_{min} \text{ and } \beta \geq \beta_{min},$$

where λ is the vector of threshold variables in g, and α_{min} and β_{min} are the minimal requirements on the input parameters. Note that, not every ground rule can lead to a meaningful rule, e.g., when all rules in the family fail to meet a minimum precision requirement due to dirty data or incomplete information. In such cases, although ground rules capture domain knowledge, they are not suitable for being used in specific applications. The weights of rules can be learned over training data using various methods [16,18]. Nonetheless, a simple but effective way is to define the weight of a rule as a linear function of the learning measure ξ. Once such a function for learning weights is determined, hard rules can be learned in terms of

$$max_\lambda \xi(\alpha, \beta) \text{ subject to } w \geq 1 - \varepsilon,$$

where ε is a fault tolerant rate, e.g., 0.005, and w is the weight of the rule $g[\lambda \mapsto a]$. To improve the efficiency of learning, the learning measure ξ should also process certain properties:

- ξ is *deterministic*, i.e., given a pair α and β, ξ always returns the same value;
- ξ is *monotonic* on α and β, i.e., $\xi(\alpha_1, \beta_1) \leq \xi(\alpha_2, \beta_2)$ iff $\alpha_1 \leq \alpha_2$ and $\beta_1 \leq \beta_2$.

4   Using Constraints

Now we discuss how to use weighted constraints to deal with the ER problem.

Given an ER model $\mathcal{M}$, and the matching result $\langle \pi, \ell, \tau \rangle$ of applying $\mathcal{M}$ over a database instance I, we use ER graphs to describe the correlation between records. An *ER graph* w.r.t. $\mathcal{M}$ and I is a triple $G = (V, E, \ell)$ consisting of a set V of vertices, each representing a record, and a set $E \subseteq V \times V$ of edges, each (u, v) being assigned a label $\ell(u, v)$. Moreover, E is the subset of π, which contains all the matches and non-matches of the same entity type, and accordingly, one ER graph corresponds to one entity type. There are two types of edges in an ER graph: a *soft edge* labelled by a real number in $(-1, 1)$, or a *hard edge* labelled by one of the symbols $\{\top, \bot\}$. Intuitively, a soft edge (u, v) represents a match or non-match, and $\ell(u, v)$ indicates the degree of confidence, i.e., if $\ell(u, v)$ is closer to 1, then (u, v) is likely a match; and if $\ell(u, v)$ is closer to -1, then (u, v) is likely a non-match.

A clustering over $G = (V, E, \ell)$, denoted as C_G, is a partition of V such that $C_G = \{V_1, \ldots, V_n\}$, where each $V_i(i \in [1, n])$ is a non-empty cluster, $\bigcup_{1 \leq i \leq n} V_i = V$ and $\bigcup_{1 \leq i \neq j \leq n} V_i \cap V_j = \emptyset$. Given an ER graph G, there are many possible clusterings over G. We are, however, only interested in valid ones. That is, a clustering C_G is *valid* iff each edge in G labelled by $\top$ is in the same cluster, and each edge in G labelled by $\bot$ are across two different clusters. As a convention, we use $C(v)$ to denote the cluster that a vertex v belongs to.

Given an ER graph $G = (V, E, \ell)$, the *ER clustering problem* over G is to find a valid clustering over G such that vertices are grouped into one cluster iff their records represent the same real-world entity. The number of clusters in such a clustering is unknown, i.e., do not know how many real-world entities the records represent.

A natural way of handling the ER clustering problem is to use existing techniques for *correlation clustering* [2,4,10]. In such cases, clustering objectives are often defined as minimizing disagreements, such as the weights of negative edges within clusters and the weights of positive edges across clusters, or maximizing agreements, which can be defined analogously. It is known that correlation clustering is a NP-hard problem [4]. For general weighted graphs, Demaine et al. and Charikar et al. [7,10] give an O(log n)-approximation algorithm w.r.t. minimizing disagreements, and Charikar et al. [7] also give a factor 0.7664 approximation w.r.t. maximizing agreements. Since our ER graphs generalize general weighted graphs by allowing negative weights and hard edges, the ER clustering problem from the viewpoint of correlation clustering is also NP-hard and finding an optimal clustering is computationally intractable.

Therefore, instead of minimizing disagreements or maximizing agreements, we propose two methods for finding ER clusterings in an ER graph: One is a variant of *pairwise nearest neighbour* (PNN) [13], and the other is called *relative constrained neighbour* (RCN). The key idea in PNN is that, given a set C of initial clusters, the pair of clusters in C which has the strongest positive evidence should be merged, unless this merge is forbidden by some hard negative evidence. Each merge would change the clustering structure. Iteratively, the next pair of two clusters that have the strongest positive evidence is picked and merged if possible, until the total weight of edges within clusters is maximized under certain specified conditions. The key idea in RCN is to consider the weights of relative constraints [15]. Given two vertices v_1 and v_2 that needs to be split. Then for another vertex v_3, if $\ell(v_3, v_1) > \ell(v_3, v_2)$, then v_3 should be clustered with v_1; otherwise v_3 should be clustered with v_2. If there is no edge between v_3 and v_2, then $\ell(v_3, v_1) > \ell(v_3, v_2)$ holds for any $\ell(v_3, v_1) > 0$. If both $\ell(v_3, v_1)$ and $\ell(v_3, v_2)$ do not exist, then v_3 can be clustered with either v_1 or v_2.

Both the PNN and RCN algorithms have three parts: Main, CutHardEdges and CutSoftEdges. The main part Main is depicted in Algorithm 1, which takes as input an ER graph G and produces as output a clustering over G. If there is no conflict in the ER graph, only Step (1) is needed, i.e., generating a cluster for each connected component in the subgraph containing only soft positive edges. Step (2) eliminates the conflicts involved hard negative edges using CutHardEdges, while Step (3) deals with the conflicts involved soft negative edges using CutSoftEdges. In a nutshell, the PNN and RCN algorithms only differ in the CutHardEdges part, i.e., the removal of hard negative edges, as presented in Algorithm 2 and Algorithm 3. To keep the discussion simple, we assume (w.l.o.g.) that ER graphs have no hard positive edges. This is because two vertices connected by a hard positive edge can always be merged into the same cluster and thus treated as one vertex.

Let $\ell(v, u) \in (-1, 1)$. We consider that $\ell(v, u) + \perp = \perp$ and $\ell(v, u) + \top = \top$, i.e., $\top$ is higher than any real number, and $\perp$ is lower than any real number. We use $G^{>a}$ to denote the subgraph of G that contains vertices of G but only soft edges whose labels are greater than a, and $G[c]$ to denote the subgraph of G that is induced by vertices in a cluster c.

Input: an ER graph $G = (V, E, \ell)$
Output: a clustering C over G

(1) Take the subgraph $G^{>0}$ and generate one cluster c_g containing all vertices in each connected component $g \in G^{>0}$, i.e., $C := \{c_g | g \in G^{>0}\}$;
(2) Check $G[c_g]$ for each $c_g \in C$ iteratively, and if there is an edge (u, v) with $\ell(u, v) = \perp$:
    (a) $C_0 := \{\langle v \rangle | v \text{ is a vertex in } G[c_g]\}$.
    (b) $C_{tmp} := \text{CutHardEdges}(G[c_g], C_0)$;
    (c) $C := C \cup C_{tmp} - \{c_g\}$.
(3) Check $G[c_g]$ for each $c_g \in C$, and if there is an edge (u, v) with $\ell(u, v) < 0$:
    (a) $C_{tmp} := \text{CutSoftEdges}(G[c_g])$;
    (b) $C := C \cup C_{tmp} - \{c_g\}$.
(4) Return C.

Algorithm 1. Main

Input: an ER graph $G = (V, E, \ell)$, and an initial
        clustering C_0 over G.
Output: a clustering C over G

(1) $C := C_0$.
(2) Perform the following iteratively until no more changes can be made on C:
    (2.1) Find two clusters $c_1, c_2 \in C$ s.t. $\sum_{u \in c_1, v \in c_2} \ell(u, v) > 0$ and is maximal among all cluster pairs in C;
    (2.2) $C := C - \{c_1, c_2\} \cup \{c_1 \cup c_2\}$.
(3) Return C.

Algorithm 2. CutHardEdges for PNN

Input: an ER graph $G = (V, E, \ell)$, and an initial
        clustering C_0 over G.
Output: a clustering C over G

(1) $C := C_0$.
(2) Select a hard negative edge $(u, v) \in E$ where $C_0(u) \neq C_0(v)$ (It can be obtained from Step (2) in Algorithm 1 in implementation).
(3) Compare $W_u := \sum_{u' \in c, v' \in C_0(u)} \ell(u', v')$ and $W_v := \sum_{u' \in c, v' \in C_0(v)} \ell(u', v')$ for each cluster c in $C_0 - \{C_0(u), C_0(v)\}$, and iteratively do the following:
    (2.1) If $W_u \geq W_v$, $C := C - \{c, C(u)\} \cup \{c \cup C(u)\}$;
    (2.2) Otherwise, $C := C - \{c, C(v)\} \cup \{c \cup C(v)\}$.
(3) Return C.

Algorithm 3. CutHardEdges for RCN

Input: an ER graph $G = (V, E, \ell)$
Output: a clustering C over G

(1) Sort the soft negative edges in G s.t. $L := [(u_1, v_1), \ldots, (u_n, v_n)]$ where $|\ell(u_i, v_i)| \geq |\ell(u_{(i+1)}, v_{(i+1)})|$.
(2) $i := 1$ and $C := \{V\}$.
(3) Do the following iteratively until $i = n + 1$:
    (3.1) If there is $c \in C$ s.t. $(u_i, v_i) \in G[c]$, then
       – Take $G^{>a}[c]$ where $a := |\ell(u_i, v_i)|$, and $C_0 := \{c_g | g \in G^{>a}[c]\}$;
       – If $C_0(u_i) \neq C_0(v_i)$, then harden $\ell(u_i, v_i)$ to be $\perp$ in $G[c]$, and $C :=$
          $C \cup \texttt{CutHardEdges}(G[c], C_0) - \{c\}$.
    (3.2) $i := i + 1$.
(4) Return C.

Algorithm 4. CutSoftEdges

5 Experimental Study

We evaluated the efficiency and effectiveness of weighted constraints for ER in three aspects. (1) *ER models*: how effectively can constraints and their weights be learned from domain knowledge for an ER model? (2) *ER clustering*: how useful can weighted constraints be for improving the ER quality, in particular, comparing with previous works on unweighted constraints such as Dedupalog [2]? (3) *ER scalability*: how scalable can our method be over large data sets?

We used two data sets in our experiments. The first one is Cora[1], which is publicly available together with its "gold standard". There are three tables in Cora, corresponding to three entity types: (a) PAPER with 1,878 records, (b) AUTHOR with 4,571 records, and (c) VENUE with 1,536 records. The second data set was taken from Scopus[2], which contains 10,784 publication records and 47,333 author records. We manually established the "gold standard" for 4,865 publication records and 19,527 author records with the help of domain experts. Our experiments were all performed on a Windows 8 (64 bit) machine with an Intel Core i5-3470 at 3.2 Ghz and 8GB RAM. The data sets and compuation results were stored in a PostgreSQL 9.2 database. We implemented ER constraints and algorithms using Java with JDBC access to the PostgeSQL database.

ER Models. Our first set of experiments evaluated how effectively weighted constraints can be learned from domain knowledge over the Cora and Scopus data sets. We chose $\xi(\alpha, \beta) = (2 * \alpha * \beta)/(\alpha + \beta)$ as the learning measure for both data sets (i.e., α is precision, β is recall and $\xi(\alpha, \beta)$ is F1-measure for positive rules) and Jaccard for measuring string similarity. In the following, we use r_g to refer to the rule learned from a ground rule g. Due to the space limitation, we will focus on discussing the ER model of the Cora data set, while omitting the one of the Scopus data set.

---

[1] http://www.cs.umass.edu/~mccallum/
[2] http://www.scopus.com/home.url

Table 2. Ground positive rules in Cora

g_1: $\text{PAPER}^*(x,y) \leftarrow \text{TITLE}(x,t), \text{TITLE}(y,t'), t \approx_{\lambda_1} t'$

No	λ_1	Precision	Recall	F1-measure
1	0.8	0.879	0.815	0.846
2	0.7	0.818	0.926	0.869
3	0.6	0.725	0.985	0.835

g_2: $\text{PAPER}^*(x,y) \leftarrow \text{TITLE}(x,t), \text{TITLE}(y,t'), t \approx_{\lambda_1} t', \text{AUTHORS}(x,z), \text{AUTHORS}(y,z'), z \approx_{\lambda_2} z', \text{YEAR}(x,u), \text{YEAR}(y,u'), u \approx_{\lambda_3} u'$

No	λ_1	λ_2	λ_3	Precision	Recall	F1-measure
1	0.5	0.5	0.5	0.990	0.640	0.778
2	0.4	0.4	0.4	0.991	0.672	0.801
3	0.3	0.3	0.3	0.978	0.677	0.800

g_3: $\text{PAPER}^*(x,y) \leftarrow \text{TITLE}(x,t), \text{TITLE}(y,t'), t \approx_{\lambda_1} t', \text{AUTHORS}(x,z), \text{AUTHORS}(y,z'), z \approx_{\lambda_2} z'$

No	λ_1	λ_2	Precision	Recall	F1-measure
1	0.7	0.7	0.849	0.741	0.792
2	0.6	0.6	0.773	0.916	0.838
3	0.5	0.5	0.711	0.944	0.811

g_4: $\text{PAPER}^*(x,y) \leftarrow \text{AUTHORS}(x,z), \text{AUTHORS}(y,z'), z \approx_{\lambda_1} z', \text{PAGES}(x,t), \text{PAGES}(y,t'), t \approx_{\lambda_2} t'$

No	λ_1	λ_2	Precision	Recall	F1-measure
1	0.6	0.6	1.000	0.650	0.788
2	0.5	0.5	1.000	0.663	0.797
3	0.4	0.4	0.965	0.700	0.811

g_5: $\text{AUTHOR}^*(x,y) \leftarrow \text{ANAME}(x,n), \text{ANAME}(y,n'), n \approx_{\lambda_1} n'$

No	λ_1	Precision	Recall	F1-measure
1	0.8	0.947	0.292	0.447
2	0.7	0.806	0.321	0.459
3	0.6	0.441	0.476	0.458

g_6: $\text{VENUE}^*(x,y) \leftarrow \text{VNAME}(x,n), \text{VNAME}(y,n'), n \approx_{\lambda_1} n'$

No	λ_1	Precision	Recall	F1-measure
1	0.8	0.339	0.720	0.461
2	0.7	0.336	0.767	0.468
3	0.6	0.281	0.786	0.414

Table 3. Ground negative rules in Cora

Data sets	Constraints	Weights
Cora	g_7: $\neg\text{PAPER}^*(x,y) \leftarrow \text{PAGES}(x,z), \text{PAGES}(y,z'), \neg z \approx_{0.2} z'$	1.00
Cora	g_8: $\neg\text{PAPER}^*(x,y) \leftarrow \text{TITLE}(x,t), \text{TITLE}(y,t'), \text{DIFFERENT}(t,t')$	1.00
Cora	g_9: $\neg\text{PAPER}^*(x,y) \leftarrow \text{HASTECH}(x,t), \neg\text{HASTECH}(y,t')$	0.96
Cora	g_{10}: $\neg\text{PAPER}^*(x,y) \leftarrow \text{HASJOURNAL}(x,t), \neg\text{HASJOURNAL}(y,t')$	0.98

We have studied 10 ground rules $g_1 - g_{10}$ over the Cora data set for three entity types: (1) $g_1 - g_6$ are positive as shown in Table 2; (2) $g_7 - g_{10}$ are negative as shown in Table 3. In general, $g_1 - g_4$ describe that two paper records are likely to represent the same paper if some of their attributes are similar, and g_5 (resp. g_6) describes that two author (resp. venue) records are likely to represent the same author (resp. venue) if their names are similar. We conducted a range of tests to learn $r_{g_1} - r_{g_6}$ from $g_1 - g_6$ over training data. Table 2 presents some of these tests. In accordance to our learning measure $\xi(\alpha, \beta)$, the "most suitable" set of similarity thresholds for each ground rule is highlighted in Table 2. For instance, the similarity threshold for title in r_{g_1} is 0.8 because it leads to the highest F1 score. The weight of each rule is determined by its precision, representing how accurately the rule can be used for ER. That is, $\omega(r_{g_1}) = 0.82$, $\omega(r_{g_2}) = 0.99$, $\omega(r_{g_3}) = 0.77$, $\omega(r_{g_4}) = 1$, $\omega(r_{g_5}) = 0.81$ and $\omega(r_{g_6}) = 0.34$. r_{g_4} is a hard rule, while the others are soft. For negative ground rules g_7 - g_{10}, they describe that two paper records likely refer to different papers in the following cases: (1) the pages of two paper records are not similar; (2) the titles of two paper records are different; (3) one paper record is a technical report but the other is not; (3) one paper record is a journal but the other is not. Among these rules, only g_7 contains a similarity threshold variable λ_1, which, according to our learning measure, is selected to be 0.2 in r_{g_7}. The weights of these rules are: $\omega(r_{g_7}) = 1$, $\omega(r_{g_8}) = 1$, $\omega(r_{g_9}) = 0.96$ and $\omega(r_{g_{10}}) = 0.98$. We consider r_{g_6} and r_{g_7} as hard rules.

ER Clustering. We compared the quality of ER over our data sets when using three different matching and clustering methods: (1) Dedupalog [2], (2) ER-PNN, and (3) ER-RCN, where ER-PNN and ER-RCN refer to the matching and clustering methods we discussed in Section 4, and they only differ in the clustering algorithms, i.e, we used the PNN algorithm in ER-PNN, and the RCN algorithm in ER-RCN.

All the experiments used the same set of rules. However, rather than using weights for soft rules, Dedupalog distinguished two kinds of soft rules: one is called *soft-incomplete rules* that provide only positive information, and the other is called *soft-complete rules* that require a cost penalty when being violated. In our experiments using the Dedupalog method, the weighted soft rules are treated as soft-incomplete rules, and conflicts were solved by using their voting, election and hardening algorithms (see Figure 4-6 in [2]). In this setting, Table 4 shows that our methods ER-RCN and ER-PNN performed equally well, and also considerably outperformed the Dedupalog method over the Cora data set. A detailed analysis over the Cora data set reveals that using the weighted rules leads to a significant improvement in precision. This is because negative rules can eliminate many false positives without increasing the number of false negatives in the matching stage. Then the clustering process resolved potential clustering conflicts, which resulted in a higher precision for both ER-RCN and ER-PNN, but a small drop in recall for ER-PNN. Our analysis over the Scopus data set indicated a similar trend. Nevertheless, because the negative rule in Scopus does not produce many conflicts with the positive rules, the effects are not obvious in this case.

ER Scalability. In order to evaluate the scalability of our methods and understand how well they compare with Dedupalog, we created four test data sets of different sizes which contain 10%, 40%, 60%, 80% and 100% of the records in the full scopus data set. Then we run the same set of rules, as previously described, three times for each test data set, and took their average runtime used in the clustering process. Table 5 presents the data characteristics of the test data sets used in our scalability tests. Figure

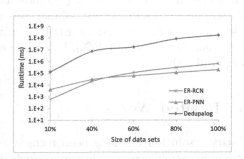

Fig. 2. Scalability tests over Scopus

Table 4. ER quality comparison

Types of rules	Methods	Cora			Scopus		
		Precision	Recall	F1-measure	Precision	Recall	F1-measure
Positive	All	0.7324	0.9923	0.8428	0.9265	0.9195	0.9230
Both positive and negative	Dedupalog	0.7921	0.9845	0.8779	0.9266	0.9196	0.9231
	ER-RCN	0.9752	0.9685	0.9719	0.9271	0.9192	0.9231
	ER-PNN	0.9749	0.9660	0.9705	0.9271	0.9193	0.9232

Table 5. Data characteristics in scalability tests

Size of data sets	#Records	#Matches and non-matches	#Clusters			#Records in maximal clusters
			Dedupalog	ER-RCN	ER-PNN	
10%	4,733	4,294	3,412	3,417	3,417	18
40%	19,527	76,984	8,804	8,813	8,810	101
60%	28,400	142,865	12,973	13,030	13,018	119
80%	37,866	250,576	15,742	15,819	15,804	213
100%	47,333	387,277	18,424	18,524	18,502	264

2.(c) presents our experimental results, in which the vertical axis indicates the runtime at a logarithmic scale, and the horizontal axis indicates the sizes of data sets. From the figure, we can see that the time required by the Dedupalog method is much longer than our methods, i.e., for the test data set at 10%, the average runtime taken by ER-PNN is roughly 3% of the average runtime taken by the Dedupalog method, while for the test data set at 100%, the average runtime taken by ER-PNN is roughly 0.1% of the average runtime taken by the Dedupalog method. The reason for such inefficiency is mainly because Dedupalog treats clustering graphs as being complete, and the edge label (positive or negative) between every two vertices thus needs to be considered when solving conflicts. As a result, the computation may become very expensive, especially when a data set has large clusters. Note that, three optimization strategies were discussed for Dedupalog [2], and one of them can make Dedupalog being executed much more efficiently in the case of having no hard rules. In order to keep the consistency of comparison, we implement all methods natively except for using implicit representation of negative edges. ER-RCN also outperforms ER-PNN over smaller data sets (i.e., 10% and 40%), but falls behind over larger data sets (i.e., 60%, 80% and 100%).

6   Related Works

Entity resolution has long been recognized as being significant to many areas of computer science [9]. Numerous works have been done in this area from different perspectives. Although many works did not discuss constraints explicitly in their ER models, constraints are implicitly used in certain form or the other. For example, the ER models in [5,6,11,12,22] can be specified by a set of soft positive constraints in our framework, together with some negative constraints to reduce certain errors. In previous studies, a variety of constraints have also been explicitly investigated in relating to ER [2,8,11,12,14,19,20]. In particular, constraints may be pairwise or groupwise. Pairwise constraints [12,11] are concerned about matching or non-matching relationships between two records, while groupwise constraints focus on aggregate properties of a group of records [8,19]. Some ER techniques handle both of them [17]. In this paper our focus is on pairwise constraints.

Dedupalog developed by Arasu et al. [2] is the closest work to ours. Distinguished from Dedupalog, our framework has the following features: (1) We generalized clustering graphs used in Dedupalog to be graphs that may not be complete, and provided efficient clustering algorithms over such graphs; (2) We considered weights for constraints, which can not only improve the ER quality but also provide a nice tool to solve conflicts among positive and negative constraints; (3) We proposed to use pairwise nearest neighbour (PNN) and relative constrained neighbour (RCN) for handling the ER clustering, departing from the traditional correlation clustering viewpoint.

7   Conclusions

In this paper, we studied weighted constraints in relation to the questions of how to properly specify and how to efficiently use weighted constraints for performing ER tasks. We developed a learning mechanism to "guide" the learning of constraints and their weights from domain knowledge. Since our framework supports both positive and negative constraints, conflicts may arise in the ER process. Our experiments showed that adding weights into constraints is helpful for conducting ER tasks, and particularly, weights allow us to leverage domain knowledge to build efficient and effective algorithms for resolving conflicts. For the future work, we plan to compare our method with other existing ER techniques in terms of quality and scalability.

References

1. Abiteboul, S., Hull, R., Vianu, V.: Foundations of Databases. Addison-Wesley (1995)
2. Arasu, A., Ré, C., Suciu, D.: Large-scale deduplication with constraints using dedupalog. In: ICDE, pp. 952–963 (2009)
3. Baldi, P., Brunak, S., Chauvin, Y., Andersen, C.A., Nielsen, H.: Assessing the accuracy of prediction algorithms for classification: an overview. Bioinformatics 16(5), 412–424 (2000)
4. Bansal, N., Blum, A., Chawla, S.: Correlation clustering. Machine Learning 56(1-3), 89–113 (2004)
5. Bhattacharya, I., Getoor, L.: Relational clustering for multi-type entity resolution. In: MRDM, pp. 3–12 (2005)
6. Bhattacharya, I., Getoor, L.: Collective entity resolution in relational data. TKDD 1(1), 5 (2007)

7. Charikar, M., Guruswami, V., Wirth, A.: Clustering with qualitative information. Journal of Computer and System Sciences 71(3), 360–383 (2005)
8. Chaudhuri, S., Das Sarma, A., Ganti, V., Kaushik, R.: Leveraging aggregate constraints for deduplication. In: SIGMOD, pp. 437–448 (2007)
9. Christen, P.: Data Matching. Springer (2012)
10. Demaine, E.D., Emanuel, D., Fiat, A., Immorlica, N.: Correlation clustering in general weighted graphs. TCS 361(2), 172–187 (2006)
11. Doan, A., Lu, Y., Lee, Y., Han, J.: Profile-based object matching for information integration. Intelligent Systems 18(5), 54–59 (2003)
12. Dong, X., Halevy, A., Madhavan, J.: Reference reconciliation in complex information spaces. In: ACM SIGMOD, pp. 85–96 (2005)
13. Equitz, W.H.: A new vector quantization clustering algorithm. IEEE Trans. Acoustics, Speech and Signal Processing 37(10), 1568–1575 (1989)
14. Lee, T., Wang, Z., Wang, H., Hwang, S.-W.: Web scale taxonomy cleansing. PVLDB 4(12) (2011)
15. Liu, E.Y., Zhang, Z., Wang, W.: Clustering with relative constraints. In: KDD, pp. 947–955 (2011)
16. Lowd, D., Domingos, P.: Efficient weight learning for markov logic networks. In: Kok, J.N., Koronacki, J., Lopez de Mantaras, R., Matwin, S., Mladenič, D., Skowron, A. (eds.) PKDD 2007. LNCS (LNAI), vol. 4702, pp. 200–211. Springer, Heidelberg (2007)
17. Shen, W., Li, X., Doan, A.: Constraint-based entity matching. In: AAAI, pp. 862–867 (2005)
18. Singla, P., Domingos, P.: Discriminative training of Markov logic networks. In: AAAI, pp. 868–873 (2005)
19. Tung, A.K., Han, J., Lakshmanan, L.V., Ng, R.T.: Constraint-based clustering in large databases. In: ICDT, pp. 405–419 (2001)
20. Wagstaff, K., Cardie, C.: Clustering with instance-level constraints. In: ICML (2000)
21. Wang, F., Wang, H., Li, J., Gao, H.: Graph-based reference table construction to facilitate entity matching. Journal of Systems and Software (2013)
22. Whang, S.E., Benjelloun, O., Garcia-Molina, H.: Generic entity resolution with negative rules. The VLDB Journal 18(6), 1261–1277 (2009)

Analogical Prediction of Null Values:
The Numerical Attribute Case

William Correa Beltran, Hélène Jaudoin, and Olivier Pivert

University of Rennes 1 – Irisa
Technopole Anticipa 22305 Lannion Cedex France
william.correa_beltran@irisa.fr, {jaudoin,pivert}@enssat.fr

Abstract. This paper presents a novel approach to the prediction of
null values in relational databases, based on the notion of analogical
proportion. We show in particular how an algorithm initially proposed in
a classification context can be adapted to this purpose. In this paper, we
focus on the situation where the relation considered may involve missing
values of a numerical type. The experimental results reported here, even
though preliminary, are encouraging as they show that the approach
yields a better precision than the classical nearest neighbors technique.

1 Introduction

In this paper, we propose a novel solution to a classical database problem that
consists in estimating null (in the sense "currently missing but existing") values in incomplete relational databases. Many approaches have been proposed to
tackle this issue, both in the database community and in the machine learning
community (based on functional dependencies [2,6], association rules [20,21,3,9],
decision trees [19], classification rules [10], clustering techniques [8], partial ordering comparison [1], etc). See also [13,14].

Here, we investigate a new idea, that comes from artificial intelligence and
consists in exploiting analogical proportions [17]. An analogical proportion is a
statement of the form "A is to B as C is to D". As emphasized in [18], analogy is
not a mere question of similarity between two objects (or situations) but rather
a matter of proportion or relation between objects. An analogical proportion
equates a relation between two objects with the relation between two other
objects. These relations can be considered as a symbolic counterpart to the case
where the ratio or the difference between two similar things is a matter of degree
or number. As such, an analogical proportion of the form "A is to B as C is to
D" poses an analogy of proportionality by (implicitly) stating that the way two
objects A and B, otherwise similar, differ is the same way as the two objects C
and D, which are similar in some respects, differ.

Up to now, the notion of analogical proportion has been studied mainly in
artificial intelligence, notably for classification purposes (see, e.g., [4]). Our objective is to exploit it in a database context in order to predict the null values
in a tuple t by finding quadruples of items (including t) that are linked by an

Y. Manolopoulos et al. (Eds.): ADBIS 2014, LNCS 8716, pp. 323–336, 2014.
© Springer International Publishing Switzerland 2014

analogical proportion. In [7], we presented a first approach that was limited to the prediction of Boolean attributes (case of a transactional database). Here, we deal with the situation where the attribute values to be estimated are of a numerical type.

The remainder of the paper is organized as follows. In Section 2, we provide a refresher on the notion of analogical proportion. Section 3 presents the general principle of the approach that we propose for estimating null values, inspired by the classification method proposed in [4,11]. Two views of the analogical prediction of numerical attributes are discussed in Section 4. Section 5 reports on an experimentation aimed at assessing the performances of the approach and at comparing its results with those obtained using other types of estimation techniques. Finally, Section 6 recalls the main contributions and outlines perspectives for future work.

2   Refresher on Analogical Proportions

The following presentation is mainly drawn from [12]. An analogical proportion is a statement of the form "A is to B as C is to D". This will be denoted by $(A : B :: C : D)$. In this particular form of analogy, the objects A, B, C, and D may correspond to descriptions of items under the form of objects such as sets, multisets, vectors, strings or trees. In the following, if the objects A, B, C, and D are tuples having n attribute values, i.e., $A = \langle a_1, \ldots, a_n \rangle, \ldots, D = \langle d_1, \ldots, d_n \rangle$, we shall say that A, B, C, and D are in analogical proportion if and only if for each component i an analogical proportion "a_i is to b_i as c_i is to d_i" holds.

We now have to specify what kind of relation an analogical proportion may mean. Intuitively speaking, we have to understand how to interpret "is to" and "as" in "A is to B as C is to D". A may be similar (or identical) to B in some respects (i.e., on some features), and differ in other respects. The way C differs from D should be the same as A differs from B, while C and D may be similar in some other respects, if we want the analogical proportion to hold. This view is enough for justifying three postulates that date back to Aristotle's time:

- (ID) $(A : B :: A : B)$
- (S) $(A : B :: C : D) \Leftrightarrow (C : D :: A : B)$
- (CP) $(A : B :: C : D) \Leftrightarrow (A : C :: B : D)$.

(ID) and (S) express reflexivity and symmetry for the comparison "as", while (CP) allows for a central permutation.

A *logical proportion* [16] is a particular type of Boolean expression $T(a, b, c, d)$ involving four variables a, b, c, d, whose truth values belong to $\mathbb{B} = \{0, 1\}$. It is made of the conjunction of two distinct equivalences, involving a conjunction of variables a, b on one side, and a conjunction of variables c, d on the other side of $\equiv$, where each variable may be negated. Analogical proportion is a special case of a logical proportion, and its expression is [12]: $(a\overline{b} \equiv c\overline{d}) \wedge (\overline{a}b \equiv \overline{c}d)$. The six valuations leading to truth value 1 are thus $(0, 0, 0, 0)$, $(1, 1, 1, 1)$, $(0, 0, 1, 1)$, $(1, 1, 0, 0)$, $(0, 1, 0, 1)$ and $(1, 0, 1, 0)$.

As noted in [18], the idea of proportion is closely related to the idea of extrapolation, i.e., to guess/compute a new value on the ground of existing values, which is precisely what we intend to do. In other words, if for whatever reason, it is assumed or known that a logical proportion holds between four binary elements, three being known, then one may try to infer the value of the fourth one.

3   General Principle of the Approach

3.1   Starting with a Classification-by-Analogy Algorithm

The approach we propose is inspired by a method of "classification by analogy" introduced in [4] where the authors describe an algorithm named FADANA. This algorithm uses a measure of *analogical dissimilarity* between four objects, which estimates how far these objects are from being in analogical proportion. Roughly speaking, the analogical dissimilarity ad between four Boolean values is the minimum number of bits that have to be switched to get a proper analogy. For instance $ad(1, 0, 1, 0) = 0$, $ad(1, 0, 1, 1) = 1$ and $ad(1, 0, 0, 1) = 2$. Thus, denoting by $\mathcal{A}$ the relation of analogical proportion, we have $\mathcal{A}(a, b, c, d) \Leftrightarrow ad(a, b, c, d) = 0$.

When dealing with four Boolean vectors in $\mathbb{B}^m$, we add the ad evaluations componentwise, which leads to an integer in the interval $[0, 2m]$. This principle has been used in [4] to implement a classification algorithm that takes as an input a training set S of classified items, a new item x to be classified, and an integer k. The algorithm proceeds as follows:

1. For every triple (a, b, c) of S^3, compute $ad(a, b, c, x)$.
2. Sort these n triples by increasing value of their ad when associated with x.
3. If the k-th triple has the integer value p for ad, then let k' be the greatest integer such that the k'-th triple has the value p.
4. Solve the k' analogical equations on the label of the class[1]. Take the winner of the k' votes and allocate this winner as the class of x.

Example 1. Let S be a training set composed of four labelled objects. The set of objects in S are showed in Table 1 (left), where the first column indicates their number or *id*, the columns A_1, A_2, and A_3 their attribute values, and the column *cl* gives the class they belong to.

Now, let $x \notin S$ be an object to be classified, defined as $A_1 = 1$, $A_2 = 0$, $A_3 = 0$.

One first has to compute the ad value between x and every possible triple of objects from S. Table 1 (right) shows the ad value obtained with the triple (1, 2, 3). Table 2 shows the list of the first seven triples (ranked according to ad).

Let $k = 5$; all the triples such that their associated ad value equals at most that of the 5th tuple (here, 1), are chosen. The triples 1 to 6 are then used to find the class of x. The six corresponding analogical equations are then solved. For instance, combination 2) yields the equation $1 : 1 :: 0 : cl$, leading to $cl=0$. Finally, the class that gets the most votes is retained for d.   ◇

---

[1] The analogical equation $a : b :: c : x$ is solvable iff $(a \equiv b) \vee (a \equiv c)$ holds true. In this case, the unique solution is given by $x = (a \equiv (b \equiv c))$

Table 1. Training set (left). Computation of **ad** (right).

id	A_1	A_2	A_3	cl		id	A_1	A_2	A_3	
1	0	0	0	0		1	0	0	0	
2	0	1	0	1		2	0	1	0	
3	0	1	1	1		3	0	1	1	
4	1	1	1	1		x	1	0	0	
						ad	1	2	1	= 4

Table 2. Triples ranked according to **ad**

Combination	a	b	c	d	ad
1)	3	1	4	x	0
2)	2	3	4	x	1
3)	3	4	2	x	1
4)	2	4	1	x	1
5)	3	1	2	x	2
6)	2	1	3	x	2
7)	4	1	3	x	2

3.2   Application to the Prediction of Missing Values

Case of Boolean Attributes. This method may be adapted to the case of null value prediction in a transactional database as follows. Let r be a relation of schema $(A_1, \ldots, A_m)$ and t a tuple of r involving a missing value for attribute A_i: $t[A_i] = null$. In order to estimate the value of $t[A_i]$ — that is 0 or 1 in the case of a transactional database —, one applies the previous algorithm considering that A_i corresponds to the class cl to be determined. The training set S corresponds to a sample (of a predefined size) of tuples from r (minus attribute A_i that does not intervene in the calculus of **ad** but represents the "class") involving no missing values. Besides, the attributes A_h, $h \neq i$ such that $t[A_h] = null$ are ignored during the computation aimed at predicting $t[A_i]$.

Case of Numerical Attributes. In the case of numerical attributes, a first solution consists in coming back to the Boolean case: a numerical attribute A is then derived into as many Boolean attributes as there are values in the domain of A. However, this solution is debatable inasmuch as binarization leads to considering cases of analogy that are rather limited (equality or non-equality). Two alternative solutions consist respectively in

1. relaxing the concept of analogical proportion by considering a gradual analogical dissimilarity measure;
2. introducing some tolerance in the comparison of values, i.e., replacing equality by an approximate equality relation.

These two approaches are presented and discussed in the following section.

4   Analogical Prediction of Numerical Attributes

In this section, we describe different aspects of the prediction of numerical attributes. The first step consists in normalizing the numerical values present in the relation so as to be able to use domain-independent measures in the next steps. A value a from $domain(A)$ is transformed into $(a - min_A)/(max_A - min_A)$ where min_A (resp. max_A) is the minimal (resp. maximal) value in the active domain of A. Hereafter, we discuss two points of view for relaxing the notion of an analogical proportion in this context (Subsections 4.1 and 4.2).

4.1   Use of a Gradual Analogical Dissimilarity Measure

We look for analogies of the type

$$(a : b :: c : d) \Leftrightarrow (a - b = c - d) \tag{1}$$

or

$$(a : b :: c : d) \Leftrightarrow (ad = bc), \tag{2}$$

the first one being called *arithmetic proportion* and the second one *geometric proportion*. These expressions may be relaxed by introducing some tolerance in analogical relations, so as to cover more cases. For instance, one may consider that $(100 : 50 :: 80 : 39)$ is *almost true* from a geometric proportion viewpoint.

A way of doing, which obviates the introduction of thresholds, is to adopt a gradual view. Then, the value of **ad** is not an integer anymore but a real number. In the case of an arithmetic proportion, one uses the formula proposed by Prade *et al.* in [15] and **ad** is then defined as:

$$w_1 = |(a - b) - (c - d)|. \tag{3}$$

In the case of a geometric proportion, we propose the following formula which is consistent with the three basic properties of analogical proportion mentioned in section 2, i.e., reflexivity, symmetry, and central permutation:

$$w_2 = \frac{\min(ad, bc)}{\max(ad, bc)}. \tag{4}$$

Notice that if $\max(ad, bc) = 0$, this formula is inapplicable and one can then only search for an arithmetic proportion.

Notice also that Formulas (3) and (4) do not make it possible to know whether the variation between a and b is of the same sign as that between c et d. Now, it seems reasonable to say that if, for instance, $a > b$ and $c < d$, there cannot exist any analogical relation of the form $a : b :: c : d$.

The analogical dissimilarity for a quadruple (a, b, c, d) of numerical values is computed by means of Algorithm 1.

Input: four numerical values a, b, c, d;

  $type\_a$: type of analogy (1: arithmetic or 2: geometric) considered

Output: the value of the analogical dissimilarity ad

begin

  if $(a \geq b$ and $c \geq d)$ or $(a \leq b$ and $c \leq d)$ **then**

    if $type\_a = 1$ **then**

      | compute ad with Formula (3) : $ad \leftarrow w_1$

    end

    else

      | compute ad with Formula (4) : $ad \leftarrow w_2$

    end

  end

  else

    | $ad \leftarrow 1;$ /* the maximal dissimilarity value */

  end

end

return ad;

Algorithm 1: Algorithm that computes **ad** for a numerical attribute

Computation of a Candidate Value. For a triple (a, b, c) in the top-k' list built by FADANA, the prediction of d is as follows:

– arithmetic proportion:

$$d = b + c - a; \tag{5}$$

– geometric proportion:

$$d = (bc)/a; \tag{6}$$

If the computed value d is smaller than 0 or greater than 1 (which corresponds to a predicted value that would be outside the active domain of the attribute considered), no candidate value d is produced and the triple (a, b, c) does not take part in the vote.

An important difficulty is that one does not know *a priori* which type of analogical proportion — geometric or arithmetic —, if any, is the most relevant for a given attribute A_i (or a set of attributes). Now, one needs to have this information for estimating the missing value in the last step of the algorithm (see Subsection 3.1). In order to overcome this difficulty, we tried two strategies.

The first one performs a preprocessing by taking a set of tuples where the sole missing values (artificially introduced) concern attribute A_i. This preprocessing consists in

1. applying FADANA using Formula (1) for predicting the values of A_i (for the other numerical attributes, the computation of **ad** is done by taking the minimum of w_1 and w_2), then
2. computing the proportion of missing values that are correctly estimated.

One does the same using Formula (2), and checks which expression yields the best precision.

The second strategy involves a dataset where null values have been artificially introduced for every attribute (scattered over the different tuples). We perfom steps 1-3 of FADANA for each missing value and we count, for each attribute A_i, the overall number of times $(n_{i,1})$ that w_1 is smaller than w_2 and the overall number of times $(n_{i,2})$ that the opposite is true when computing the ad values. For a given attribute A_i, if $n_{i,1} > n_{i,2}$ then we decide that arithmetic proportion will be the most suited to predict the missing values for A_i, otherwise we conclude that geometric proportion will be the most appropriate.

Unfortunately, none of these strategies appeared conclusive. In each case, and for each dataset considered, we observed that it was always either the arithmetic analogy or the geometric one that won *for all attributes*. In other terms, as surprising as this may be, it seems that the type of analogical proportion that yields the best result depends on the dataset, not on the attributes considered individually. This, of course, makes the choice easier, but raises some questions. What makes one kind of analogy perform better than the other is still unclear and must be investigated further. In the following section, we define another way to relax the notion of an analogical proportion, which does not imply to make a choice between two types of relations.

4.2   Use of an Approximate Equality Relation

The idea here is to introduce some tolerance in the comparison of the values involved in an analogical proportion:

$$a : b :: c : d \Leftrightarrow (((a \approx b) \wedge (c \approx d)) \vee ((a \approx c) \wedge (b \approx d))). \tag{7}$$

We interpret $x \approx y$ as $|x - y| \leq \lambda$ where $\lambda \in [0, 1]$ ($\lambda = 0$, $\lambda = 0.05$, and $\lambda = 0.1$ have been used in the experimentation).

Computation of a Candidate Value. For a triple (a, b, c) in the top-k' list built by FADANA, the prediction of d is as follows:

$$\text{if } |a - b| \leq \lambda \text{ then } d \leftarrow c \text{ else if } |a - c| \leq \lambda \text{ then } d \leftarrow b. \tag{8}$$

If one has neither $|a - b| \leq \lambda$ nor $|a - c| \leq \lambda$, then the triple (a, b, c) cannot be used to predict any candidate value d and it does not take part in the vote.

4.3   Computation of the Final Value

Let us denote by $v_1, \ldots, v_n$ the number of votes obtained respectively by the different predicted candidate values d_i in the FADANA algorithm ($\sum_{i=1}^{n} v_i = k'$ and $n \leq k'$ since different predicted values may be equal). Let us assume that a ranking has been performed so that $v_1 \geq v_2 \geq \ldots \geq v_n$. There are of course many different ways to compute the final value d, among which:

− by an arithmetic mean:

$$d = (\sum_{i=1}^{n} d_i)/n \tag{9}$$

– by a weighted mean:

$$d = (\sum_{i=1}^{n} v_i \times d_i)/k' \tag{10}$$

– by keeping only the candidate value with the most votes:

$$d = d_i \text{ such that } v_i = \max_{j=1..n} v_j$$

(but then, ties are problematic).

However, it is somewhat risky to predict a precise value whereas the prediction process is fundamentally uncertain. A more cautious solution is to return a probability distribution of the form:

$$\{\frac{v_1}{k'}/d_1, \ldots, \frac{v_n}{k'}/d_n\}, \tag{11}$$

Example 2. Let us assume that $k' = 10$ and that the pairs (v_i, d_i) obtained are:

$$(4, 17), (4, 12), (1, 8), (1, 15).$$

Using the arithmetic mean, one gets $d = 13$. With the weighted mean, the result is: $d = 14$ (rounding 13.9). On the other hand, the associated probability distribution is $\{0.4/17, 0.4/12, 0.1/8, 0.1/15\}$.   ◇

Note that Equation (11) implicitly assumes that all the other values that could be predicted using a larger k' are considered totally unlikely (their probability degree is set to zero).

4.4   Evaluating the Precision of the Method

So as to evaluate the precision of the prediction method m for a dataset D, one may use, in the three first cases considered above, the measure:

$$prec(m, D) = \frac{\sum_{null\ values\ x\ in\ D} 1 - |x_{actual} - x_{predicted}|}{|null\ values\ in\ D|} \tag{12}$$

where $x_{predicted}$ is the value estimated for x using m. It is of course assumed that the null values have been artificially introduced in D, i.e., one replaced by *null* some values that were initially precisely known (the precise value of x is denoted by x_{actual} in the formula).

Let us now consider the case where $x_{predicted}$ is a probability distribution. Let us denote by $cand(x_{predicted})$ the crisp set of candidate values c appearing in the distribution, and by $pr(c)$ the probability degree associated with candidate c in $x_{predicted}$. The precision measure is defined as:

$$prec(m, D) = 1 - \frac{\sum_{null\ values\ x\ in\ D} penalty(x)}{|null\ values\ in\ D|} \tag{13}$$

where $penalty(x) = \sum_{d_i \in cand(x_{predicted})} |x_{actual} - d_i| \times pr(d_i)$.

5   Preliminary Experimentation

The main objective of the experimentation was to compare the results obtained using this technique with those produced by other approaches (in particular the "nearest neighbors" technique), thus to estimate its relative effectiveness in terms of *precision* (i.e., of percentage of values correctly predicted).

Let us first emphasize that the performance aspect (in terms of execution time) is not so crucial here, provided of course that the class of complexity remains reasonable. Indeed, the prediction of missing values is to be performed offline. However, this aspect will be tackled in the conclusion of this section, and we will see that different optimization techniques make it possible to significantly improve the efficiency of the algorithm.

5.1   Experimental Results

In order to assess the effectiveness of the approach, four datasets from the UCI machine learning repository[2], namely *adult, blood, cancer*, and *energy* have been used. For each dataset, a sample M of 50 tuples has been modified, i.e., a 40% of values of its tuples has been replaced by *null*. Then, the FADANA algorithm has been run so as to predict the missing values: for each tuple d involving at least a missing value, a random sample D of $E - M$ (thus made of complete tuples) has been chosen. This sample D is used for running the algorithm inspired from FADANA, detailed in Section 3. Each time, we tested the following approaches:

- arithmetic
  - aritmean: Formulas 5 and 9
  - aritweight: Formulas 5 and 10
  - aritprob: Formulas 5 and 11
- geometric
  - geomean: Formulas 6 and 9
  - geoweight: Formulas 6 and 10
  - geoprob: Formulas 6 and 11
- tolerant with $\lambda = 0$, $\lambda = 0.05$, $\lambda = 0.1$
  - t0mean, t005mean, t01mean: Formulas 8 and 9
  - t0weight, t005weight, t01weight: Formulas 8 and 10
  - t0prob, t005prob, t01prob: Formulas 8 and 11
- kNN
  - knnmean: kNN technique and Formula 5
  - knnweight: kNN technique and Formula 6
  - knnprob: kNN technique and Formula 11.

Tables 3 and 4 show how precision evolves with the value of k, using a training set made of 40 tuples. A remarkable result is that the value of k does not have a strong impact on the precision of each approach. The best performances

---

[2] http://http://archive.ics.uci.edu/ml/datasets.html

were obtained with *geoweight* (84.5%), *geoprob* (90.8%), *knnmean* (86.9%), and *t005mean* (89.6%) for the datasets *adult, blood, cancer,* and *energy* respectively.

Table 5 shows how precision evolves with the size of the training set, using $k = 40$. The best performances were obtained with *geoweight* (84%), *aritmean* (88.9%), *t01mean* (85.6%), and *aritmean* (88.2%) for *adult, blood, cancer,* and *energy* respectively. Notice that all of the approaches have a poor precision when the size of the training set is around 10, but as soon as this size gets around 20, the precision of the analogical approaches considerably increases, which is not the case of kNN.

Table 3. Results with $ts = 40$ and $k \in [10\text{-}40]$ (datasets *adult* and *blood*)

k value	adult 10	20	30	40	blood 10	20	30	40
aritmean	70.17	73	73.07	70.69	87.76	**88.69**	**88.8**	**88.09**
aritweight	69.75	71.97	72.08	70.38	85.95	84.71	84.95	83.62
aritprob	70.39	71.67	71.72	69.4	86.32	86	86.15	85.36
geomean	81.53	83.71	83.71	83.14	84.7	85	86.45	83.35
geoweight	**82.14**	**84.3**	**84.5**	**84.18**	83.14	80.35	82.5	77.5
geoprob	70.51	72.2	71.6	69.22	**90.8**	86.22	84.69	82.3
t0mean	67.19	68.35	68.48	66.14	85.09	87.38	87.6	86.9
t0weight	72.29	74.3	74.25	71.63	86.46	84.8	86.1	84.5
t0prob	73.38	75.3	75	72.3	86.82	86.2	86.6	85.4
t005mean	72.17	75.8	75.75	72.3	88.21	88.22	88.12	88
t005weight	72	75.6	75.6	76.2	88.2	85.3	85.8	85.3
t005prob	71.63	74.8	74.7	71.6	86.3	86.7	86	86
t01mean	71.3	74.2	74.5	71.1	87.3	87.47	88	88
t01weight	71.2	73.9	74.2	71	87.3	84.8	86.2	85.8
t01prob	70.9	73.2	73.5	70.4	84.8	84.7	85.3	86
knnmean	72.5	74.8	73.8	68.6	87.8	87	87.4	86.2
knnweight	72.3	74.7	73.8	68.6	87.7	86.1	87	86
knnprob	71.3	73.6	72.6	67.5	86	84.3	85	83

An interesting result is that, even though the analogical approach based on approximate equality (Formula 7) is usually outranked by either the arithmetic or the geometric proportion-based one, it is in general better than the worst among these two *and* than kNN. We thus can consider it a good compromise. From these experimental results, it seems that $\lambda = 0.05$ is a good choice for the threshold, and that the best way to determine the final predicted values with this method is either the arithmetic mean (Formula 5) or the use of a probability distribution (Formula 11).

However, it appears that the best method overall is either the arithmetic or the geometric proportion-based analogical approach (depending on the dataset) and let us recall that a preprocessing such as that described at the end of Subsection 4.1 makes it possible to determine which one among the two is the most

Table 4. Results with $ts = 40$ and $k \in [10\text{-}40]$ (datasets *cancer* and *energy*)

k value	cancer 10	20	30	40	energy 10	20	30	40
aritmean	84.98	85.23	85.29	84.38	**89.29**	**89.12**	89.05	**89.04**
aritweight	84.98	85.23	85.29	84.38	**89.29**	**89.12**	89.05	**89.04**
aritprob	82.97	82.57	82.7	81.9	86.95	86	85.6	85.7
geomean	85.57	**86.12**	85.5	84.5	88.1	87.4	88.1	87.5
geoweight	85.6	**86.12**	85.5	84.5	88.1	87.38	88.1	87.5
geoprob	83	83.22	82.77	81.22	85.84	84.26	84.91	84.3
t0mean	78	81.3	83.5	83.32	72.71	71.5	70.4	74
t0weight	75.2	74.9	75.13	74.46	66.3	64.74	68.2	66.2
t0prob	84.25	84	84.25	83.36	86.2	85	86.86	85.09
t005mean	85	85.6	**85.6**	**85.11**	89.12	87.9	**89.6**	88.7
t005weight	75.72	75.2	75.2	74.5	72.6	71.3	75.7	72
t005prob	83.8	83.8	84	83.2	87.2	86	87.4	86.5
t01mean	85.1	85.6	**85.6**	85.1	88.6	87.5	88.6	87.6
t01weight	75.8	75.3	75.25	74.6	72.3	71	75	71
t01prob	83.8	83.9	84	83.3	86.12	85	86	85
knnmean	**86.9**	86	82.4	75.4	84.9	81.7	75.2	69.6
knnweight	86.9	86	82.4	75.4	84.9	82	75	69.6
knnprob	84	82.7	78.6	70	81.26	76.8	69	64.2

suitable for a given dataset. For the arithmetic (resp. geometric) proportion-based analogical method, the experimental measures reported here tend to show that the best way to compute the final values is to use an arithmetic (resp. weighted) mean. Of course, these are just preliminary results that need to be confirmed on larger and more diverse datasets, but they are very encouraging as to the relevance of applying analogy in this context.

5.2 Optimization Aspects

As mentioned in the preamble, temporal performances of the approach are not so crucial since the prediction process is to be executed offline. However, it is interesting to study the extent to which the calculus could be optimized. With the base algorithm presented in Section 3, complexity is in $\theta(N^3)$ for the prediction of a missing value, where N denotes the size of the training set TS (indeed, an incomplete tuple has to be associated with every triple that can be built from this set, the analogical relation being quaternary). An interesting idea consists in detecting *a priori* the triples from TS that are the most "useful" for the considered task, i.e., those the most likely to take part in a sufficiently valid analogical relation. For doing so, one just has to run the algorithm on a small subset of the database containing artificially introduced missing values, and to count, for each triple of the training set, the number of k-lists in which it appears as well as the average number of cases in which the prediction is correct. One can

Table 5. Results with k=40 and ts ∈ [10-40]

	adult			blood			cancer			energy		
ts size	10	20	30	10	20	30	10	20	30	10	20	30
aritmean	68.24	71.54	71.3	84.7	**88.4**	**88.9**	79	84.2	85	**82.17**	**86.5**	**88.2**
aritweight	67.2	70.8	71	82.5	83.7	84.5	79.2	**84.3**	85	**82.17**	**86.5**	**88.2**
aritprob	66.7	69.9	69.7	82.4	86	86	76	80.8	82.3	77.2	82.2	84.1
geomean	80.4	83	83	75.6	84	85.9	**80.4**	83.2	85	80.4	84.44	86.5
geoweight	**81.95**	**84**	**83.9**	70	79	79.2	80.3	83.2	85	80.4	84.4	86.5
geoprob	66.7	70.5	69.8	76	81.8	83.2	77.6	79.8	82.2	75.3	80	82.6
t0mean	61.5	67	67	84.5	86.56	87	74	76.6	80	60.1	66.5	74
t0weight	69.3	73.2	72	84.5	85.6	85.9	71	69	69	61.2	64.8	65.8
t0prob	71.58	74.1	72.8	82.86	85.7	85.3	77.6	81.2	83.58	75.8	82.5	83.4
t005mean	70.3	74.2	73	82.2	87	86.7	79.8	83.5	85	78.8	85.3	87.6
t005weight	70.1	73.86	73.34	80.9	85.4	84.8	71.3	68.7	69.3	68.5	72	71
bt005prob	69.8	73.1	72.2	80.7	85	84.8	77	79.7	83.2	74.7	82	85
t01mean	69.5	72.5	72.2	83.6	86.4	87.4	79.7	82.5	**85.6**	80.5	86.2	87
t01weight	69	721	72.2	81.7	84.8	86	71.2	67.3	69.2	69	72.3	70.9
t01prob	69	71.3	71.2	81.4	84.1	84.5	77	80	83.4	76.6	82.5	84.5
knnmean	68	70.8	68.8	**85.7**	86	86.6	76.5	74.8	77	69.2	70.2	70
knnweight	68	70.8	68.8	**85.7**	86	86.6	76.5	74.8	77	69.2	70.2	70
knnprob	67.3	69.6	67.7	83.6	83.8	84	72.9	69.9	72.5	64.3	63.8	63.9

then keep the sole N' triples that appear the most frequently with a good rate of success, and use them to predict the missing values in the entire database. Complexity is then in $\theta(N')$ for estimating a given missing value.

We ran this optimized algorithm on several Boolean datasets, with k varying between 20 and 40, the size of the training set between 20 and 40, and N' between 100 and 1000. For a total of 3000 incomplete tuples, the basic algorithm was run over the first 500, treating the others with the optimized method.

While the precision of the regular FADANA algorithm was 91% on average, that of the optimized method was about 84%, i.e., there was a difference of about 7 percents (whereas the precision of the kNN method over the same dataset was about 85%). On the other hand, the optimized method is much more efficient: it is 1300 times faster than the regular FADANA algorithm when the size of the training set equals 40, and 25 times faster when it equals 20.

These results show that this method does not imply a huge loss of precision, but leads to a very significant reduction of the overall processing time. Further experiments and analyses are needed, though, in order to determine which properties make a triple more "effective" than others.

Let us mention that another optimization axis would consist in parallelizing the calculus on the basis of a vertical partitioning of the relation involved, which would make it possible to assign a subset of attributes to each processor, the intermediate results being summed in order to obtain the final value of the analogical dissimilarity ad.

6   Conclusion

In this paper, we have presented a novel approach to the estimation of missing values in relational databases, that exploits the notion of analogical proportion. This study is a follow-up to a previous work which showed that analogical prediction is an effective technique in the case of Boolean values. The main aim of the present work was to extend it to the numerical attribute case. We have investigated different ways to relax the notion of an analogical proportion in the context of numerical values, and we have evaluated their pros and cons. Then, we have shown how an algorithm proposed in the context of classification could be adapted to a prediction purpose. The experimental results obtained, even though preliminary, appear very encouraging since the different variants of the analogical-proportion-based approach always yield a better precision than the classical nearest neighbors technique as soon as the training set is not too small.

Among the many perspectives opened by this work, let us mention the following ones. Future work should notably i) investigate how to deal in a sophisticated way with categorical attributes by taking into account notions such as synonymy, hyponymy/hypernymy, etc. ii) study the way predicted values must be handled, in particular during the database querying process. This will imply using an uncertain database model (see e.g. [22] for a survey of probabilistic database models and [5] about a model based on the notion of possibilistic certainty) inasmuch as an estimated value remains tainted with uncertainty, even if the prediction process has a good level of reliability.

References

1. Abraham, M., Gabbay, D.M., Schild, U.J.: Analysis of the talmudic argumentum a fortiori inference rule (kal vachomer) using matrix abduction. Studia Logica 92(3), 281–364 (2009)
2. Atzeni, P., Morfuni, N.M.: Functional dependencies and constraints on null values in database relations. Information and Control 70(1), 1–31 (1986)
3. Bashir, S., Razzaq, S., Maqbool, U., Tahir, S., Baig, A.R.: Using association rules for better treatment of missing values. CoRR abs/0904.3320 (2009)
4. Bayoudh, S., Miclet, L., Delhay, A.: Learning by analogy: A classification rule for binary and nominal data. In: Veloso, M.M. (ed.) IJCAI, pp. 678–683 (2007)
5. Bosc, P., Pivert, O., Prade, H.: A model based on possibilistic certainty levels for incomplete databases. In: Godo, L., Pugliese, A. (eds.) SUM 2009. LNCS, vol. 5785, pp. 80–94. Springer, Heidelberg (2009)
6. Chen, S.M., Chang, S.T.: Estimating null values in relational database systems having negative dependency relationships between attributes. Cybernetics and Systems 40(2), 146–159 (2009)
7. Correa Beltran, W., Jaudoin, H., Pivert, O.: Estimating null values in relational databases using analogical proportions. In: Laurent, A., Strauss, O., Bouchon-Meunier, B., Yager, R.R. (eds.) IPMU 2014, Part III. CCIS, vol. 444, pp. 110–119. Springer, Heidelberg (2014)
8. Fujikawa, Y., Ho, T.-B.: Cluster-based algorithms for dealing with missing values. In: Chen, M.-S., Yu, P.S., Liu, B. (eds.) PAKDD 2002. LNCS (LNAI), vol. 2336, pp. 549–554. Springer, Heidelberg (2002)

9. Kaiser, J.: Algorithm for missing values imputation in categorical data with use of association rules. CoRR abs/1211.1799 (2012)
10. Liu, W.Z., White, A.P., Thompson, S.G., Bramer, M.A.: Techniques for dealing with missing values in classification. In: Liu, X., Cohen, P., Berthold, M. (eds.) IDA 1997. LNCS, vol. 1280, pp. 527–536. Springer, Heidelberg (1997)
11. Miclet, L., Bayoudh, S., Delhay, A.: Analogical dissimilarity: Definition, algorithms and two experiments in machine learning. J. Artif. Intell. Res. (JAIR) 32, 793–824 (2008)
12. Miclet, L., Prade, H.: Handling analogical proportions in classical logic and fuzzy logics settings. In: Sossai, C., Chemello, G. (eds.) ECSQARU 2009. LNCS, vol. 5590, pp. 638–650. Springer, Heidelberg (2009)
13. Myrtveit, I., Stensrud, E., Olsson, U.H.: Analyzing data sets with missing data: An empirical evaluation of imputation methods and likelihood-based methods. IEEE Trans. Software Eng. 27(11), 999–1013 (2001)
14. Nogueira, B.M., Santos, T.R.A., Zárate, L.E.: Comparison of classifiers efficiency on missing values recovering: Application in a marketing database with massive missing data. In: CIDM, pp. 66–72. IEEE (2007)
15. Prade, H., Richard, G., Yao, B.: Enforcing regularity by means of analogy-related proportions — a new approach to classification. International Journal of Computer Information Systems and Industrial Management Applications 4, 648–658 (2012)
16. Prade, H., Richard, G.: Reasoning with logical proportions. In: Lin, F., Sattler, U., Truszczynski, M. (eds.) KR. AAAI Press (2010)
17. Prade, H., Richard, G.: Homogeneous logical proportions: Their uniqueness and their role in similarity-based prediction. In: Brewka, G., Eiter, T., McIlraith, S.A. (eds.) KR. AAAI Press (2012)
18. Prade, H., Richard, G.: Analogical proportions and multiple-valued logics. In: van der Gaag, L.C. (ed.) ECSQARU 2013. LNCS, vol. 7958, pp. 497–509. Springer, Heidelberg (2013)
19. Quinlan, J.R.: Induction of decision trees. Machine Learning 1(1), 81–106 (1986)
20. Ragel, A.: Preprocessing of missing values using robust association rules. In: Żytkow, J.M. (ed.) PKDD 1998. LNCS, vol. 1510, pp. 414–422. Springer, Heidelberg (1998)
21. Shen, J.J., Chen, M.T.: A recycle technique of association rule for missing value completion. In: AINA, pp. 526–529. IEEE Computer Society (2003)
22. Suciu, D., Olteanu, D., Ré, C., Koch, C.: Probabilistic Databases. Synthesis Lectures on Data Management. Morgan & Claypool Publishers (2011)

Observations on Fine-Grained Locking
in XML DBMSs

Martin Hiller, Caetano Sauer, and Theo Härder

University of Kaiserslautern, Germany
{hiller,csauer,haerder}@cs.uni-kl.de

Abstract. Based on XTC, we have redesigned, reimplemented, and reoptimized BrackitDB, a native XML DBMS (XDBMS). Inspired by "optimal" concurrency gained on XTC using the taDOM protocol, we applied an XML benchmark on BrackitDB running now on a substantially different computer platform. We evaluated important concurrency control scenarios again using taDOM and compared them against an MGL-protocol. We report on experiments and discuss important observations w.r.t. fine-grained parallelism on XML documents.

1   Introduction

In the past, we have addressed—by designing, implementing, analyzing, optimizing, and adjusting an XDBMS prototype system called XTC (XML Transactional Coordinator)—all these issues indispensable for a full-fledged DBMS. To guarantee broad acceptance for our research, we provided a *general solution* that is even applicable for a spectrum of XML language models (e. g., XPath, XQuery, SAX, or DOM) in a multi-lingual XDBMS environment [5]. At that time, all vendors of XML(-enabled) DBMSs supported updates only at document granularity and, thus, could not manage highly dynamic XML documents, let alone achieve ambitious performance goals. For this reason, we primarily focused on locking mechanisms which could support fine-grained, concurrent, and transaction-safe document modifications in an efficient and effective way.

The outcome of this research was the *taDOM* family of complex lock protocols [7] tailor-made for fine-grained concurrency in XML structures and guaranteeing ACID-quality transaction serializability. Correctness of taDOM [13] and its superiority against about a dozen of competitor protocols were already experimentally verified using our (disk-based) XTC (XML Transaction Coordinator) [6]. Changes and developments of computer and processing architectures (e.g., multi-core processors or use of SSDs) also impact the efficacy of concurrency control mechanisms. Therefore, we want to review taDOM in a—compared to the study in [6]—substantially changed environment. Because computer architectures provided fast-growing memories in recent years, we want to emphasize this aspect—up to main-memory DBMS—in our experimental study. Furthermore, based on XTC, we have redeveloped our testbed system, called BrackitDB as a disk-based XDBMS [1,3]. To gain some insight into the concurrency control

Y. Manolopoulos et al. (Eds.): ADBIS 2014, LNCS 8716, pp. 337–351, 2014.

behavior of these reimplemented and improved system, we run an XML benchmark under various system configurations and parameter settings [8]. Because
the conceivable parameter space is so huge, we can only provide observations on
the behavior of important XML operations.

To build the discussion environment, we sketch the system architecture of
BrackitDB, the efficacy of important taDOM concepts, and critical implementation issues in Sect. 2. In the following section, we describe the test database
and the XML benchmark used for our evaluation, before we report on our measurements obtained and, in particular, the most important observations made in
Sect. 4. Finally we summarize our results and conclude the study.

2 Environment of the Experimental Study

To facilitate the understanding of our study, we need at least some insight into
the most important components influencing the concurrency control results.

2.1 Hierarchical DBMS Architecture

Using a hierarchical architecture model,
we have implemented BrackitDB as a
native XDBMS aiming at efficient and
transaction-safe collaboration on shared
documents or collection of documents.
Hence, it provides fine-grained isolation
and full crash recovery based on a native
XML store and advanced indexing capabilities.

Fig. 1 gives a simplified overview of its
architecture where its layers describe the
major steps of dynamic abstraction from
the storage up to the user interface.

File Layer, Buffer Management, and
the file formats (Container, Log, Metadata)—taken from XTC— are not XML-
specific and similar to their counterparts
in relational DBMSs. The components of
the Storage Layer are more important for
our study. They manage XML documents
and related index structures and provide
node-oriented access which has to be isolated in close cooperation with the Lock

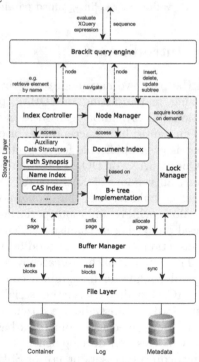

Fig. 1. Architecture of BrackitDB

Manager (see Sect. 2.3). The document index—a B*-tree where each individual XML node is stored as a data record in one of the tree's leaf pages—is the
core structure. To identify nodes, a prefix-based labeling scheme is applied[1].

---

[1] The node labeling scheme is *the key to efficient and fine-grained XML processing* [5].

A DeweyID, for example, 1.5.7 identifies a node at level 2, where its parent has DeweyID 1.5 at level 1 and its grandparent DeweyID 1 at level 0. The DeweyIDs serve as keys in the document index and easily allow to derive sibling and ancestor relations among the nodes. The latter is particularly important for locking; if, e.g., a node is accessed via an index, intention locks have to be acquired on the ancestor nodes first. Because DeweyIDs also serve as index references, they contain the ancestor path and, hence, intention locks can be set automatically.

Efficient XML processing requires the *Path Synopsis* which represents in a tiny memory-resident repository all different path classes of a document. Using a path synopsis, each existing path in the document can be indexed by the so-called *Path Class Reference* (or PCR). By storing the DeweyID together with the PCR as an index reference, the entire path of the indexed node to the document root can be reconstructed without document access. This technique also enables virtualization of the inner nodes, i.e., it is sufficient to store only the leaf nodes of an XML document [11]. Moreover, name index, content index, path index, and CAS index (content and structure)—all using the B*-tree as base structure—contribute to the efficiency of XML query processing.

The top layer in Fig. 1 is embodied by BrackitDB's XQuery engine which compiles and executes high-level XQuery expressions. Because we circumvent here the query engine to have more precise control over node operations and locking aspects, we do not need to discuss this engine.

2.2 Lock Concepts of taDOM

Due to its complexity (with 20 lock modes and sophisticated lock conversion rules [7]), taDOM cannot even be sketched here completely for comprehension. Based on some examples, we try to convince the reader of taDOM's concepts and their potential. For this reason, we visualize its very fine granularity and compare it to the well-known MGL-protocol (Multi-Granularity Locking) [4].

taDOM uses the MGL intention locks IR and IX, renames R, U, and X to SR, SU, and SX (subtree read, update, and exclusive). Furthermore, it introduces new lock modes for *single nodes* called NR (node read), NU (node update), and NX (node exclusive), and for *all siblings under a parent* called LR (level read). NR and LR allow, in contrast to MGL, to read-lock only a node or all nodes at a level (under the same parent), but not the corresponding subtrees.

To enable transactions to traverse paths in a tree having (levels of) nodes already read-locked by other transactions and to modify subtrees of such nodes, a new intention mode CX (child exclusive) had to be defined for a context (parent) node. It indicates the existence of an SX or NX lock on some *direct child* nodes and prohibits inconsistent locking states by preventing LR and SR locks. It does not prohibit other CX locks on a context node *cn*, because separate child nodes of *cn* may be exclusively locked by other transactions (compatibility is then decided on the child nodes). Altogether these new lock modes enable serializable schedules with read operations on inner tree nodes, while concurrent updates may occur in their subtrees. An important and unique feature is the optional variation of the *lock depth* which can be dynamically controlled by a parameter.

Lock depth n determines that, while navigating in documents, individual locks are acquired for existing nodes down to level n. If necessary, nodes below level n are locked by a subtree lock (SR, SX) at level n. Fig. 2 summarizes the lock compatibilities among taDOM's core modes.

Fig. 3 highlights taDOM's flexibility and tailor-made adaptations to XML documents as compared MGL. Assume transaction $T1$—after having set appropriate intention locks on the path from the root—wants to read-lock context node cn. Independently of whether or not $T1$ needs subtree access, MGL only offers an R lock on cn, which forces a concurrent writer ($T2$ in Fig. 3(a)) to wait for lock release in a lock request queue (LRQ). In the same situation, node locks (NR and NX) would al-

	IR	IX	SR	SU	SX	NR	LR	CX
IR	+	+	+	-	-	+	+	+
IX	+	+	-	-	-	+	+	+
SR	+	-	+	-	-	+	+	-
SU	+	-	+	-	-	+	+	-
SX	-	-	-	-	-	-	-	-
NR	+	+	+	-	-	+	+	+
LR	+	+	+	-	-	+	+	-
CX	+	+	-	-	-	+	-	+

Fig. 2. Core modes of taDOM

low greatly enhance permeability in cn's subtree (Fig. 3(b, c)). As the only lock granule, however, node locks would result in excessive lock management cost and catastrophic performance behavior, especially for subtree deletion [6]. Scanning of cn and all its children could be done node by node in a navigational fashion (Fig. 3(b)). A special taDOM optimization—using a tailor-made LR lock for this frequent read scenario—enables stream-based processing and drastically reduces locking overhead; in huge trees, e.g., the DBLP document, a node may have millions of children. As sketched in Fig. 3(c), LR also supports write access to deeper levels in the tree. The combined use of node, level, and subtree locks gives taDOM its unique capability to tailor and minimize lock granules.

Document entry from any secondary index [4] involves the risk of phantoms. For instance, retrieving a list of element nodes from the *name index* [11] causes every returned node to be NR-locked. While this protects the elements from being renamed or deleted, other transactions might still insert new elements with matching names. Hence, opening the name index again at a later time with the same query parameters (i.e., given element name), might fetch an extended list of element nodes, including phantom elements that have been inserted concurrently. To prevent this kind of anomalies, BrackitDB implements Key-Value Locking (or KVL [12]) for all B*-trees employed as secondary indexes. In a nutshell,

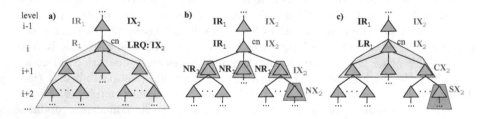

Fig. 3. Locking flexibility and effectivity: MGL (a) vs. taDOM (b, c)

KVL aims not only for locking single index entries but also protects key ranges traversed by index scans. Therefore, KVL effectively prevents phantoms.

2.3 Lock Management

Additional flexibility comes from dynamic lock-depth variations and *lock escalation*, which help to find an adequate balance between lock overhead (number of locks managed) and parallelism achieved (concurrency-enabling effects of chosen lock granules). An illustrative example for this locking trade-off is given above: *T1* started to navigate the child set of node *cn* in a low-traffic subtree (Fig. 3(b)). If no interfering transaction is present, lock escalation could be performed (Fig. 3(c)) to reduce lock management cost. In BrackitDB, two kinds of lock escalation are distinguished. Using *static lock depth*, locks are acquired only up to a pre-specified level *n* in the document tree. If a node is accessed at a deeper level, its corresponding ancestor will be locked instead. Keep in mind that escalating a lock request to an ancestor potentially widens the lock mode as well. *Dynamic lock escalation* works at runtime to overcome the inflexibility of the static approach which handles each subtree in the same way. Some parts of the document might not even be accessed at level 1, while others might accommodate hotspot regions deeper than level *n*. Tracking locality of nodes to collect "escalation" information is cheap, because ancestor locks have to be acquired anyway. While performing intention locking, an internal lock request counter is incremented for the parent of requested nodes, indicating that one of its children is accessed. The following heuristics determines the escalation threshold:

$$threshold = \frac{maxEscalationCount}{2^{level} * escalationGain}$$

The deeper a node is located in the document, the lower is the threshold for performing lock escalation. The parameter from the numerator, `maxEscalationCount`, constitutes an absolute basis for the number of locks that need to be held at most before escalation is applied. This value is not changed throughout different escalation strategies, whereas the denominator parameter, `escalationGain`, affects how the threshold changes from one level to the next, i.e., the higher this value is set, the more aggressive is the escalation policy for deeper nodes. Throughout the benchmark runs, the following escalation policies are applied:

	maxEscalationCount	escalationGain
moderate	1920	1.0
eager	1920	1.4
aggressive	1920	2.0

The concept of *meta-locking* realized in BrackitDB provides the flexibility to exchange lock protocols at runtime. Hence, such dynamic adaptations are a prerequisite to achieve workload-dependent optimization of concurrency control and to finally reach autonomic tuning of multi-user transaction processing [6]. This concept enables us to execute all benchmark runs under differing lock protocols in an *identical system environment*.

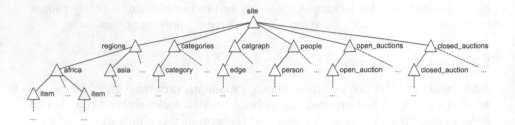

Fig. 4. Fragment of the XML document used in the benchmark

Lock management internals are encapsulated in so-called lock services with a lock table as their most important data structure. Lock services provide a tailored interface to various system components, e.g., for DB buffer management [7]. A lock table must not be a bottleneck for performance and scalability reasons. Therefore, it has to be traversed and modified by a multitude of transaction threads at a time. Hence, preservation of its physical consistency becomes a major challenge for an implementation viewpoint. Furthermore, frequent blocking situations must be avoided when lock table operations (look-up, insertion of entries) or house-keeping operations are performed. Therefore, the use of latches on individual entries of a hash table is mandatory for lock table access. As compared to a single monitor for a hash table, such a solution avoids hot spots and guarantees physical consistency under concurrent thread access [4].

Another major component of BrackitDB's lock management is the *deadlock detector (DD)*. Our DD component runs periodically in a separate thread so that deadlock resolution is still possible, even when all transaction workers are already blocked. Every time the DD thread wakes up, its task is to crawl through the lock table and to construct a *wait-for* graph on its way. While building the wait-for graph, the DD thread has to comply with the same latching protocol as every other thread accessing the lock table. If a cycle is detected, one of the participating transactions is to be aborted. The decision heuristics for a suitable abort candidate is based on the minimal number of locks obtained.

3   Benchmark Document and Workload

The benchmark data (see Fig. 4) was created by the XMark Benchmark Data Generator. Its command line tool `xmlgen` produces documents that model auction data in a typical e-commerce scenario. Words for text paragraphs are randomly picked from Shakespeare's plays so that character distribution is fairly representative for the English language. Further, the XML output includes referential constraints across the document through respective ID/IDREF pairs, thus providing a basis for secondary index scenarios.

But the most outstanding feature is arguably the *scaling factor*. Varying this factor allows for generating documents from the KB range up to several GBs. The documents still preserve their characteristics under scaling so that bottlenecks found on smaller documents would apply for larger documents proportionally. It is worth noting that scaling works only horizontally, where document depth and complexity remains untouched, while scaling appends new elements of different types w.r.t. a probability distribution. The benchmark is based on an XMark document with a scaling factor of 10, resulting in a file of approximately 1.1 GB in size (stored on HDD) and specific document cardinalities (see Fig. 5).

Items:	217,500
Categories:	10,000
Catgraph edges:	10,000
Persons:	250,000
Open auctions:	120,000
Closed auctions:	97,500

Fig. 5. Cardinalities

The XMark document models an auction application where *items* coming from a certain *region* are put up for auction. Each item is thereby associated with one or more *categories*, which is realized by a sequence of child elements referring to category elements by their respective ID. Furthermore, every item contains a *mailbox* element surrounding a list of *mails*. Following on the item declarations, the *categories* element accumulates the available item categories in the system which are simply characterized by a name and a description text. The *catgraph* element defines a graph structure on top of the categories, whereas each *edge* ties two categories together. The *person* elements gathered under the *people* element act as users of the virtual auction platform. A few personal information like the name or the email address are exposed at this place, but their main usage is to be referenced by *auction* elements as bidders or sellers, respectively. In terms of auctions, XMark distinguishes between open and closed auctions and store them separately as either *open_auction* or *closed_auction* element in the corresponding subtree. But in fact, their XML structure is quite similar. Both kinds of auctions contain a description, a reference to the traded item and its seller.

The workload that is putting stress on BrackitDB is a mix of eight different transactions. Some of these are read-only, others perform both read and write accesses. Moreover, transactions are picked randomly from this pool according to predefined weights (see Table 1), which are supposed to reflect a distribution we might find in a realistic auction-based application where the focus of activity is certainly on placing bids. Here, we can sketch the workload transactions (TX) only by their names to give some flavor of the benchmark evaluated.

The transactions need a way to randomly jump into the document to begin their processing. For instance, placing a bid requires the transaction to pick a random auction element, while another operation might start on a random person element as context node. Another feature needed by some operations is the possibility to follow ID references within the document, e.g., jump from the auction to the corresponding item element. Although both requirements (random entry and navigation via references) could be met without utilizing any secondary indexes by accessing the root node and scanning for the desired node,

Table 1. Mix of transactions with their types and weights

Transaction	Type	Weight	Transaction	Type	Weight
Place Bid	r-w	9	Check Mails	r-o	7
Read Seller Info	r-o	4	Read Item	r-o	5
Register	r-w	1	Add Mail	r-w	4
Change User Info	r-w	4	Add Item	r-w	1

system performance could drastically suffer from this unfavorable access plan. For that reason, we provide two secondary indexes:

- A name index maps element names to a set of DeweyIDs enabling node reconstructions or direct document access to the name occurrences.
- A CAS index [11] allows for content-based node look-ups, i.e., attribute and text nodes can be retrieved based on their string value.

When fetching random elements from the name index, the elements are selected w.r.t. some predefined *skew*. Locking-related contention must also be artificially enforced. Higher skew implies denser access patterns and increased contention. Skew is defined as $1 - \sigma$, while σ denotes the relative standard deviation of the normal distributions. Hence, a skew of 99% (or a relative σ of 1%) implies that only 1% of the records in the page are picked with a probability of approximately 68%, effectively producing a hotspot.

Note, we report numbers (tpm, ms) for each experiment averaged over > 10 runs. Transaction order varied from run to run due to random elements fetched and, more influential, randomly picked transactions. Hence, distribution of locks and, in turn, blocking delays might have strongly differed from run to run.

4   Measurements

All benchmark runs performed were executed in the following runtime environment. Note that benchmark suite (BenchSuite) as well as BrackitDB server were both running concurrently on the same node during the measurements.

CPU: 2x Intel Xeon E5420 @ 2.5 GHz, totaling in 8 cores
Memory: 16 GB DDR2 @ 667 MHz
Storage: RAID-5 with 3 disks (WD1002FBYS), 1 TB each, 7200 rpm (HDD)
Log: Samsung SSD 840 PRO, 256 GB (SSD)
Operating System: Ubuntu 13.04 (GNU/Linux 3.8.0-35 x86-64)
JVM: Java version 1.7.0_51, OpenJDK Runtime Environment (IcedTea 2.4.4)

4.1   System Parameters

We need to sketch the system parameters (not discussed so far) which strongly influence the benchmark results. As buffer size, we distinguish two corner cases: A *small buffer* keeps 1% of the XMark document incl. secondary indexes, thereby provoking lots of page replacements to HDD. In turn, a *large buffer* holds the

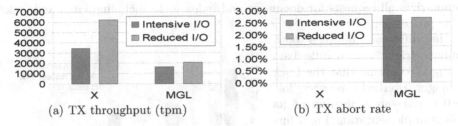

Fig. 6. Baseline experiments for X- and MGL-protocols

entire document in main memory; hence, I/O is only needed for logging. Every experiment starts on a warm buffer.

We run *ACID* transactions, where BrackitDB has to perform ARIES-style logging and recovery [4] based on *HDD* or *SSD*. To reveal special effects, logging (and, in turn, ACID) may be *disabled*. The skew parameter was set to 50% (*low*) or 99% (*high*). The *number of threads* was always 8. Scheduling is insofar *optimal* that a transaction is only initiated, if a worker thread is available. Hence, no queuing effects occur, apart from blocking delays in front of the Lock Manager. Note, the assignment of threads to processing cores is encapsulated by the OS, such that applications (e.g., DBMS) cannot exert influence.

4.2 Baseline Experiments

We started running the same XMark workload under pure X- and MGL-protocols to reveal especially the influence of buffer size or I/O activity and lock management overhead. To obtain the largest spread of transaction throughput, configurations with small buffer/low skew/log on HDD (Intensive I/O) and large buffer/high skew/log on SSD (Reduced I/O) were chosen (Fig. 6). Reducing I/O to the minimum necessary more than doubled throughput in case of the X-locking. It always exclusively locked the entire document—only a single worker thread was accessing the document at a time, while all others waited for the global X-lock to be released—and thereby prevented lock conflicts (no aborts), whereas MGL handled 8 threads concurrently. Lock conflicts together with multi-threading [14] obviously led to substantial throughput loss. For *Reduced I/O*, MGL only achieved one third of the X-lock throughput, i.e., throughput of serial workload executions profits much more from a larger DB buffer than that of parallel executions. Further, up to 3% of the transactions had to be aborted.

The result of Fig. 6 is seductive and points to *degree of parallelism* = 1 as the seemingly best case. But Fig. 7 reveals the downside of this approach. The response times dramatically grow even in case of very short transactions, although optimal scheduling was provided. A mix of long and short transactions would make *parallelism* = 1 definitely unacceptable for most applications.

TX timings in Fig. 7 are avg. times over the benchmark—all runs of all transaction types. *Block time* is the aggregated wait time for a transaction in front of the Lock Manager for lock releases. *Lock request time* containing the block time summarizes all requests for document and index locks including latch and lock table processing.

The difference between both timings (right and middle bars in Fig. 7 confirms that the Lock Manager overhead is very low (~ 0.1 ms)—an indication of its salient implementation. Fig. 7 further reveals the critical role of I/O on the *total runtime*—even in the case of MGL. Analyzing the deviation between the total transaction runtime (left bar) and the accumulated lock request time

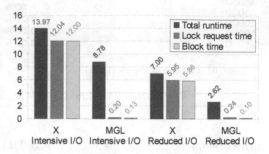

Fig. 7. TX timings (ms)

(middle bar) makes clear that the non-locking-related overhead is strongly dependent on I/O. This overhead includes TCP interaction between BenchSuite and BrackitDB, look-ups in meta-data, , accessing storage and evaluating navigation steps, logging every modification, fixing pages at buffer level and much more—these fixed costs make up a major portion of the total query runtime. In case of X-locking, only (parts of) communication and look-ups can be done in parallel leaving the major share of non-locking-related work for serial execution, whereas, in case of MGL, all work (except latch synchronization) could be concurrently done in all threads making the I/O dependency even stronger. Based on X-locking (mostly serial execution), we can approximately calculate the the total query runtime from Fig. 6(a): Intensive I/O: $\sim 33,000$ tpm = ~ 550 tps) $\equiv 1.9$ ms/TX and Reduced I/O: $\sim 62,000$ tpm = $\sim 1,030$ tps) $\equiv 0.97$ ms/TX. These values are confirmed by the corresponding TX timings in Fig. 7. In contrast, waits for I/O can be partially masked in MGL runtimes, however, latch waits (for buffer pages and lock table), lock conflicts, and multi-threading loss cannot be fully amortized.

Note, due to our "optimal" BenchSuite scheduling, queuing delays before TX initiation are not considered in the given TX timings. For user-perceived *response times*, we have to add a variable time component for queuing, which is dependent on traffic density of TX requests and TX execution time variance. Similar TX timings as shown for the MGL-protocol were obtained in the subsequent experiments, such that we will not repeat this kind of figure.

4.3 Lock Depth and Lock Escalation

Because taDOM acts as an extension to MGL, providing the same set of core lock modes plus some new taDOM-specific locks, we can expect this protocol to behave similar to MGL. Because we want to focus on lock conflicts, we have chosen a configuration with large buffer and high skew.

The first observation we can make from examining Fig. 8(a) is the huge impact on the transaction throughput when varying the lock depth in the range between 0 and 3. While a lock depth of 0 basically escalates every lock request to the

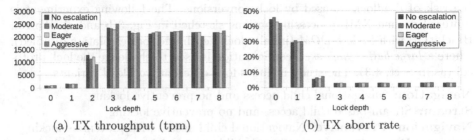

(a) TX throughput (tpm)   (b) TX abort rate

Fig. 8. taDOM with variation of lock depth and lock escalation policy

root node and thereby achieves a dreadful result in terms of system performance, taDOM seems to reach its climax when restricting the lock depth to 3. In fact, this pattern could be well explained by the XMark document structure and the workload. Increasing throughput values for lock depths up to 3 indicate that most transactions clash at the fourth level of the XML tree, where most of the document modifications take place. If the focus of transaction contention is not deeper in the XML tree, higher (static) lock depths only increase the lock management overhead. Fig. 8(b) is somewhat complementary to Fig. 8(a). High locking contention poses a higher risk of deadlocks which again leads to high abort rates and eventually a drastically reduced transaction throughput.

As compared to MGL (with $\sim 20,000$ tpm in Fig. 6(a)), taDOM variations are constantly superior and achieve an throughput gain of up to 20%. In this benchmark, the lock escalation policy applied seems to have only limited effect on throughput or abort rate whatsoever. A lock escalation count essentially tells us that only escalation "Aggressive" leads to actually performed lock escalations and exhibits some throughput gain of $\sim 10\%$ at lock depth 3. In this experiment, the conditions it takes for escalating were hardly satisfied for weaker escalation policies. Hence, *No escalation*, *Moderate* and *Eager* are not only coincidentally producing similar results, but they literally constitute the same scenario. On the other hand, scenarios with varying high-traffic contention in subtrees could certainly profit from a well-chosen escalation policy. Because escalation checking did hardly increase lock management overhead, the proposed policies should be offered as valuable optimization options for fine-grained XML lock protocols.

Although not comparable one-to-one, we obtained throughput values (tpm) for BrackitDB's SSD-based benchmarks that are up to 100% higher than the former HDD-based benchmarks using XTC ([6,2]. Due to space limits, we abstain from reporting our experiment using a small DB buffer, i.e., enforcing *Intensive I/O* with a small buffer, because it did not exhibit new aspects.

4.4 Workload Variants

To figure out whether there are still optimizations for taDOM left, we prepared different implementations of the same XMark transactions. Semantically, these workload variants are equivalent, but they utilize slightly different node operations to fulfill their task, or explicitly acquire locks here and there to reduce the risk of deadlocks caused by lock conversions. The following experiment is centered around XML processing options sketched in Fig. 3(b and c). We run BrackitDB with *basic taDOM* (infinite lock depth and lock escalation disabled), where a *large buffer/high skew* configuration is used to challenge the lock protocol (as in Sect. 4.3). The study includes four different workload variants:

Navigation: Navigational child access and no preventive locking
Stream: Stream-based child access and no preventive locking
Navigation+Preventive: Navigational child access with preventive locking
Stream+Preventive: Stream-based child access with preventive locking.

The first workload variant, where the transactions always perform navigation steps like `First-Child` and `Next-Sibling` whenever child traversal is necessary, is the same as used in the experiment in Sect. 4.3. The other alternative is to call `getChildren()` to retrieve a child stream using a single LR lock on the parent node. Although stream-based access might sound like the clear winner, there are situations where single navigation steps make more sense, e.g., when only a few of the first children are actually accessed. If indeed deadlocks were the limiting factors for transaction throughput in the previous benchmark, we should also notice a difference when applying *preventive locking*, since this strategy should drastically reduce deadlocks or even prevent them altogether. Preventive locking is a strategy where the strongest necessary lock mode is obtained before any particular node is accessed for the first time; thus, lock upgrades as the major source of deadlocks are avoided. It is clear that such a proceeding needs substantial application knowledge. Because automatic conversion is currently out of reach, we did it manually per transaction type. Hence, the results obtained are somewhat "fictitious", marking future optimization potential.

The navigational workload achieves a throughput of slightly more than 20,000 tpm. Modifying the workload implementation by substituting single navigation steps by child streams (*stream-based*) does not have a striking impact on performance either (see Fig. 9). The results for the non-preventive variants are pretty stable. In fact, the stream-based strategy consistently exposes lower lock request times throughout all benchmark repetitions. Also the total query runtime turns out to be lower if child streams are utilized instead of navigation steps, which might indicate a slight performance benefit resulting from storage-related aspects (e.g., more efficient sibling traversal).

What improves the throughput significantly, however, is reducing the risk of deadlocks, achieved by application-specific locking optimizations in the workload variants with preventive locking. As a result, the abort rate could be brought down from 2.8% (non-preventive variants) to literally 0%. In case of the navigational approach, the *preventive* locking strategy roughly

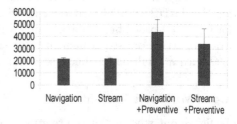

Fig. 9. Different workload variants

doubles the throughput, while it also improves the stream-based implementation

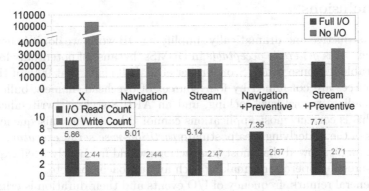

Fig. 10. TX throughput (tpm) for the I/O edge cases and I/O counts for *Full I/O*

considerably, yet not to the same extent. The higher throughput of *Navigation + Preventive* might result from frequent, incomplete child-set traversals. The reason why this diagram displays an additional error bar is to emphasize the high standard deviation for the two measurements involving preventive locking. No matter how often the experiment was repeated, these high fluctuations in the transaction throughput remain.

4.5 I/O Edge Cases

Eventually, due to the strong throughput dependency on the available I/O configuration, we want to review the four different workload variants employing taDOM under extreme I/O setups. So far, we gained the high-skew throughput results in Fig. 9 for "Reduced I/O" (I/O only for log flushes to SSD). Now, we look at the two edge cases concerning I/O activity. *Full I/O* is obtained by a tiny buffer covering only 0.1% of the document—enforcing very frequent page replacements to HDD—, rather slow log flushes to HDD, and low-skew access of the benchmark transactions. *No I/O*, in contrast, uses a large buffer (no replacements needed) and disables log flushes completely. Note, this edge case is achievable in the near future, when persistent *non-volatile RAM (NVRAM)* (enabling log writes in the sub-μsec range [10]) can replace part of DRAM.

With a minimum of DB buffer size, the avg. number of I/O events per transaction for *Full I/O* is similar in all experiments. Note, *No I/O* has neither read-nor write-I/O. The X-case with its throughput spectrum from ~23,000 tpm to >100,000 tpm is only present in Fig. 10 to highlight again the critical role of I/O, at least for serial execution. In the other experiments, based on multi-threaded processing under taDOM, the I/O impact is obvious, but far less critical. Obviously, the only moderate throughput gain by reduced I/O is a consequence of processing frictions caused by multi-threading and latch/lock/deadlock conflicts. Hence, there is future optimization potential for multi-core scheduling in the OS domain and for adjusted concurrency control in the DBMS domain.

5 Conclusions

Although seductive and dramatically simplifying DB work, we don't believe that *parallelism* = 1 is *generally acceptable* in DBMSs, because of potentially long and unpredictable response times. Stonebraker et al. [14] claim that 90% of the run-time resources are consumed by four sources of overhead, namely buffer pool, multi-threading, record-level locking, and an ARIES-style [4] write-ahead log. Even if this is correct, most applications cannot tolerate the consequences. Refinements of the underlying idea postulate *partition-wise serial execution*. Again, Larson et al. [9] showed that most transactions would frequently need expensive partition-crossing operations, making such an approach obsolete.

As a general remark, frequency of I/O events and their duration is critical for high-performance transaction processing. Large main memories and use of SSDs (in near future also NVRAM)—at least for logging tasks—help to substantially increase the level of concurrency and throughput. Yet, multi-threading and lock conflicts diminish this gain considerably, as all result figures confirm. Because scheduling and multi-core mapping are OS tasks, improvements can not be obtained by the DBMS alone—an application program from the OS perspective.

With appropriately chosen taDOM options, we observed in our study a through-put gain of ~ 20% compared to the widely used MGL-protocol. Optimal lock depth values strongly depend on document and workload; if wrongly chosen, they may have a suboptimal impact on concurrency. Well-chosen (or unlimited) lock depth combined with lock escalation may be a better solution with acceptable costs, because it dynamically cares for a balance of appropriate concurrency-enabling lock granules and lock management overhead. As a rule of thumb, setting the lock depth value *too high is much better than too low*.

Using application knowledge, processing XML documents could be significantly improved thereby reducing the risk of deadlocks. Scanning the child set of a node is a frequent operation, where—as a kind of context knowledge—preventive locking can be applied. As a result, it brought the abort rate down to 0% and, in turn, substantially enhanced the transaction throughput.

References

1. Bächle, S., Sauer, C.: Unleashing XQuery for Data-Independent Programming. Datenbank-Spektrum 14(2), 135–150 (2014)
2. Bächle, S., Härder, T.: The Real Performance Drivers Behind XML Lock Protocols. In: Bhowmick, S.S., Küng, J., Wagner, R. (eds.) DEXA 2009. LNCS, vol. 5690, pp. 38–52. Springer, Heidelberg (2009)
3. BrackitDB – Google Project Hosting (2014), http://code.google.com/p/brackit/wiki/BrackitDB
4. Gray, J., Reuter, A.: Transaction Processing: Concepts and Techniques. Morgan Kaufmann (1993)
5. Haustein, M.P., Härder, T.: An Efficient Infrastructure for Native Transactional XML Processing. Data & Knowl. Eng. 65(1), 147–173 (2008)
6. Haustein, M.P., Härder, T., Luttenberger, K.: Contest of XML Lock Protocols. In: Proc. VLDB, pp. 1069–1080 (2006)

7. Haustein, M.P., Härder, T.: Optimizing Lock Protocols for Native XML Processing. Data & Knowl. Eng. 65(1), 147–173 (2008)
8. Hiller, M.: Evaluation of Fine-grained Locking in XML Databases. Master Thesis, University of Kaiserslautern (2014)
9. Larson, P.-A., et al.: High-Performance Concurrency Control Mechanisms for Main-Memory Databases. PVLDB 5(4), 298–309 (2011)
10. Pelley, S., et al.: Storage Management in the NVRAM Era. PVLDB 7(2), 121–132 (2013)
11. Mathis, C., Härder, T., Schmidt, K.: Storing and Indexing XML Documents Upside Down. Computer Science - R&D 24(1-2), 51–68 (2009)
12. Mohan, C.: ARIES/KVL: A Key-Value Locking Method for Concurrency Control of Multiaction Transactions Operating on B-Tree Indexes. In: Proc. VLDB, pp. 392–405 (1990)
13. Siirtola, A., Valenta, M.: Verifying Parameterized taDOM+ Lock Managers. In: Geffert, V., Karhumäki, J., Bertoni, A., Preneel, B., Návrat, P., Bieliková, M. (eds.) SOFSEM 2008. LNCS, vol. 4910, pp. 460–472. Springer, Heidelberg (2008)
14. Stonebraker, M., Weisberg, A.: The VoltDB Main Memory DBMS. IEEE Data Eng. Bull. 36(2), 21–27 (2013)

Multi-dialect Workflows*

Leonid Kalinichenko, Sergey Stupnikov,
Alexey Vovchenko, and Dmitry Kovalev

Institute of Informatics Problems, Russian Academy of Sciences, Moscow, Russia
{leonidandk,itsnein,dm.kovalev}@gmail.com, ssa@ipi.ac.ru

Abstract. The results presented in this paper contribute to the techniques for conceptual representation of data analysis algorithms as well as processes to specify data and behavior semantics in one paradigm. An investigation of a novel approach for applying a combination of semantically different platform independent rule-based languages (dialects) for interoperable conceptual specifications over various rule-based systems (RSs) relying on the rule-based program transformation technique recommended by the W3C Rule Interchange Format (RIF) is extended here. The approach is coupled also with the facilities for heterogeneous information resources mediation. This paper extends a previous research of the authors [1] in the direction of workflow modeling for definition of compositions of algorithmic modules in a process structure. A capability of the multi-dialect workflow support specifying the tasks in semantically different languages mostly suited to the task orientation is presented. A practical workflow use case, the interoperating tasks of which are specified in several rule-based languages (RIF-CASPD, RIF-BLD, RIF-PRD) is introduced. In addition, OWL 2 is used for the conceptual schema definition, RIF-PRD is used also for the workflow orchestration. The use case implementation infrastructure includes a production rule-based system (IBM ILOG), a logic rule-based system (DLV) and a mediation system.

Keywords: conceptual specification, workflow, RIF, production rule languages, database integration, mediators, PRD, multi-dialect infrastructure.

1 Introduction

This work keeps on the intention of developing the facilities for conceptual declarative problem specification and solving in data intensive domains (DID). In [1] it was claimed that conceptual data semantics alone (e.g., formalized in ontology languages based on description logic) are insufficient, so that conceptual representation of data analysis algorithms as well as processes for problem solving are required to specify data and behavior semantics in one paradigm.

---

* This research has been done under the support of the RFBR (projects 13-07-00579, 14-07-00548) and the Program for Basic Research of the Presidium of RAS.

Y. Manolopoulos et al. (Eds.): ADBIS 2014, LNCS 8716, pp. 352–365, 2014.
© Springer International Publishing Switzerland 2014

The results presented in this paper extend the research [1] aimed at the definition and implementation of the facilities for conceptually-driven problems specification and solving in DID aiming at ensuring eventually the following capabilities for expressing the specifications:

1. an ability to provide complete and precise specification of the abstract structure and behavior of the domain entities, their consistency, relationship and interaction;
2. well-grounded diversity of semantics of the modeling facilities providing for the best attainable expressiveness, compactness and precision of the definition of the problem solving algorithm specifications;
3. arrangements for the extensions of the modeling facilities satisfying the changing technological and practical needs;
4. specification independence from implementation platforms (languages, systems);
5. specification independence from concrete information resources (databases, services, ontologies and others) combined with facilities for their semantic integration and interoperability;
6. built-in methodologies for creation of unifying specification languages providing for construction of semantics-preserving mappings of conceptual specifications into their implementations in specific platforms.

The research reported in [1] investigated the conceptual modeling facilities for DID applying rule-based declarative logic languages possessing different, complementary semantics and capabilities combined with the methods and languages for heterogeneous data mediation and integration. Two fundamental techniques were combined: (1) constructing of the unifying extensible language providing for semantics-preserving mapping into it of various information resource (IR) specification languages (e.g., such as DDL and DML for databases); (2) creation of the unified extensible family of rule-based languages (dialects) and a model of interoperability of the programs expressed in such dialects.

The first technique is based on the experience obtained in course of the SYNTHESIS language development [2]. The kernel of the SYNTHESIS language is based on the object-frame data model used together with the declarative rule-based facilities in the logic language similar to a stratified Datalog with functions and negation. The extensions of the kernel are constructed in such a way that each extension together with the kernel is a result of semantic preserving mapping of some IR language into the SYNTHESIS [2]. The canonical information model is constructed as a union of the kernel with such extensions defined for various resource languages. Canonical model is used for development of mediators positioned between the users, conceptually formulating problems in terms of the mediators, and distributed resources. A schema of a subject mediator for a class of problems includes the specification of the domain concepts defined by the respective ontologies.

Another, multi-dialect technique for rule-based programs interoperability applied is based on the RIF standard [3] of W3C. RIF introduces a unified family

of rule-based languages together with a methodology for constructing of semantic preserving mappings of specific languages used in various Rule-based
Systems (RS) into RIF dialects. Examples of RS include *SILK, OntoBroker,
DLV, IBM Websphere ILOG JRules*, and others (more examples can be found
at http://www.w3.org/2005/rules/wiki/Implementations). From the RIF
point of view an IR is a program developed in a specific language of some RS.

In [1] the first results obtained were presented including the description of an
approach and an infrastructure supporting:

- application domain conceptual specification and problem solving algorithms
  definitions based on the combination of the heterogeneous database mediation technique and the rule-based multi-dialect facilities;
- interoperability of distributed multi-dialect rule-based programs and mediators integrating heterogeneous databases;
- rule delegation approach for the peer interactions in the multi-dialect environment.

The proof-of-concept prototype of the infrastructure based on the SYNTHE-
SIS environment and RIF standards has been implemented. The approach for
multi-dialect conceptualization of a problem domain, rule delegation, rule-based
programs and mediators interoperability were explained in detail and illustrated
on an use-case in the finance domain [1]. For the conceptual definition of the
use-case problem the OWL was used for the domain concepts definition and
two RIF logic dialects RIF-BLD [4] and RIF-CASPD [5] were used and mapped
for implementation into the SYNTHESIS formula language and the ASP-based
DLV [6] language respectively.

The results obtained so far are quite encouraging for future work: they show
that the mentioned in the beginning capabilities (1-6) sought for conceptual
modeling become feasible. This paper reports the results of extending the research in the direction of modeling of the processes for the problem solving
following the approach briefly outlined above. These results include extensions
of the infrastructure and specification languages considered in [1] to the workflow level keeping the same approach and paradigm as well as aiming at the
capabilities of the conceptualization (1-6) that were stated in [1] and mentioned
in the beginning of the introduction.

For investigation of such extension w.r.t. the choice of rule-based languages
it was decided not to go outside the limits of the existing set of the published
RIF dialects. Such decision would allow to retain well-defined semantics of the
conceptual rule-based languages with a possibility to check preservation of their
semantics by various languages of the implementing systems.

The production rule dialect RIF PRD [7] has been chosen as the language for
the workflow modeling in such a way that the tasks of the workflow can have
multi-dialect rule-based representation (as defined in [1]). This paper reporting
the results of such investigation is structured as follows. To make the paper
self-contained, the next section provides a brief overview of the infrastructure
supporting multi-dialect programming defined in details in [1]. Here we stress

that this infrastructure is suitable for the workflow tasks specification. Workflow-oriented extension of the multi-dialect infrastructure is considered in the section 3. Use case implementation in the proof-of-concept prototype is given in the section 4. Related works are reviewed in the section 5. Conclusion summarizes contributions of the research.

2 Basic Principles of the Workflow Tasks Representation in the Multi-dialect Infrastructure

Every workflow task (besides those that for pragmatic reasons are defined as externally specified functions) is assumed to be represented in the novel infrastructure defined in details in [1]. Conceptual programming of tasks is performed using the RIF dialects (now not only logic but also production rule dialects can be used). Conceptual tasks are implemented by their transformation into the rule-based programs of the respective RSs and mediation systems (MSs). *Conceptual specification of a task* is defined in the context of a subject domain and consists of a set of RIF-documents (document is a specification unit of RIF). The *conceptual schema* of the domain is defined using OWL 2 [8] ontologies. Such usage of ontology is analogous to [9], however it is specifically important in the multidialect environment due to the formally defined compatibility between RIF and OWL. The ontologies contain entities of the domain and their relationships (Fig. 1, right-hand part). Conceptual specification of a task is defined over conceptual schema. Ontologies are imported into the RIF-documents specifying an import profile, for instance, *OWL Direct*. Documents *import* other documents having the same semantics (the *Import* directive), *link* documents defined using other dialects and having different semantics (remote module directive *Module*) or *refer* to entities contained in other documents using *external terms*.

Semantics of a conceptual task definition in such setting become a multidialect one. The specification modules of a task are treated as peers. Mediation modules are assumed to be defined in RIF-BLD for representation of the mediator rules (to be interpreted in SYNTHESIS) supporting schema mapping and semantic integration of the information resources. Multi-dialect task is implemented by means of transformation of conceptual specifications into modular, component-based P2P program represented in the languages of the mediation (MSs) and rule-based systems (RSs) with the respective semantics. Interoperability of logic rule components of such distributed program is carried out by means of the delegation technique ([1] section 3.3). Production rule components are considered as external functions, interoperability is achieved through the mechanism of external terms.

A schema S_R of a peer R is a set of entities (classes or relations and their attributes) corresponding to extensional and intensional predicates of the resource implementing the peer R. The RS or the mediation system (MS) of every peer R should be a conformant D_R consumer, where D_R is a respective RIF dialect (Fig. 1, left-hand part). Conformance is formally defined using formula entailment and language mappings [3].

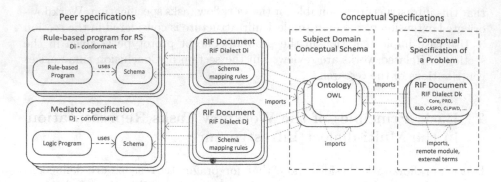

Fig. 1. Conceptual schema and peer specifications

The peer R is relevant to a RIF-document d of a conceptual specification of a problem (Fig. 1, right-hand part) if (a) D_R is a subdialect of the document d dialect (subdialect is a language obtained from some dialect by removing certain syntactic constructs and imposing respective restrictions on its semantics [4]; every program that conforms with the subdialect also conforms with the dialect) and (b) entities of the peer schema S_R (if they exist) are *ontologically relevant* to entities of the conceptual schema the names of which are used in d for extensional predicates.

The schema of a relevant peer is mapped into the conceptual schema. The mapping establishes the correspondence of the conceptual entities referred in the document d to their expressions in terms of entities of the schema S_R using rules of the D_R dialect. These schema mapping rules constitute separate RIF-document (Fig. 1, middle part).

Peers communicate using a technique for distributed execution of the rule-based programs. The basic notion of the technique is *delegation* – transferring facts and rules from one peer to another. A peer is installed on a node of the multi-dialect infrastructure. A node is a combination of a wrapper, an RS or an MS, and a peer (to save space, we refer a reader for the details to the paper [1], Fig. 3). A wrapper transforms programs and facts from the specific RIF dialect into the language of the RS or MS and vice versa. A wrapper also implements the delegation mechanism. Transferring facts and rules among peers is performed in the RIF dialects.

A special component (*Supervisor*) of the architecture defined in [1] stores shared information of the environment, i.e. conceptual specifications related to the domain and to the problem, a list of the relevant resources, RIF-documents combining rules for the conceptual specification and a resource schema mapping.

Implementation of the conceptual specification includes the following steps:

- Rewriting of the conceptual documents into the RIF-programs of the peers performed by the *Supervisor*. The rewriting includes also (1) replacing the document identifiers (used to mark predicates) by peer identifiers and (2) adding schema mapping rules to programs (Fig. 1, middle part).

- A transfer of the rewritten programs to nodes containing peers relevant to the respective conceptual documents. The transfer is performed by the *Supervisor* by calling the method *loadRules* of the respective node wrappers.
- A transformation of the RIF-programs into the concrete RS or MS languages. The transformation is performed by the *NodeWrapper* or by the RS or MS itself (if the RS or MS supports the respective RIF dialect).
- An execution of the produced programs in P2P environment.

During the process of rewriting of the conceptual schema into the resource programs the relationships between RIF-documents of the conceptual schema defined by remote or imported terms are replaced by relationships between peers also defined by remote or imported terms. To implement remote and imported terms a rule delegation mechanism is used to transfer facts and rules from one peer to another. The details of *rule delegation* approach including description of the related algorithms are provided in [1].

3   Workflow-Oriented Extension of the Multi-dialect Infrastructure

The aim of the infrastructure proposed is a conceptual programming of problems in the RIF-dialects and an implementation of conceptual specifications using rule-based languages of the RSs and MSs. One of the objectives of this particular paper is to introduce an extension of the existing multi-dialect infrastructure [1] aiming at the conceptual specification of rule-based workflows.

Conceptual specification of a problem (class of problems) is defined in the context of a subject domain and consists of a set of RIF-documents. Besides the documents expressed in the logic dialects of RIF, the documents expressed in the production rule dialect (RIF-PRD) also can be a part of conceptual specification of a problem. In particular, these documents are aimed to express a process of solving the problem as the production rule-based workflow.

A workflow consists of a set of tasks orchestrated by specific constructs (*workflow patterns* [10], for instance *sequence, split, join*) defining the order of tasks execution. The specification of such orchestration is called here a *workflow skeleton*. A skeleton is defined using RIF-PRD production rules. Workflows and workflow patterns can be represented using production rules in various ways, e.g. as in [10][19]. The approach applied in this paper to represent workflows requires the extension of RIF-PRD dialect by several built-in predicates (they are considered to be a part of *wkfl* namespace referenced by http://www.w3.org/2014/rif-workflow-predicate# URI similarly to *func* and *pred* namespaces defined in [23] for built-in functions and predicates of RIF).

Predicates *wkfl:variable-definition* and *wkfl:variable-value* allow to specify workflow variables and their values and thus to organize the data flow within a workflow. Predicates *wkfl:parameter-definition* and *wkfl:parameter-value* allow to specify workflow parameters and their values and thus to define the interface of a workflow in terms of input and output parameters. Using of workflow parameters and variables is illustrated in the next section.

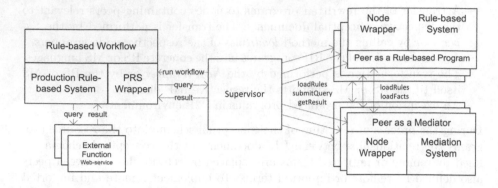

Fig. 2. Extended multi-dialect infrastructure

Predicate *wkfl:end-of-task(?arg)*, where *?arg* is an identifier of a task, turns into true if a task *?arg* has been completed. The predicate allows to orchestrate the order of execution of workflows tasks using conditions and actions of production rules. For instance, *AND-Split* workflow pattern is represented in RIF-PRD by the following production rule template using *wkfl:end-of-task* predicate:

```
If Not(External(wkfl:end-of-task(A)))
Then Do (Act(A)  Assert(External(wkfl:end-of-task(A))))
If And(Not(External(wkfl:end-of-task(B))) External(wkfl:end-of-task(A)))
Then Do (Act(B)  Assert(External(wkfl:end-of-task(B))))
If And(Not(External(wkfl:end-of-task(C))) External(wkfl:end-of-task(A)))
Then Do (Act(C)  Assert(External(wkfl:end-of-task(C))))
```

The template includes three rules for tasks A, B and C respectively. $Act(A)$, $Act(B)$ and $Act(C)$ denotes actions associated with tasks A, B and C. Orchestration (tasks B and C are executed concurrently right after task A is completed) is specified using *wkfl:end-of-task* predicate in conditions and *Assert* actions of rules. More complicated patterns like OR-, XOR- splits and joins, structured loops, subflows and others are represented in RIF-PRD similarly.

Workflow tasks can be specified as:

- separate RIF-documents in various logic RIF-dialects (this is the way how multi-dialect infrastructure [1] is extended with workflow capabilities);
- separate RIF-documents in the RIF-PRD dialect;
- set of production rules embedded into the workflow skeleton;
- external functions treated as "black boxes".

Semantics of tasks specified as multi-dialect logic programs are defined in accordance with the RIF-FLD [3] standard and standards for the respective RIF-dialects. Semantics of tasks specified as production rule programs are defined in accordance with the RIF-PRD standard. Semantics of external functions "are assumed to be specified externally in some document" [3].

All kinds of tasks (except those that are embedded into a workflow skeleton) are referenced in the workflow skeleton as *external terms* [3] like *External(t)* where term t is defined by an external resource identified by internationalized resource identifier (IRI) [3].

Workflows defined in the conceptual specification are implemented in the environment shown on Fig. 2. P2P environment [1] intended to implement logic programs is extended with a production rule-based system — PRS in short (for instance, a production system compliant with the OMG Production Rule Representation [18]) and with external functions, implemented as web-services. Implementation of the conceptual specification includes the following steps:

- Transfer of the conceptual RIF-documents constituting a workflow skeleton to the production rule-based system node (performed by the *Supervisor* component).
- Transformation of the conceptual RIF-documents constituting a workflow skeleton into the language of the production rule-based system (performed by the *PRS Wrapper* component).
- Transferring RIF logic programs related to tasks to the relevant nodes of the environment and transformation of the RIF-programs into the concrete RS or MS languages [1].
- Execution of the workflow.

The interface of the *Supervisor* includes methods for submitting and executing a workflow represented as a set of RIF-documents, and for getting the result of the workflow execution.

4   Multi-dialect Workflow Use-Case

Motivation of the use case that illustrates the proposed approach comes from the finance area. The use case extends the investment *portfolio diversification problem* defined in ([1], Appendix) by adding workflow orchestration applying the production rules dialect RIF-PRD. The idea of the portfolio diversification problem is as follows. The portfolio is a collection of securities of companies, and its size is the number of securities in the portfolio. The problem is to build a diversified portfolio of maximum size. Diversification means that the prices of the securities in portfolio should be almost independent of each other. If the price of one security falls, it will not significantly affect the prices of other. Thus the risk of a portfolio sharp decrease is reduced.

The input data for the problem is a set of securities and respective time series of indicators of the security price for each security. Time series for each security is a set of pairs (d, v) where d is a date and v is an indicator of the security price (for instance, closing price). The financial services Google Finance (https://www.google.com/finance) and Yahoo! Finance (http://finance.yahoo.com/) are considered. They include various indicators of the security price for all trading days of the last decades. For the diversified portfolio the securities having non-correlated time series should be used. Non-correlation of the time series means that their correlation is less than some predetermined price correlation value. The output data for the problem is a set of subsets of securities of the maximum size, for which the pairwise correlation will be less than the predetermined one.

The maximum satisfying subset of securities is calculated in the following way. Let G be a graph where the vertices are the securities. An edge between two securities exists if absolute value of their correlation is less than a specified number. So any two securities connected by an edge are considered as non-correlated. In such case, the problem of finding the portfolio of the maximum size is exactly the problem of finding a maximum clique in an undirected graph. A maximal clique is a maximal portfolio. Note that several different maximal portfolios can be found.

The conceptual specification of the use case [1] used two RIF-dialects: RIF-BLD and RIF-CASPD. The use case was implemented in the environment containing a mediation system used as a platform for RIF-BLD [4] and ASP-based DLV system [6] – a platform for RIF-CASPD. The RIF-BLD was used to specify the problem of data integration, and RIF-CASPD – the problem of finding a maximum clique in an undirected graph.

The portfolio use case is extended in this work in the following way. The goal is not only to build a set of diversified portfolios, but to choose the "best" of them according to some criteria. There are several approaches to choose the most appropriate portfolio.

The most recognized one is based on the Markovitz portfolio theory [11]. The idea is to choose the portfolio, which has the maximum risk/return ratio. The most well-known metric to operate with risk/return is Sharpe-ratio [12]: $(r_p - r_f)/\sigma^2$. Here r_p denotes the expected return of the portfolio, r_f denotes a risk free rate, σ^2 denotes a portfolio standard deviation (risk). The more the Sharpe-ratio, the better the investment is.

Another approach is based on an idea that with the advent of social networks, it became possible to monitor ideas, sentiments, actions of people and lots of available information has to do with the markets and investments. In [13] Bollen et al. draw the connection between the mood of investor tweets and the move of Dow Jones Index, stating that correlation between them is more that 80%. The idea of using tweets to assess market movements has been implemented in several hedge funds.

Combining these two strategies could provide benefits of both of them, which leads to the following problem statement: having S&P500 (a stock market index maintained by the Standard & Poor's, comprising 500 large-cap American companies) list of companies, compute the diversified portfolio of maximum size with the best riskreturn and sentiment ratios. Fig. 3 demonstrates the workflow of the extended portfolio problem. It contains six tasks:

- *getPortfolios.* A set of diversified portfolio candidates is computed. The multi-dialect task specification consists of two RIF-documents in BLD and CASPD dialects ([1], Appendix). Portfolios received as a result contain only security tickers, they have to be augmented by financial and sentiments ratios.

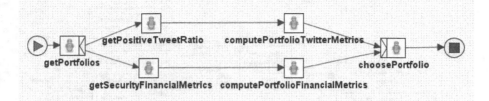

Fig. 3. Portfolio workflow

- *getPositiveTweetRatio.* This task is responsible for computing a sentiment ratio of tweets for every security. Every tweet is assessed to be positive, negative or neutral. The task is specified as a call of external function.
- *computePortfolioTwitterMetrics.* The portfolio sentiment ratio is computed as the average of its securities sentiment ratio. The task is specified using RIF-PRD.
- *getSecurityFinancialMetrics.* For every security in a portfolio the financial rates (the *expected return* and the *standard deviation*) are calculated on the basis of historical rates of securities specified as an OWL 2 class of the ontology of the application domain. The task is specified using RIF-BLD dialect.
- *computePortfolioFinancialMetrics.* The computation of the portfolio expected return, risk, and Sharpe-ratio is done within this task. The task is specified using RIF-PRD dialect.
- *choosePortfolio.* The best portfolio is chosen according to maximizing the (*Sharpe ratio* ∗ *sentiment ratio*) product. The task is specified using RIF-PRD dialect.

Workflow skeleton is specified as a RIF-PRD document importing the ontology of the application domain. To save space we provide below the orchestration rules only for the task *getPortfolios*:

```
Document( Dialect(RIF-PRD)
 Import(<http://synthesis.ipi.ac.ru/portfolio/ontology#>
        <http://www.w3.org/ns/entailment/OWL-Direct>)
 Prefix(ont  <http://synthesis.ipi.ac.ru/portfolio/ontology#>)
 Prefix(svc  <http://synthesis.ipi.ac.ru/portfolio/services#>)
Group 2 ( Do(
Assert(External(wkfl:parameter-definition(startDate xsd:string IN)))
Assert(External(wkfl:parameter-definition(endDate xsd:string IN)))
Assert(External(wkfl:variable-definition(ps List<ont:Portfolio> IN))))
Group 1 (
Forall ?sd ?ed such that( External(wkfl:parameter-value(startDate  ?sd))
 External(wkfl:parameter-value(endDate  ?ed))  )
( If Not(External(wkfl:end-of-task(getPortfolios)))
  Then
  Do( Modify(External(
        wkfl:variable-value(ps External(svc:getPortfolios(?sd ?ed))))
  Assert(External(wkfl:end-of-task(getPortfolios)))  )  )  )
```

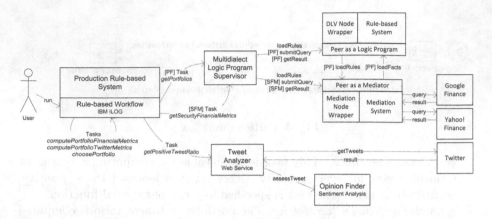

Fig. 4. Portfolio problem implementation structure

Production rules of the document are divided into two groups. The first group with priority 2 contains rules defining workflow parameters and variable. Parameters are *start date* and *end date* of historical rates used for calculation of *portfolio metrics*. Workflow variable *ps* denotes a set containing *portfolio candidates*.

The second group with priority 1 contains the orchestration rules – workflow skeleton. The only orchestration rule provided in the example above corresponds to the task *getPortfolios*. The external function *getPortfolios* encapsulates a multi-dialect logic program calculating portfolio candidates ([1], Appendix). A *Modify* action is used to call the function and to put the returned result into the *ps* variable.

The implementation structure of the use case is shown on Fig. 4.

The RIF-PRD workflow skeleton was transformed into a program in the ILOG [14] language combining production rules and workflow facilities (like *fork* and *sequence*). The ILOG program was executed in the *IBM Operational Decision Manager* tool.

The *computePortfolioTwitterMetrics*, *computePortfolioFinancialMetrics*, and *choosePortfolio* tasks are implemented as production rules in ILOG.

The *getSecurityFinancialMetrics* task uses the same instance of the mediation system as the *getPortfolios* task. The reason is that financial metrics are calculated using the historical rates of the securities. This is exactly the information that is extracted by the mediation system from *Google Finance* and *Yahoo! Finance*. The difference between two tasks is that the *getPortfolios* is implemented as a submission of a query to the DLV node, but the *getSecurityFinancialMetrics* is implemented as a submission of a different query to the Mediation Node.

The *getPositiveTweetRatio* task is implemented as a Java-program wrapped by a web service. First it collects tweets using the *Twitter Streaming API*. After that a sentiment analysis is done by the *Polarity Classifier* of the *OpinionFinder* tool [15] which assesses if tweet is positive, negative or neutral. Finally the sentiment ratio for every security in a portfolio is calculated.

Detailed specifications of the use case including ontologies, logic programs, production rules, workflow specification and implementation are provided in Appendix of [1] and in a technical report [16]. The technical report includes also results obtained by one of the workflow runs.

5   Related Work

Two types of workflow models, namely abstract and concrete were identified [17]. In the abstract model, a workflow is described in an abstract form, without referring to specific resources. In this paper we propose workflow representation in abstract and platform independent (PIM) form.

A classification model for scientific workflow characteristics [10] contributes to better understanding of scientific workflow requirements. The list of structural patterns discovered during this analysis (including sequential, parallel, parallel-split, parallel-merge, mesh) influenced our choice of the workflow patterns.

The OMG standard [18] reflects an attitude to production rules from the industrial side providing an OMG MDA PIM model with a high probability of support at the PSM (Platform-Specific Model) level from the rule engine vendors. Similar capabilities though formally defined are used as the basis for the production rule dialect RIF-PRD [7].

Some vendors of such production rule engines have extended their languages with the workflow specification capabilities. IBM has extended ILOG to provide the ruleflow capability. Microsoft supports Windows Workflow Foundation as a platform providing the workflow and rules capabilities. Examples of specific formalisms for PIM rule-based process specifications are provided also in [19].

Comparing to the known variants of the PIM production rule representations, selection of the RIF-PRD production rule dialect we consider well-grounded: (1) the RIF-PRD is formally defined; (2) RIF ensures support of interoperability of modules written in different rule-based dialects with different semantics; (3) RIF provides foundations for PIM to PSM semantic preserving transformation; (4) RIF also provides an ability for specification of the concepts in application domain terms combining rule-based specifications with the OWL ontologies.

Importance of providing the inter-dialect interoperation is advocated in [20] for combining the functionalities of production systems and logic programs for abductive logic programming (ALP). The ALP framework gives a model-theoretic semantics to both kinds of rules and provides them with powerful proof procedures, combining backward and forward reasoning.

Papers related to RIF-PRD experimentations are focused mainly on the issue of the PRD programs transformation to an implementation system. In [21] a case study of bridging the ILOG Rule Language (IRL) to RIF-PRD and vice versa is considered. In [22] implementation of RIF-PRD in three different paradigms: Answer Set Programming, Production Rules and Logic Programming (XSB) is investigated.

The contribution of this paper w.r.t. previous works of the authors [1] consists in extensions of the infrastructure and specification languages considered in [1] to the workflow level.

6   Conclusion

Progress in the investigation of the infrastructure [1] for the conceptual multi-dialect interoperable programming in the abstract, rule-based, platform independent notations is reported. We present an extension of the coherent combination of the multi-dialect rule-based programming technique recommended by the W3C RIF with the approach for unifying modeling of heterogeneous data bases for their semantic mediation. The extension of the infrastructure and specification languages considered in [1] in the direction of the workflow modeling is presented.

Sticking to the limits of the existing set of the published RIF dialects, we present a capability of the multi-dialect workflow support with the tasks specified in semantically different languages mostly suited to the task orientation. We present a realistic problem solving use case containing the interoperating tasks specified in several platform independent rule-based languages: RIF-CASPD, RIF-BLD, RIF-PRD. In addition, OWL 2 is used for the conceptual schema definition, RIF-PRD is applied for the workflow orchestration. The platforms selected for implementation of the tasks include: DLV, SYNTHESIS, IBM ILOG. Such approach retains well-defined semantics of the platform independent rule-based languages with a possibility to check preservation of their semantics by various languages of the implementing systems. The principle of independence of tasks from the specific IRs is carried out by the heterogeneous database mediation facilitates contributing to the re-use of tasks and workflows. Alongside with the further extension of the approach, in the future work we plan to apply the conceptual multi-dialect programming philosophy for support of the experiments in data intensive sciences. In particular, we plan to investigate modeling hypotheses in astronomy representing them as a set of rules applying the multiplicity of the dialects required.

References

1. Kalinichenko, L.A., Stupnikov, S.A., Vovchenko, A.E., Kovalev, D.Y.: Conceptual Declarative Problem Specification and Solving in Data Intensive Domains. Informatics and Applications 7(4), 112–139 (2013),
   http://synthesis.ipi.ac.ru/synthesis/publications/13ia-multidialect
2. Kalinichenko, L.A., Stupnikov, S.A., Martynov, D.O.: SYNTHESIS: A language for canonical information modeling and mediator definition for problem solving in heterogeneous information resource environments, p. 171. IPI RAN, Moscow (2007)
3. Boley, H., Kifer, M. (eds.): RIF Framework for Logic Dialects. W3C Recommendation, 2nd edn. (February 5, 2013)
4. Boley, H., Kifer, M. (eds.): RIF Basic Logic Dialect. W3C Recommendation, 2nd edn. (February 5, 2013)
5. Heymans, S., Kifer, M. (eds.): RIF Core Answer Set Programming Dialect (2009),
   http://ruleml.org/rif/RIF-CASPD.html
6. Leone, N., Pfeifer, G., Faber, W., Eiter, T., Gottlob, G., Perri, S., Scarcello, F.: The DLV System for Knowledge Representation and Reasoning. ACM Transactions on Computational Logic 7(3), 499–562 (2006)

7. de Sante Marie, C., Hallmark, G., Paschke, A. (eds.): RIF Production Rule Dialect. W3C Recommendation, 2nd edn. (February 5, 2013)
8. Bock, C., et al. (eds.): OWL 2 Web Ontology Language Structural Specification and Functional-Style Syntax. W3C Recommendation, 2nd edn. (December 11, 2012)
9. Calvanese, D., et al.: Ontology-based database access. In: Proceedings of the Fifteenth Italian Symposium on Advanced Database Systems, pp. 324–331 (2007)
10. Ramakrishnan, L., Plale, B.: A Multi-Dimensional Classification Model for Scientific Workfow Characteristics. In: Proceedings of the 1st International Workshop on Workflow Approaches to New Data-centric Science. ACM, New York (2010)
11. Markowitz, H.M.: Portfolio Selection: Efficient Diversification of Investments. Wiley (1991)
12. Sharpe, W.F.: Mutual Fund Performance. J. Business 39(S1), 119–138 (1966)
13. Bollen, J., Maoa, H., Zeng, X.: Twitter mood predicts the stock market. J. Comp. Sci. 2(1) (2011)
14. IBM WebSphere ILOG JRules Version 7.0. Online documentation, http://pic.dhe.ibm.com/infocenter/brjrules/v7r0/index.jsp
15. Wilson, T., Wiebe, J., Hoffmann, P.: Recognizing Contextual Polarity in Phrase-Level Sentiment Analysis. In: Proceedings of the Conference on Human Language Technology and Empirical Methods in Natural Language Processing, pp. 347–354. Association for Computational Linguistics, Stroudsburg (2005)
16. Kalinichenko, L.A., Stupnikov, S.A., Vovchenko, A.E., Kovalev, D.Y.: Multi-Dialect Workflows: A Use Case. Technical Report. IPI RAN, Moscow (2014), http://synthesis.ipi.ac.ru/synthesis/projects/RuleInt/Multidialect-Workflows-Use-Case.pdf
17. Yu, J., Buyya, R.: A taxonomy of scientific workflow systems for grid computing. ACM SIGMOD Records 34(3), 44–49 (2005)
18. Production Rule Representation (PRR), Version 1.0. OMG Document Number: formal/2009-12-01 (2009), http://www.omg.org/spec/PRR/1.0
19. Boukhebouze, M., Amghar, Y., Benharkat, A.-N., Maamar, Z.: A rule-based approach to model and verify flexible business processes. Int. J. Business Process Integration and Management 5(4), 287–307 (2011)
20. Kowalski, R., Sadri, F.: Integrating Logic Programming and Production Systems in Abductive Logic Programming Agents. In: Polleres, A., Swift, T. (eds.) RR 2009. LNCS, vol. 5837, pp. 1–23. Springer, Heidelberg (2009)
21. Cosentino, V., Del Fabro, M.D., El Ghali, A.: A model driven approach for bridging ILOG Rule Language and RIF. In: Proceedings of the 6th International Symposium on Rules, RuleML 2012. CEUR-WS.org, vol. 874, pp. 96-102 (2012)
22. Veiga, F.D.J.: Implementation of the RIF-PRD. Master thesis. Universidade Nova de Lisboa (2011)
23. Polleres, A., Boley, H., Kifer, M. (eds.): RIF Datatypes and Built-Ins 1.0 W3C Recommendation, 2nd edn. (February 5, 2013)

Context-Aware Adaptive Process Information Systems: The Context-BPMN4V Meta-Model

Imen Ben Said[1], Mohamed Amine Chaabane[1], Eric Andonoff[2], and Rafik Bouaziz[1]

[1] MIRACL/FSEG, Route de l'aéroport, BP 1088, 3018 Sfax, Tunisia
{Imen.Bensaid,MA.Chaabane,Raf.Bouaziz}@fsegs.rnu.tn
[2] IRIT/UT1-Capitole, 2 rue du doyen Gabriel Marty, 31042 Toulouse, France
andonoff@univ-tlse1.fr

Abstract. This paper introduces Context-BPMN4V, an extension of BPMN for modeling variability of processes using versions and also considering their contextual dimension. More precisely, it shows how we extend BPMN meta-models to support version modeling to deal with process adaptation, along with context modeling to characterize the situations in which instances of processes are executed. Because of space limitation, this paper only focuses on intra-organizational processes.

Keywords: Context, Adaption, Process Information Systems, Versions, BPMN.

1 Introduction

In the last two decades, there has been a shift from data-aware information systems to Process-Aware Information Systems [1]: processes play now a fundamental role in Enterprise Information Systems (EIS) and they are the support of the alignment between EIS and business strategies of enterprise actors [2,3]. A Process (aware) Information System (PIS) is a software system that manages and executes operational processes involving people, applications, and/or information sources on the basis of process schemas (models). Examples of PISs are workflow management systems (*e.g.* YAWL), case-handling systems (*e.g.* Flowers), enterprise resource planning systems, Business Process Management (BPM) suites *(e.g.* BizAgi) and service oriented architecture-based process implementation [1].

Adaptation is a major challenge for PIS, before their definitive acceptance and use in enterprises [4]. This issue is fundamental, as the economic environment in which enterprises are involved is more and more dynamic, competitive, and open [5,6]: enterprises frequently change their processes in order to meet, as quickly and efficiency as possible, new operational, organizational or customer requirements, or new law regulations. Thus, the economic success of enterprises is closely related to their ability to integrate changes happening in their environment and to make evolve their processes accordingly [7].

This need for adaptive PIS has led BPM researchers to intensively investigate process adaptation issue. Several typologies have been introduced to classify process

Y. Manolopoulos et al. (Eds.): ADBIS 2014, LNCS 8716, pp. 366–382, 2014.
© Springer International Publishing Switzerland 2014

adaptation, and even if they are different, they all agree to distinguish two different times for process adaptation (design-time and run-time), two abstraction levels for process adaptation (the schema –model– level and the instance –case– level), and three types of process adaptation [8–11]: (i) *adaptation by design*, for handling foreseen changes in processes where strategies are not necessarily defined at design-time to face these changes and must be specified at run-time by process users (*e.g.* late modelling and late biding [12]), (ii) a*daptation by deviation*, for handling occasional unforeseen changes and where the differences with initial process are minimal, and (iii) *adaptation by evolution*, for handling unforeseen changes in processes, which require occasional or permanent modifications in process schemas.

We distinguish several approaches to deal with process adaptation issue: activity-driven approach [13–21], constraint-driven approach [22,23], data-driven approach [24,25], case-driven approach (case handling) [26], and more recently, social-driven approach [27]. Activity-driven approach is based on the explicit representation of process schemas (models): the activities of the process and the way there are synchronized are modelled along with information handled and resources involved in it. On the other hand, constraint-driven, case-driven, social-driven and even date-driven approaches avoid describing the way activities are synchronized: in these approaches, process schemas are not explicitly and a priori defined. Even if these approaches are promising, this paper focuses on activity-driven adaptive PIS, as activity-oriented models are used in the majority of (service-oriented) process management systems. Consequently, BPM community has to provide solutions to deal with activity-oriented process adaptation [7].

When dealing with adaptation in activity-driven PIS, in addition to the behavioural, organisational and informational dimensions usually considered for processes and which respectively define the activities of the process and their synchronization, the involved resources in the realization of these activities and the data they produce or consume, we also have to consider the contextual dimension of processes in order to characterize the situations in which instances of processes are executed [7,28–30].

This paper focuses on context-aware adaptation in activity-driven PIS. More precisely, it presents Context-BPMN4V, an extension of BPMN to model context of versions of processes. Versions have been introduced in activity-driven PIS to deal with adaptation issue as they facilitate adaptation by design, by deviation and by evolution [7]. They also facilitate the migration of process instances from an initial schema to a final one, allowing cases in which this migration is impossible and enabling the execution of a same process according to different schemas [11,13–16]. BPMN4V (BPMN for Versions) is an extension of BPMN to model process variability using versions. Basic concepts for process versions have been introduced in [7,21] and BPMN4V has been presented in [31]. In this paper, we extend BPMN4V in order to consider context of process versions. Because of space limitation, we particularly focus on intra-organizational processes [32].

This paper is organized as follows. Section 2 focuses on related works about adaptation in activity-driven PIS and context in BPM area. Section 3 presents BPMN4V to deal with process adaptation using version of processes. Section 4 shows how we extend BPMN4V in order to take into account the contextual dimension of processes. Finally, section 5 concludes the paper and gives some directions for future works.

2 Related Works

Adaptation in activity-driven PIS is a highly investigated issue since the end of the nineties. Even if existing contributions are significant [12–21], considering run-time and design-time adaptation, both at the schema and the instance level, we can observe that they mainly focus on adaptation of process behavior, leaving aside the organizational, informational and contextual dimensions of processes. Moreover, proposed notations are not standards and are unlikely to be used by process designers, who are in charge of modeling variability of processes.

In order to take into account the second previous remark, some contributions have extended BPMN, which is now known as the standard notation for processes, to deal with adaptation issue. For instance, [33] has extended BPMN including concepts of a generic meta-model to support process goals and process performance. [34] has investigated more deeply this issue, proposing goals models, depicted as tree-graphs, to represent process goals and their possible variations. However, these works did not adopt a comprehensive approach, considering all the dimensions of processes at the same time, *i.e.* the behavioral, organizational, informational and contextual dimensions of processes.

The notion version has been recognized as a key notion to deal with adaptation issue. On the one hand, handling version of processes facilitate the migration of instances from an initial schema to a final one, allowing if the migration is not possible, two different instances of a same process to run according to two different schemas [7,13–16,21]. On the other hand, as defended in [7], versions are appropriate to deal with the three types of adaptations identified in the main topologies of the literature ([8–11]), *i.e.* adaptation by design, adaptation by deviation and adaptation by evolution. Consequently, in a previous proposition, and with respect to the concepts of [21], we introduced BPMN4V (BPMN for Versions), an extension of BPMN to support intra-organizational and inter-organizational process version modeling, both considering behavioral (what, how), organizational (who) and informational (when) dimensions of processes [31]. Thus, we extended in this proposition the main contributions about versions in BPM [13–18,20]. We continue our effort in adaptive in activity-driven PIS area extending BPMN4V to integrate the contextual dimension of processes in order to characterize the situations in which instances of processes are executed and thus to define *why* version of processes are defined/used instead another according to the context.

Context-awareness has been investigated in several domains of computer science (*e.g.* natural language, human computer interaction, mobile application, or web system engineering). BPM area has also taken advantages from context-awareness. A process context is defined in [28] as *the minimum set of variables containing all relevant information that impact the design and execution of a process*. This contribution also introduced a taxonomy for contextual information by distinguishing four natures of context: (i) *immediate context*, which covers information on process components, *i.e.* context of activities, events, control flow, data and organization, (ii) *internal context*, which covers information on the internal environment of an organization that impacts the process (*e.g.* process goals, organization strategies and policies), (iii) *external context*, which covers information related to external stakeholders (*e.g.* customers, suppliers, government) of the organization, and finally (iv) *environmental context*, which covers information related to external factors (*e.g.*, weather, time, workforce, economic data). Other works (*e.g.* [21], [34]) only distinguish two types of information to represent process context: functional information related to process components (activities, events, resources...) and non-functional information related to quality of process (safety, security, cost, time...). In this work, we rather adopt the taxonomy introduced in [28], which is more comprehensive.

In addition to these taxonomies, several contributions have been done to support context awareness in process modeling. For instance, [29,30,35] proposed a rule-based approach to define contextual information used to configure process instances to particular situations. On the other hand, [34,36] proposed goals models, depicted as tree-graph, to represent process goals and their possible variations. In these works, a goal model is used to adapt process definition according to the context. But only considering goal is not enough to describe context of processes. Finally, [21] also investigated the contextual dimension of processes, but as indicated in the previous section, the underlying context taxonomy is rather incomplete, and the proposed notation is unlikely to be used by process designers.

To sum up, in this paper we advocate the modeling of process versions using BPMN4V (BPMN for Versions) to deal with adaptation in activity-driven PIS [31]. We also define Context-BPMN4V, an extension of BPMN4V in order to take into account context for (process) versions and considering the taxonomy introduced in [28]. This contextual dimension is fundamental to help PIS users (i) at design-time, to indicate why a process version is defined (different variables and conditions are specified for the considered process version), and (ii) at run-time, to instantiate a particular version of a process according to a concrete situation, *i.e.* a set of values given to the different variables specified in the context of the considered process version.

3   Modeling Process Versions: The BPMN4V Meta-Model

As indicated before, this paper focuses on intra-organizational processes. Such processes are modelled in BPMN2.0 through private processes, which are internal to a specific organization.

This section first introduces the notion of version and then present BPMN4V, an extension of BPMN for modeling versions of private process.

3.1   Notion of Version

As illustrated in Fig. 1 below, a version corresponds to one of the significant states a process may have during its life cycle. So, it is possible to describe the changes of a process through its different versions. These versions are linked by a derivation link; they form a version derivation hierarchy. When created, a process is described by only one version. The definition of every new version is done by derivation from a previous one: such versions are called derived versions. Of course, several versions may be derived from the same previous one: they are called alternatives or variants. The derivation hierarchy looks like a tree if only one version is directly created from a process entity, and it looks like a forest if several versions are directly created from the considered process entity.

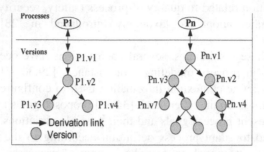

Fig. 1. Versions to Model Process Variability

We defend that this version notion subsumes the notion of variant as it is defined in [17,19,20,29] as, when considering versions, we both model process evolution and process alternative (*i.e.* variant) to describe process variability and deal with adaptation by design, adaptation by deviation and adaptation by evolution [7].

3.2   BPMN4V: BPMN2.0 for Versions

BPMN4V meta-model for private processes results from the merging of BPMN2.0 meta-model for private processes and a versioning pattern used to make classes of BPMN2.0 meta-model versionnable, *i.e.* able to handle versions. We present briefly these two layers.

Versioning Pattern. The versioning pattern is very simple: it includes only two classes: *Versionable* class and *Version of Versionable* class, and two relationships: *Is_version_of* and *Derived_From* as illustrated in Fig. 2. A versionable class is a class for which we would like to handle versions. In addition we define a new class which contains versions, called *Version of Versionable*.

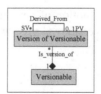

Fig. 2. Versioning Pattern

The *Is_version_of* relationship links a class to its corresponding versions. The *Derived_From* relationship allows for building version derivation hierarchies (cf. Fig. 1). This latter relationship is reflexive and the semantic of both relationship sides is the following: (i) a version (*SV*) succeeds another one in the derivation hierarchy, and (ii) a version (*PV*) precedes another one in the derivation hierarchy. Regarding properties of a *Version of Versionable* class, we introduce the classical version properties, *i.e.* version number, creator name, creation date and status.

Main Concepts of BPMN 2.0 Meta-Model [37]. These concepts are presented in Fig. 3 hereafter. The three main dimensions of processes are considered in BPMN2.0 meta-model.

The behavioral dimension of processes supports the description of process activities and their synchronization along with events happening during process execution through the notions of *FlowElementContainer* which contains *SequenceFlow*, *FlowNode* (*Gateway*, *Event*, and *Activity*), and *Data Object*. A *SequenceFlow* is used to show the order of *FlowNode* in a process. A *SequenceFlow* may refer to an *Expression* that acts as a gating condition. The *Expression* class is used to specify a condition using natural-language text. A *Gateway* is used to control how *SequenceFlow* interact within a process. An *Event* is something that happens during the course of a process. It can correspond to a trigger, which means it reacts to something (*catchEvent*), or it can throw a result (*throwEvent*). An *Event* can be composed of one or more *EventDefinitions*. There are many types of *Event Definitions*: *ConditionalEventDefinition*, *TimerEventDefinition*... An *Activity* is a work performed within a process. An *Activity* can be a *Task* (*i.e.* an atomic activity) or a *Sub Process* (*i.e.* a non-atomic activity). A *Task* is used when the work is elementary (*i.e.* it cannot be more refined). BPMN2.0 identifies different types of tasks: *Service Task*, *User Task*, *Manual Task*, *Send Task* and *Receive Task*.

Regarding the organizational dimension of processes, an activity is accomplished by a *ResourceRole*. A *ResourceRole* can refer to a *Resource*. A *Resource* can define a set of parameters called *ResourceParameters*. A *ResourceRole* can be a *Performer*, which can be a *HumanPerformer*, which can be in turn a *PotentialOwner*.

Regarding the informational dimension of processes, an *ItemAwareElement* references element used to model the items (physical or information items) that are created, manipulated and used during a process execution. An *ItemAwareElement* can be a *DataObject*, a *DataObjectReference*, a *Property*, a *DataStore*, a *DataInput* or a *DataOutput*.

BPMN4V Meta-Model. The idea is to use the versioning pattern introduced before (cf. Fig. 2) to make some classes of the BPMN2.0 meta-model versionable, i.e. able to handle versions. Fig. 3 below presents the resulting meta-model. White rectangles and relationships correspond to classes and relationships of BPMN2.0, while grey rectangles and blue relationships correspond to classes and relationships involving versions, and resulting from the introduction of the versioning pattern.

We propose to handle versions for seven classes: *Process*, *Sub Process*, *Event*, *Activity*, *ItemAwareElement*, *Resource*, and *ResourceRole* in order to support adaptive activity-driven PIS. Different instances (versions) of these classes can be created, each one representing a significant state (alternative or derived) of the considered element (*e.g.* a process). A new version of an element (*e.g.* a process or a resource) is defined according to the changes occurring to it: these changes may correspond to the adding of new information (property or relationship) or to the modification or the deletion of an existing one. Actually, these changes can affect all the dimensions of a process. The general idea is to keep track of changes occurring to components participating to the description of the way business is carried out.

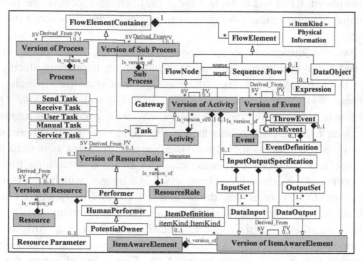

Fig. 3. BPMN4V Meta-model for Modeling Process Versions

Regarding the organizational dimension of a process, we create a new version of *Resource* when we change its parameters. For instance, a *Manager* resource may be defined using two parameters: name and experience. A new version of *Manager* may be defined if it becomes necessary to consider another parameter (*e.g.* region of the manager) and this definition can lead to the definition of a new process in which this resource is involved [21]. We also propose to create versions of *ResourceRole* when there is a change in its privileges. For instance, an *Employee* is a *HumanPerformer* resource that performs three activities. Because some activities of the process in which this employee is involved become automatic, the employee can perform anymore only two activities. A new version of the *ResourceRole* employee has then to be defined.

Regarding the informational dimension of processes, and more particularly *ItemAwareElement*, we consider that changes in the structure and/or the type of an *ItemDefinition* results in the creation of a new version. For example, if *Report* is an *ItemAwareElement* corresponding to a paper data (*Itemkind* is a *Physical* data), and if after technical changes it becomes an electronic data (*Itemkind* becomes an *Information* data), then a new version of *Report* has to be created.

Regarding the behavioral dimension of processes, several classes are versionable: *Event*, *Activity*, *Sub-Process* and, of course, *Process*. More precisely regarding activities, we create a new version of an activity when there are changes in the type of the activity (a manual activity becomes a service one), in the involved resources, or in the required or produced data. Regarding events, we create a new version of an event when there is change in the associated *EventDefinition*. For instance, if an *Alert* is a signal event (*i.e.* it has a *SignalEventDefinition*), and if, after technical changes, it becomes a message event (*i.e.* it has a *MessageEventDefinition*), then a new version of *Alert* has to be created. Regarding sub processes and processes, we create new versions when there are changes in the involved activities and events or in the way they are linked together (used patterns, *i.e.* gateways, are changed).

3.3   Example

We illustrate in Fig. 5 the instantiation of BPMN4V meta-model according to the damage compensation process of an insurance company. This process is shown in Fig. 4 below. Basically, it contains five activities: *Receive Request, Review Request, Send reject letter, Calculate claim amount,* and *Financial settlement*. For simplification reasons, we only focus on the behavioral dimension of this process.

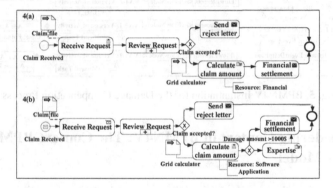

Fig. 4. Versions of the Damage Compensation Process

Two versions of this process are described in Fig. 4. The first one is given in Fig. 4(a). This version starts when the client files a claim. After checking the claim, a reject letter is sent, by mail, if the request is not accepted. Otherwise, the claim amount is calculated by the insurance manager using *GridCalculator*, and the financial service prepares and sends the financial settlement. On the other hand, further to an increase in the number of its customers, the insurance agency has modeled a second version of this process. Fig. 4(b) illustrates this version introducing an *Expertise* activity (a new activity, used when the damage amount exceeds 1000$) and both modifying the start *ClaimReceived* event and the *ReceiveRequest* and *CalculateClaimAmount* activities (their type have changed). To sum up, regarding the damage compensation process we have two versions of the process itself, two versions of *ClaimReceived* event and two versions of the *ReceiveRequest* and *CalculateClaimAmount* activities: the first version of both *ReceiveRequest* and *CalculateClaimAmount* activity hold for the first version of the process while the second ones hold for the second version of the process. In addition, the sequence flows and patterns have been modified in the second version of the process. Finally, regarding the *ClaimReceived* event, in the first version of the process, it is a *None Event*, used to indicate that this version starts when the client presents its *ClaimFile* (a paper data). However, in the second process version, it becomes a *Message Event*, indicating that the client sends the *ClaimFile* (an electronic data) as a message via the insurance web site.

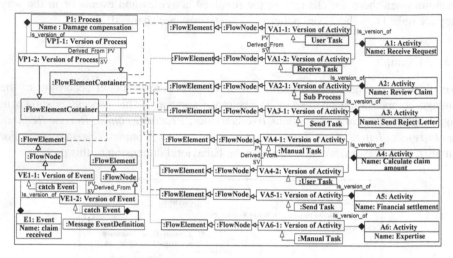

Fig. 5. BPMN4V Instantiation for the Damage Compensation Process

4   Considering Context in BPMN4V: The Context-BPMN4V Meta-Model

We advocate using the version notion to support adaptation in activity-driven PIS. Thus, several versions can be defined for a process: versions of the process itself, but also versions of activities, sub-processes, events, data produced or consumed by activities (*i.e.* ItemAwareElement) and versions for the organizational dimension of processes (*i.e.* Resource and ResourceRole). Each of the defined versions is required in a specific context, *i.e.* has to be used in a given situation. Therefore, it becomes crucial to consider the contextual dimension of versions in order to characterize the situations in which these versions have to be used. Indeed, this contextual dimension is fundamental to help PIS users (i) at design-time, to indicate why a (process) version is defined –different variables and conditions are specified for the considered (process) version–, and (ii) at run-time, to instantiate a particular version of a process according to a concrete situation, *i.e.* a set of values given to the different variables specified in the context of the considered process version.

We present below Context-BPMN4V, an extension of BPMN4V considering the contextual dimension of processes. We first introduce a Context meta-model for context description in PIS. Then, we present Context-BPMN4V which results from the merging of this Context meta-model and BPMN4V to model context for (process) versions. Finally, we illustrate context definition for process versions using the damage compensation process introduced previously.

4.1   Context Modeling

The Context meta-model given in Fig. 6 below allows the definition of a *Context Model* as the aggregation of a set of context parameters. A *Context Parameter* corresponds to a variable characterizing a situation, and to which a condition will be defined. A context parameter has a *Context Nature*, which can be immediate, internal, external or environmental, according to the taxonomy given in [28]. This taxonomy is used to specify the source of each parameter that composes a *Context Model*. In addition to this taxonomy, we also consider the type of a *Context Parameter*, which refers to the dimension to which it belongs to. Thus, we consider *Behavioral Parameters*, *i.e.* variables related to the behavioral dimension of processes (*e.g.* activity execution mode, activity duration), *Role Parameters*, *i.e.* variables related to the organizational dimension of processes (*e.g.* availability of a resource, experience of a human performer), and *Data Parameters*, *i.e.* variables related to the informational dimension of processes (*e.g.* data type, data structure). We finally consider *Goal Parameters*, *i.e.* variables describing objectives to be achieved (*e.g.* quality, cost, quantity); such type of context parameters belongs to the intentional dimension of processes [35,38].

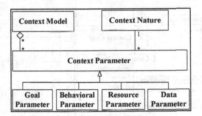

Fig. 6. Context Meta-model

4.2   Extending BPMN4V Meta-Model to Consider Context for (Process) Versions: The Context-BPMN4V Meta-Model

Context-BPMN4V, visualized in Fig. 7, results from the merging of BPMN4V and Context meta-models. It defines the necessary concepts for modeling context of process versions. Of course, context can be defined for each versionable component of the process; thus, in addition to process versions, contexts can be defined for versions of activities, sub processes, events, resource roles, resources and ItemAwareElements.

The proposed meta-model links a process to a context model aggregating a set of context parameters, corresponding to variables from the different dimensions of the considered process: goal parameters from the intentional dimension of the process, behavior parameters from the behavioral dimension of the process, resource parameters from the organizational dimension of the process, and data parameters from the informational dimension of the process. Thus, each versionable component of the process can be linked to one or several context parameters of its corresponding context model, in order to define conditions on these parameters.

More precisely, goal parameters specify objectives of versionable concepts through goal conditions. A *Goal Condition* is a boolean expression defining why a version is created (cf. section 4.3 for an goal condition example). As indicated before, each versionable component of a process can be linked to a parameter of its corresponding context model. As a consequence, a goal parameter, used in the definition of goals of a versionable component of a process, has to be part of the set of context parameters that form the context model of this process. In the proposed meta-model, we also use OCL constraints to define restrictions. We give a textual definition of these restrictions in Fig. 7.

Regarding behavioral, data and role parameters, we specify conditions, called *Assignment conditions* that allow the assignment of versionable components. These conditions, described as boolean expressions, define for a specific version of process, the situation in which versions of activities, of events, of resource role and of ItemAwareElement involved in this process have to be used.

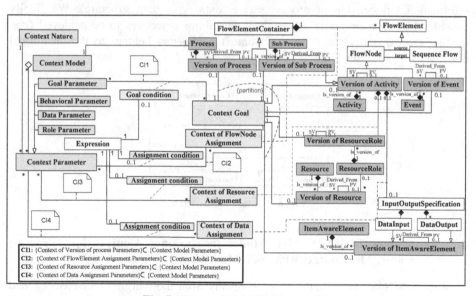

Fig. 7. Context-BPMN4V Meta-model

Context-BPMN4V extends BPMN4V adding three classes used to define assignment conditions within relationships between BPMN components:

- *Context of FlowNode Assignment* attached to the relationship between *FlowElementContainer* and *FlowElement*, defines conditions indicating in which situation a version of activity (or a version of event) has to be used in a version of process or a version of sub-process.

- In the same vein, *Context of Resource Assignment*, attached to the relationship between *Version of Activity* and *Version of ResourceRole*, defines conditions indicating in which situation a version of activity is performed by a version of *ResourceRole*.
- Finally, *Context of Data Assignment*, attached to the relationship between *Version of Activity* and *InputOutputSpecification*, defines conditions indicating in which situation a version of activity consumes or produces versions of *ItemAwareElement*.

Finally, a context parameter used in the definition of an assignment condition must belong to the set of context parameters that form the context model of the corresponding process. Constraints *CI2*, *CI3* and *CI4* express these restrictions with respect to *Context of FlowNode Assignment*, *Context of Resource Assignment* and *Context of Data Assignment* classes.

4.3   Defining Context in the Damage Compensation Process

To better illustrate our proposition, we refer to the damage compensation process previously presented (cf. section 3.3), and gives the context of each version of this process. Even if this example is rather simple, it illustrates suitably this context notion for processes. Table1 below summarizes context parameters involved in the first and the second version context definition of this process.

Table 1. Context Parameters and Conditions in the Damage Compensation Process

Context Nature	Parameter/ (Parameter type)	Context of the first version	Context of the second version
External Context	NumberofDailyClaims / (Goal parameter)	<50	≥50
Immediate Context	ClaimFile / (Data parameter)	is-a paperData	is-a ElectronicData
	GridCalculator / (Data parameter)	is-a paperData	is-a ElectronicData
	CalculateClaimAmount / (Behavioral parameter)	is-a ManualActivity	is-a ElectronicActivity
	CalculateClaimAmountRole / (Role parameter)	is-performed-by Human	is-performed-by ACA
	Expertise / (Behavioral parameter)		is-a ManualActivity

The first version of the damage compensation process is defined at the beginning, *i.e.* when insurance agency is created: only few clients and thus only few daily claims, rather simple, to deal with, and only few IT investments. In this process version, the insurance manager calculates the claim amount using a grid calculator. Parameters that characterize this situation are the number of daily claims to deal with, the claim file modeled as a paper data, and the way the claim amount is calculated: manually from a grid calculator. The second version of the process is defined to face the increasing number of clients, and thus the increasing number of daily claims to deal with. The insurance agency invested in IT software: clients no longer fill in claim files manually but rather declare their claims via the website of the insurance agency, and specific software is used to calculate claim amounts. In addition, a new role and a new activity are introduced in the process when specific claim require expertise. Parameters that characterize this second version of the damage compensation process are the number of daily claims, the claim file which is modeled as an electronic data, the presence of the expertise activity, and the way claim amounts are calculated: automatically using specific software, or manually using grid calculator.

More precisely, *NumberofDailyClaims* is a goal parameter whose nature is external: the two conditions *NumberofDailyClaims<50* and *NumberofDailyClaims>=50* define the possible values for this parameter in the two versions of the damage compensation process. *ClaimFile* and *GridCalculator* are immediate data parameters. The conditions, *is-a paperData* and *is-a ElectronicData*, define the possible values for these parameters in the two process versions. Note that *ClaimFile* and *GridCalculator* refer to data handled by the process activities. In the same vein, we refer to existing activities with the immediate behavioral parameters *CalculateClaimAmount* and *Expertise*. Regarding this latter, it is not involved in the first version of the process but it is a manual activity in the second version of the process (*is-a ManualActivity* condition). Regarding *CalculateClaimAmount* parameter, two conditions are defined, respectively in the context of the first and the second process versions: it is a manual activity in the first version (*is-a ManualActivity*) while it is an electronic activity in the second version (*is-a ElectronicActivity*). Finally, immediate resource parameters are defined for both *CalculateClaimAmount* and *Expertise* activities in order to define how these activities are performed. Thus, *CalculateClaimAmountRole* is performed by a human in the first process version (*is-performed-by Human*), while it is performed by a software application in the second process version (*is-performed-by ACASoftwareApplication*). In the same way, *ExpertiseRole* is performed by a human in the second process version. Note that classical and specific operators are used to specify these conditions (*e.g.* is-a, is-performed-by, has-experience) [21]. Fig. 8 below gives an extract of the instantiation of damage compensation process context.

In this figure, *CG1* and *CG2* are context goals (*i.e.* instances of the ContextGoal class) of the first and the second version of this process (*VP1-1* and *VP1-2*). These goals are defined using *NumberofDailyClaims* context parameter and <50 (Gc1) and >=50 (Gc2) goal conditions. In addition, *CFA1, CFA2, CDA1, CDA2, CRA1* and *CRA2* define why a version of the calculate claim amount activity is used in the two damage compensation process versions. Actually, two versions of this activity are created: *VA4-1* and *VA4-2*, each involved in one of the two process versions. *VA4-1* holds for the first version of this process: in this version, *CalculateClaimAmount* is a *ManualActivity*, *CalculateClaimAmountRole* is performed by *Human*, and *GridCalculator* is a *paperData*. Regarding *VA4-2*, it holds for the second version of the damage compensation process: in this case, *CalculateClaimAmount* is a *ElectronicActivity*, *CalculateClaimAmountRole* is performed by *ACA*, and *GridCalculator* is an *electonicData*.

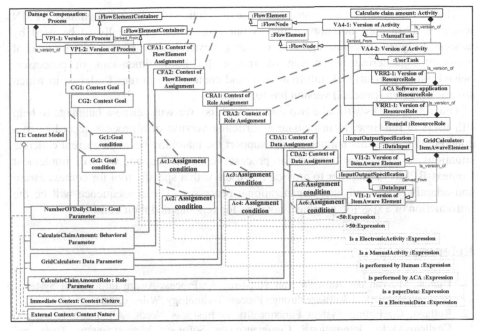

Fig. 8. Extract of the damage compensation process context instantiation

5   Conclusion

This paper has presented Context-BPMN4V, an extended BPMN meta-model to support context-aware process variability modeling in activity-driven PIS. It results from the merging of BPMN4V and a Context meta-model. Context-BPMN4V supports the modeling of adaptive processes using versions. Indeed, versions are a powerful mechanism to model process variability: (i) they facilitate the migration of process instances from an initial schema to a final one, allowing cases in which this migration is impossible and enabling the execution of a same process according to different schemas, and (ii) they are appropriate to deal with the three types of adaptation identified in the different typologies of literature: adaptation by design, adaptation by deviation and adaptation by evolution.

In addition, Context-BPMN4V also considers the contextual dimension of processes, which rather focuses on the why dimension of a process. This dimension is considered for each versionable component of Context-BPMN4V: thus, context can be defined for versions of processes of course, but also for versions of activities, sub-processes, events, resource roles, and data consumed and produced by activities or sub-processes. This contextual dimension is fundamental to help PIS users (i) at design-time, to indicate why a process version is defined (different variables and conditions are specified for the considered process version), and (ii) at run-time, to instantiate a particular version of a process according to a concrete situation, *i.e.* a set of values given to the different variables specified in the context of the considered process version.

The advantages of our contribution are the following. First, it extends BPMN which is a standard notation: it makes our proposition potentially to be used by process designers. Second, regarding versions, it extends the different contributions of literature considering the notion of version for both dimensions of processes: behavioral, organizational, informational, and contextual in order to define in which situation a given (process) version has to be used.

Our future works will take two directions. First, we will define a language to help PIS users at run-time, to instantiate a particular version of a process according to a concrete situation: this language must support the matching between current concrete situations and contexts defined for process versions. Second, we will implement Context-BPMN4V in order to provide PIS users with a specific tool for context-aware and adaptive intra and inter organizational processes. The consequence will be the introduction of a specific graphical notation for versions and their context.

References

1. Dumas, M., van der Aalst, W., ter Hofstede, A.: Process-Aware Information Systems: Bridging People and Software through Process Technology. Wiley (2005)
2. Rolland, C.: Fitting System Functionality to Business Needs: Alignment Issues and Challenges. In: International Conference on Software Methodologies, Tools and Techniques, Yokohama City, Japan, pp. 137–147 (September 2010)
3. Simonin, J., Nurcan, S., Barrios, J.: Evolution organisationnelle fondée sur la cohérence des relations entre acteurs avec les buts métier. In: National Conference on Informatique des Organisations et des Systèmes d'Information et de Décision, Paris, pp. 225–240 (May 2013)
4. Smith, H., Fingar, P.: Business Process Management: the Third Wave. Megan-Kiffer Press (2003)
5. Reijers, H.: Workflow Flexibility: the Forlon promise International Workshop on Enabling Technologies: Infrastructure for Collaborative Enterprises, Manchester, United Kingdom, pp. 271–272 (June 2006)
6. Weske, M.: Business Process Management: Concepts, Languages, Architectures. Springer (2007)
7. Chaâbane, M.A., Andonoff, E., Bouaziz, R., Bouzguenda, L.: Versions to Address Business Process Flexibility Issue. In: Grundspenkis, J., Morzy, T., Vossen, G. (eds.) ADBIS 2009. LNCS, vol. 5739, pp. 2–14. Springer, Heidelberg (2009)
8. Nurcan, S.: A Survey on the flexibility Requirements related to Business Process and Modelling Artifacts. In: International Conference on System Sciences, Waikoloa, Big island, Hawaii, USA, pp. 378–387 (January 2008)
9. Schonenberg, H., Mans, R., Russel, N., Mulyar, N., van der Aalst, W.: Process Flexibility: a Survey of Contemporary Approaches. In: Dietz, J.L.G., Albani, A., Barjis, J. (eds.) CIAO! 2008 and EOMAS 2008. LNBIP, vol. 10, pp. 16–30. Springer, Heidelberg (2008)
10. Weber, B., Sadiq, S., Reichert, M.: Beyond Rigidity – Dynamic Process Lifecycle Support: a Survey on Dynamic Changes in Process-Aware Information Systems. International Journal on Computer Science, Research and Development 23(2), 47–65 (2009)
11. Andonoff, E., Nurcan, S., Hanachi, C.: Adaptation des processus d'entreprise. In: Lopisteguy, P., Rieu, D., Roose, P. (eds.) L'adaptation dans tous ses états, ch. 3, Cepadues (2012)

12. Adams, M., ter Hofstede, A.H.M., Edmond, D., van der Aalst, W.M.P.: Worklets: a Service-Oriented Implementation of Dynamic Flexibility in Workflows. In: Meersman, R., Tari, Z. (eds.) OTM 2006. LNCS, vol. 4275, pp. 291–308. Springer, Heidelberg (2006)

13. Casati, F., Ceri, S., Pernici, B., Pozzi, G.: Workflow Evolution. In: Thalheim, B. (ed.) ER 1996. LNCS, vol. 1157, pp. 438–455. Springer, Heidelberg (1996)

14. Kammer, P., Bolcer, G., Taylor, R., Bergman, R.: Techniques for supporting dynamic and adaptive workflow. International Journal on Computer Supported Cooperative Work 9(3/4), 269–292 (1999)

15. Kradolfer, M., Geppert, A.: Dynamic workflow schema evolution based on workflow type versioning and workflow migration. In: International Conference on Cooperative Information Systems, Edinburgh, Scotland, pp. 104–114 (September 1999)

16. Reichert, M., Rinderle, S., Dadam, P.: ADEPT workflow management system: flexible support for enterprise-wide business processes. In: van der Aalst, W.M.P., ter Hofstede, A.H.M., Weske, M. (eds.) BPM 2003. LNCS, vol. 2678, pp. 370–379. Springer, Heidelberg (2003)

17. Lu, R., Sadiq, S.K.: Managing process variants as an information resource. In: Dustdar, S., Fiadeiro, J.L., Sheth, A.P. (eds.) BPM 2006. LNCS, vol. 4102, pp. 426–431. Springer, Heidelberg (2006)

18. Zhao, X., Liu, C.: Version Management in the Business Change Context. In: Alonso, G., Dadam, P., Rosemann, M. (eds.) BPM 2007. LNCS, vol. 4714, pp. 198–213. Springer, Heidelberg (2007)

19. Lu, R., Sadiq, S., Governatori, G., Yang, X.: Defining adaptation constraints for business process variants. In: Abramowicz, W. (ed.) BIS 2009. LNBIP, vol. 21, pp. 145–156. Springer, Heidelberg (2009)

20. Hallerbach, A., Bauer, T., Reichert, M.: Capturing Variability in Business Process Models: the Provop Approach. Journal of Software Maintenance 22(6-7), 519–546 (2010)

21. Chaâbane, M., Andonoff, E., Bouaziz, R., Bouzguenda, L.: Modélisation multidimensionnelle des versions de processus. Journal on Ingénierie des Systèmes d'Information 15(5), 89–114 (2010)

22. Pesic, M., Schonenberg, H., Sidorova, N., van der Aalst, W.: DECLARE: full support for Loosely-Structured Processes. In: International Conference on Enterprise Distributed Object Computing, Annapolis, Maryland, USA, pp. 287–300 (October 2007)

23. Pesic, M., Schonenberg, M.H., Sidorova, N., van der Aalst, W.M.P.: Constraint-based workflow models: Change made easy. In: Meersman, R., Tari, Z. (eds.) OTM 2007, Part I. LNCS, vol. 4803, pp. 77–94. Springer, Heidelberg (2007)

24. Müller, D., Reichert, M., Herbst, J.: Data-driven Modeling and Coordination of Large Process Structures. In: Meersman, R., Tari, Z. (eds.) OTM 2007, Part I. LNCS, vol. 4803, pp. 131–149. Springer, Heidelberg (2007)

25. Müller, D., Reichert, M., Herbst, J.: A New Paradigm for the Enactment and Dynamic Adaptation of Data-Driven Process Structures. In: Bellahsène, Z., Léonard, M. (eds.) CAiSE 2008. LNCS, vol. 5074, pp. 48–63. Springer, Heidelberg (2008)

26. van der Aalst, W., Weske, M., Grünbaur, D.: Case Handling: a New Paradigm for Business Process Support. International Journal on Data Knowledge Engineering 53(2), 129–162 (2005)

27. Bruno, G., Dengler, F., Jennings, B., Khalaf, R., Nurcan, S., Prilla, M., Sarini, M., Schmidt, R., Silva, R.: Key challenges for enabling agile BPM with social software. Journal of Software Maintenance and Evolution: Research and Practice 23(4), 297–326 (2011)